The Basics of Physics

The Basics of Physics

Rusty L. Myers

Basics of the Hard Sciences
Robert E. Krebs, Series Editor

GREENWOOD PRESS
Westport, Connecticut • London

Library of Congress Cataloging-in-Publication Data

Myers, Rusty [Richard] L.
   The basics of physics / Rusty L. Myers.
      p. cm. — (Basics of the hard sciences)
   Includes bibliographical references and index.
   ISBN 0-313-32857-9 (alk. paper)
   1. Physics—Textbooks. I. Title. II. Series.
   QC21.3.M94   2006
   530—dc22   2005030786

British Library Cataloguing in Publication Data is available.

Library of Congress Catalog Card Number: 2005030786
ISBN: 0-313-32857-9

First published in 2006

Greenwood Press, 88 Post Road West, Westport, CT 06881
An imprint of Greenwood Publishing Group, Inc.
www.greenwood.com

Printed in the United States of America

The paper used in this book complies with the
Permanent Paper Standard issued by the National
Information Standards Organization (Z39.48–1984).

10 9 8 7 6 5 4 3 2 1

While the physics experiments ("activities") listed in chapter 16 of this volume have been exhaustively tested for safety, the author and publisher cannot be held liable for any injury or damage that may occur during the performance of these experiments.

Contents

Preface

The word *physics* comes from the Greek word *physika* meaning pertaining to nature. Physics can be considered the most basic of the sciences. Ernest Rutherford, a prominent physicist of the early twentieth century, who ironically received the 1908 Nobel Prize in chemistry for his work on the structure of the atom, said all science is either physics or stamp collecting. While other scientific disciplines have developed in the last three hundred years, physics is the foundation of all science.

Each of us is subject to the laws of physics and physics attempts to explain basic phenomena such as gravity, electricity, sound, light, and heat. Physics applications constantly surround us as we go about our daily activities. Every time we buckle a seat belt, flip a switch, use a tool, or ride a bike the principles of physics are used. Basic physical activities such as walking, speech, hearing, and lifting are governed by physics. Life experience has given each of us a basic understanding of physics and we constantly use this knowledge, often unconsciously. Every time we speed up to make a green light, use a jack to change a flat tire, or place ice in a glass to cool a drink we are applying the laws of physics. In this sense, we all have an intuitive knowledge of physics, yet many of us have never studied physics formally and may even dread the thought.

The Basics of Physics is written for students beginning a formal study of physics. These are primarily high school and college students enrolled in their first physics course or anticipating enrolling in such a course. Individuals not enrolled in a physics course, but wanting a general overview of the subject should find this book helpful. Teachers of all grades may use *The Basics of Physics* as a general reference on the subject. Librarians can use *The Basics of Physics* as a ready reference as a starting point for both scientific and historical information.

The Basics of Physics is a general reference book that presents the basic scientific concepts of physics in addition to providing the historical and biographical context of the individual responsible for their development. Chapters progress through several areas. The first two chapters focus on physics' history as a modern science. Most of the chapters in *The Basics of Physics* are devoted to explaining basic physics concepts. Chapters are devoted to principles such as mechanics, motion, energy, matter, light, electricity, and modern physics. Building on these general concepts, additional chapters explore several areas of physics and present additional concepts especially important to these areas. Chapters on quantum physics, nuclear physics, and relativity are included. Chapter 16 presents the experimental method and 25 physics activities suitable for students beginning a study of physics. Chapter 17 gives an overview of the physics profession, possible careers in physics information about pursuing a physics education.

The Basics of Physics includes ample illustrations, tables, and analogies to clarify the basic physics concepts and highlight important physics information. Important terms are typed in boldface when they first appear in the text and summarized in a glossary. Year of birth and death of important individuals are given in parentheses following the first appearance of a person's name. Important physical constants, comprehensive conversion tables, and physics Nobels are presented in appendices.

In writing a book called *The Basics of Physics*, the most difficult task is deciding what material to include. What may be "the basics" for one person may not be to another. Rather than present a list of basic facts, *The Basics of Physics* presents physics from a broad perspective. An introductory book on such a broad subject should give the student a glimpse of the discipline, provide basic concepts and information, and provide a foundation for further study. In this respect, an introductory book is like wiping the moisture from a fogged window. Many objects become clear through the transparent opening, but the rest of the glass remains foggy. Objects viewed through the fog are unclear. Even after wiping away all of the moisture, only a small portion of the outside world can be seen. My hope is that *The Basics of Physics* provides the student with a small cleared opening into the world of physics. Once a glimpse is gained through this opening, the student can continue to enlarge the opening.

Acknowledgments

Many individuals helped in the preparation of this book. Greenwood Press initiated this book as part of a series of general reference books in the sciences. I would like to thank the staff at Greenwood for the opportunity to write this book. Greenwood's Debbie Adams shepherded the project through completion with timely e-mails and gentle reminders. Mark Kane put the finishing touches on the work. Rae Dejúr diligently completed artwork while dodging hurricanes in Florida's panhandle. A special thanks goes to Robert Krebs who provided valuable feedback with his reviews and encouragement. Apex publishing did an excellent editing job transforming the material into a coherent work.

Physics and Its Early Foundations

Introduction: What Is Physics?

Physics is experience, arranged in economical order.

Ernst Mach (1836–1915),
Austrian philosopher and physicist

Throughout life, each of us has been conditioned to accept, without thought, regular patterns as we conduct our daily business. The Sun rises in the morning and sets at night, dropped objects fall to the ground, water flows downhill, and an ice cube melts when taken out of the freezer. We know that a window will shatter when stuck by a baseball hit by a line drive, and if we live in a glass house it is best to throw nerf balls. The universe constantly reveals its behavior as we go about our daily lives. Conversely, we accept other facts based not on direct experience, but on the authority of others. At times, these facts even contradict our experience and direct observation. For example, it is difficult to comprehend that as you read this book you are sitting on a roughly spherical object rotating at a speed of approximately 800 miles per hour. (This assumes you are reading in the continental United Sates. In

Alaska, the speed is only about 500 miles per hour, while at the equator it is just over 1,000 miles per hour.) In addition to its rotation, the Earth is also revolving around the Sun at a speed close to 67,000 miles per hour, and the entire solar system is moving around the center of our Milky Way galaxy at more than 500,000 miles per hour.

The several examples cited above illustrate the science of physics. Physics involves the study of matter and energy in its different forms, and the transformation of matter and energy. This same definition might also apply to chemistry, and the two disciplines are closely related. Chemists tend to focus more on the specific characteristics of matter and how different forms of matter are transformed into other forms of matter. Chemists tend to treat matter and energy as separate entities. Physicists are concerned with the general properties that govern all of matter and energy, and in this sense a clear distinction between the two is unnecessary.

As the quote at the beginning of this chapter states, physics involves trying to explain everyday experience in the simplest terms. The science of physics is a continual quest to explain the behavior of the universe using relatively few basic principles. These basic principles should be applicable to all scales of matter ranging from fundamental particles (quarks) and atoms to galaxies. In a sense, physics can be thought of as the search for a general explanation for everything. During physics' history there have been periods when physicists have claimed they were on the verge of knowing everything needed to explain the universe. For instance, the confidence in Newtonian mechanics at the end of the nineteenth century caused some physicists to claim that little remained to be discovered and this heralded the end of physics. Shortly thereafter, relativity and quantum mechanics gave rise to modern physics. Today the quest for simplicity continues in the form of string theory, but a **general unified theory** also termed a **Theory of Everything** doesn't exist.

Divisions of Physics

When discussing physics it is helpful to distinguish between the different branches of physics. Two broad areas that can be used to categorize physics are **theoretical physics** and **experimental physics.** Theoretical physicists use the laws of physics to refine theories, suggest experiments, and predict the results of experiments. Albert Einstein can be thought of as the quintessential theoretical physicist. Einstein's laboratory was his mind. Einstein didn't conduct experiments in the classical sense of using instruments to collect data, but used the body of physics knowledge available at the beginning of the twentieth century to introduce new ideas to explain both old and new observations. As the name implies, experimental physicists design and conduct experiments to test hypotheses and verify theories. Theoretical and experimental physics should not be viewed as distinct areas, but as different strategies

for understanding the universe. The interaction between theoretical and experimental physics is critical to the advancement of the science as a whole. Theoretical physicists use the results of experiments to refine their ideas and at the same time suggest new experiments. Experimental physicists, likewise, use refined theories to modify techniques and develop new experiments.

Applied physics is concerned with using physics to meet societal needs. The twenty-first century's modern life is largely a product of applied physics. Modern society uses applied physics to produce a plethora of simple products and integrated engineered systems. A toaster applies the concept of electrical resistance. Cars are the result of applying thermodynamics (internal combustion engine), aerodynamics (design), material science (body, tires, windshield), electronics (ignition and diagnostics), and mechanics (brakes, transmission). Air transportation exists in its present form due to the application of fluid mechanics, in the form of the Bernoulli principle, to achieve lift. The modern information age is intimately connected to applied physics. This is exemplified by prestigious physics labs operated by large companies such as Bell and IBM. While applied physics is concerned with practical uses of the principles of physics, a great deal of basic research and development occurs in the process of applying physics principles to societal needs.

Classical physics refers to the body of knowledge that began to define physics as a discipline starting in the Renaissance. Galileo Galilei (1564–1642) and Isaac Newton (1643–1727) were the two key figures who established the foundation of classical physics. Classical physics is often defined as physics developed up to the year 1900. Classical physics deals primarily with familiar phenomena and macroscopic objects. Classical physics is perfectly adequate for describing everyday observations. If we want to know how fast to drive to make a green light or how long it takes for a dropped object to strike the floor, equations developed hundreds of years ago can be used to make accurate calculations. The laws of classical physics were questioned starting in the late nineteenth century, as paradoxical results were obtained from experiments exploring light and atomic structure. These paradoxes led to the development of **modern physics.** Modern physics allows physicists to address questions of nature outside the realm of everyday experience. **Relativistic** and **quantum** principles from modern physics are needed to explain observations on objects approaching the speed of light or the behavior of matter in an extreme gravitational field, such as a black hole.

Classical and modern physics should not be viewed as distinct branches of physics but as different levels for understanding natural phenomena. In fact, classical physics can be considered to be modern physics applied to conditions applicable to everyday life. Since we exist in the macroscopic world of the Earth's gravity where objects move at a snail's pace compared to the speed of light, the laws of classical physics suffice quite well. Results using the equations and laws of classical physics will, for all practical purposes, be the same as using the laws of modern physics. Much more will be said about this in the chapter on relativity. Although classical physics adequately describes common experiences, modern physics provides writers and directors of futuristic material ample plots. Episodes of *Star Trek* and its numerous spin-offs are prime examples of the modern physics genre. Theories incorporating the equivalency of matter and energy, the speed of light as the ultimate speed limit in the universe, and quantum processes demonstrate how physics has the ability to stretch the imagination and provide a springboard for speculation about the future. Much of what is presented in science fiction is beyond twenty-first century physics, but modern physics provides a keyhole through which to peer into the future. For example, **holography** lets us imagine aliens being projected on the deck of a starship in the year 2005.

Classical physics can be divided into several main branches. These include **mechanics, thermodynamics, electromagnetism, sound,** and **optics.** Mechanics is the study of motion (or lack of motion). The pure description of motion is termed **kinematics. Dynamics** is the study of changes in motion and the forces associated with this change. Mechanics has enabled scientists and engineers to put men on the Moon, design intricate mechanical devices, and populate all areas of the Earth. The tem "mechanics" also appears with other terms to further delineate mechanics. Fluid mechanics deals with the motion of liquids and gases. **Quantum mechanics** is an area of modern physics that enables physicists to describe the structure and behavior of atoms.

Thermodynamics is the study of various forms of energy and how energy is transformed. It involves the study of heat. The **first law of thermodynamics** states that energy cannot be created or destroyed, while the **second law** dictates the natural flow of energy in a system. These basic laws can be used to calculate the amount of work that can be obtained from a given amount of energy and to determine the efficiency of a process. By applying the laws of thermodynamics, humans continually convert both nonrenewable (fossil fuels) and renewable (hydroelectric, solar, biomass, etc.) resources into energy to fuel modern society. The industrial revolution in the nineteenth century owes its start to the ability to apply thermodynamics to develop industrial machines.

Electromagnetism is the study of the related phenomena of electricity and magnetism. The movement of electric charge produces a magnetic field, and the movement of a conductor, such as wire, through a magnetic field produces an electric current. The study of electromagnetism deals with the propagation of electromagnetic waves through space. Society changed drastically as humans built vast networks to deliver electricity throughout developed countries at the beginning of the twentieth century. The modern information age with its plethora of electronic devices and global communication networks is based on harnessing electromagnetism for society's use.

Sound is the vibration of air and the propagation of sound waves through space. The study of sound, how it's created, and its interaction with the environment is termed **acoustics**. Human communication—speech and hearing—is based on sound. Sound waves, in the form of sonar, can be used to probe beneath the sea. Acoustics is used to design musical instruments and the grand concert halls that house the musicians that play in them.

Optics is the study of light and, in this sense, may be considered a branch of electromagnetism. **Geometrical optics** deals with the tracing of light rays, how they travel, and how objects affect them. The use of corrective lenses (glasses or contacts) is probably the most common example of geometrical optics. The behavior of light as an electromagnetic wave is referred to as **physical optics.**

Physics combined with other disciplines produces unique areas of study. Geophysics is physics of the Earth. Geophysicists use physics to study both the external and internal structure of the Earth. Biophysics applies the principles of physics to the study of life processes, for example, using thermodynamics to study metabolism. Astro-physics applies physics to the study of the universe. The evolution of stars, the propagation of electromagnetic waves through interstellar space, and the structure of the universe are just a few areas that astrophysics addresses. Many distinct areas of science are highly dependent on physics. For example, meteorologists use physics to study the motion of the atmosphere and to understand weather patterns. Oceanographers apply physics to study wave motion and water circulation throughout the oceans. Engineers constantly apply physics to carry out tasks such as building bridges, designing cars, and refining oil. Numerous other disciplines utilize physics. In fact, it's difficult to think of any science or technical area that does not depend on physics to some degree.

The Philosophical Foundation of Physics

The development of modern science can be traced backed to ancient Greece and the desire to explain natural phenomena using rational logic. Before the fifth century B.C.E., Greek mythology served to explain phenomena such as seasons, floods, droughts, plagues, earthquakes, lightning, meteors, eclipses, and tides. Gods and goddesses in Greek mythology were the personification of natural phenomena. One area of Greek mythology that survives today is the observation of the night sky and the identification of familiar constellations. Constellations such as Ursa Major (the Great Bear), Orion, Gemini, Leo, and Pegasus have individual stories, often several, associated with how they were formed. For example, Ursa Major, the Great Bear, lives in the northern skies. According to Geek myth, Callisto was the daughter of King Lyacon. Callisto was chosen to be a maiden of Artemis. Artemis was the sister of Apollo and the protector of babies. Artemis demanded fidelity and chastity of her maidens, but Callisto became pregnant by Zeus. Artemis, who was an avid hunter, exacted her punishment on Callisto by turning her into a bear that she would hunt down. Zeus took pity on Callisto and gave her permanent refuge in the heavens. Later her son, Arcas, who formed the constellation Ursa Minor, joined her.

Around the sixth century B.C.E., Greek philosophers proposed alternative explanations for natural phenomena based on their observations, reason, and logic. The roots of physics, as well as other scientific disciplines, can be traced back to this period of the pre-Socratic Greek philosophers. Miletus, located on Turkey's western shore of the Aegean Sea, was the center of early Greek philosophy. Miletus was ideally situated as a trading port where Babylonian and Egyptian learning could be integrated with Greek thought. It was in Miletus that pre-Socratic philosophers started to integrate rational logic into traditional Greek mythology to explain their world. Gods and supernatural causes were removed from natural explanations. A major area the pre-Socratic philosophers focused upon was the relationship between change and life. Life was interpreted in a holistic sense by the early Geeks. In contrast to the modern view of the world consisting of living and nonliving entities, objects such as stones, water, and fire were seen as biotic components of the world. For example, the Greeks believed the Earth gave birth to rocks deposited on the Earth's surface through volcanic eruptions. Thus, a volcano was similar to a woman in labor.

Coupled with the idea of life and change, was the belief of a primordial substance that formed all matter. Thales of Miletus (624–546 B.C.E.) believed water was this primordial substance. His thoughts on water as the essence of life were probably influenced by Egyptian and Babylonian contacts he met in his travels throughout the region. The idea of water as the basic element was appealing to the early Miletian philosophers. After all, water could exist in three forms—solid, liquid, and gas—that readily changed from one to another. The deposition of sediments out of water in deltas gave evidence that Earth itself sprang forth from water. Although none of his writings survive, most of our knowledge on Thales comes from Plato, who considered Thales to be the first true natural philosopher.

Other philosophers dismissed water as the primordial substance. Anaximander (610–546 B.C.E.), a pupil of Thales, introduced a universal aperion as the source of matter. Anaximander's aperion was a universal infinite abstraction that continually mixed to form substances and dissolved back into aperion. In this sense, Anaximander's reality might be compared to images that fade in and out on a screen. Anaximenes (550–474 B.C.E.) believed air was the basic substance of matter. Life depended on a steady supply of air, as did fire. Anaximenes considered fire to be rarefied air, and condensed air to give rise to water and earth. In his philosophy, Anaximenes viewed changes in air as producing other forms of matter. Similar to Thales and other philosophers, Anaximenes found evidence of his philosophy in the world around him. For example, air would rise and form clouds that would eventually produce water falling to the ground as rain.

Heraclitus (550–480 B.C.E.) lived in Ephesus, which was located just north of Miletus. Heraclitus observed that the world was constantly changing and attributed this change to fire. Fire itself was constantly in a state of flux, and Heraclitus saw this as representative of life itself. Heraclitus' philosophy also considered reality in terms of opposites. The way up and the way down were one and the same, depending only on one's perspective.

A contemporary of Heraclitus and the founder of another school of philosophy was Pythagoras (582–500 B.C.E.). The Pythagoreans established themselves in southern Italy, and their natural philosophy was based on numbers. For the Pythagoreans, numbers governed the harmony of the universe and had symbolic significance that revealed the secrets of nature. Music had a major influence on Pythagoras' philosophy. He discovered that harmonics on a stringed instrument could be produced with a specific whole-number ratio of string lengths. According to Pythagoras, the harmony of music also applied to nature. He believed the position and motion of the Earth and heavens mimicked a musical scale. The Earth, planets, and Sun were considered to be perfect spheres that moved in perfect circles around the central substance of fire.

The Eleatic school of Greek philosophy was yet another group and was centered in the southern Italian city-state of Elea, which flourished at the same time as the Pythagoreans. The Eleatics directly opposed the ideas of Heraclitus and believed the universe did not change. They felt knowledge acquired through sensory perception was faulty because reality was distorted by our senses. The founder of the Eleatics was Parmenides (515–450 B.C.E.), who viewed the world as a unity and believed change was an illusion. In order to make their ideas about the absence of change conform to their observations, the Eleatics expanded upon the idea of one primary substance. Rather than positing water, air, or fire as the primary substance, the Eleatic philosopher Empedocles of Acragas (492–432 B.C.E.) proposed four primary elements: air, earth, fire, and water.

The atomist school developed in response to the Eleatics. The atomists were led by Leucippus (circa 450–370 B.C.E.) and his student Democritus (460–370 B.C.E.). The atomists refuted the idea that change was an illusion and developed a philosophy that supported the sensory observations of the physical world. The atomists proposed that the world consisted of an infinite void (vacuum) filled with atoms. According to Diogenes Laertius, a third-century biographer of great philosophers, Democritus's view of nature was summarized in the belief that nothing exists except atoms and empty space, everything else is opinion. Atoms were eternal, indivisible, invisible, minute solids that comprised the physical world. Atoms were homogeneous but existed in different sizes and shapes. The idea of atoms developed on logical grounds. It was reasoned that if a piece of gold could be continually divided into smaller and smaller units, that the gold would eventually reach a point where it was indivisible. The ultimate unit of gold, or any other substance, would be an atom. According to Democritus, an infinite number of atoms moved through the void and interacted with each other. The observable physical world was made of aggregates of atoms that had collided and coalesced. The properties of matter depended on the size and shape of the atoms. Change was explained in terms of the motion of the atoms as they moved through the void or arranged themselves.

Plato

Plato (427–347 B.C.E.), pupil of Socrates, integrated the work of his master and other predecessors into a comprehensive worldview. As was the case with Socrates, who was content to address political and moral issues, natural philosophy did not hold a central place in Plato's philosophy. Plato held a Pythagorean view of the universe, believing that mathematics was the key to understanding nature. He reasoned that the four basic elements consisted of geometrical shapes. Fire consisted of tetrahedral particles, air particles were octahedral, water particles were icosahedra, and earth particles consisted of cubes. The heavens consisted of a fifth element, quintessence, which came from dodecahedra. Plato accepted the view that the Earth and planets were perfect spheres and moved in uniform circular motion. Because mathematics held a central place in Plato's philosophy, arithmetic, geometry, and astronomy were studied as branches of mathematics. In this regard, astronomy served to develop mathematical constructs, rather than to explain observation of the heavens. Astronomical observations made by Plato and his students served to reveal the basic truths contained in the mathematics. Plato's simple model of the universe described in his *Timeaus* consisted of two crystalline spheres. An outer sphere that contained the fixed stars that rotated around the earth once every 23 hours 56 minutes, and a solar sphere that rotated once every 24 hours and contained the Sun. Beyond the crystalline sphere containing the stars was an infinite void. Between the Earth and the stars, the planets—the term itself means wanderers—moved.

Eudoxus (408–355 B.C.E.) from Cnidus, in present-day Turkey, was a well-traveled contemporary of Plato and a philosopher who studied mathematics and astronomy in Egypt, Athens, Italy, and Asia Minor. He was greatly influenced by the Pythagorean philosophers and was exposed to the teaching of Plato during his trips to Athens. Eudoxus is best remembered for his development of an astronomical system based on a series of homocentric spheres. Like the Pythagoreans and Plato, Eudoxus believed the planets and stars rotated around the Earth in circular motion. He observed that the planets did not follow a uniform circular motion. A planet's speed was not constant through the heavens, and some planets even stopped and reversed direction, a phenom-

enon referred to as retrograde motion. To account for these observations and preserve perfect circular motion, Eudoxus described a planetary body's motion as the result of movement on several spheres. The axis of each sphere revolved around the Earth in perfect spherical motion. In this manner, Eudoxus could preserve the idea of perfect circular motion and at the same time account for the apparent movement of heavenly bodies.

Eudoxus' universe was Earth centered. The fixed stars revolved around the Earth in the most distant sphere. Each of the five planets' (Mercury, Venus, Mars, Jupiter, and Saturn) motion could be described using four spheres. The Sun and the Moon required three spheres each to describe their motions. Therefore, a total of 27 spheres described the motion of the planets in terms of circular motion. It should be emphasized that Eudoxus, as a mathematician, developed a geometrical description of the universe. There is no evidence that Eudoxus presented his homocentric spherical model as physical reality. The methods Eudoxus employed represent those of modern science. Eudoxus' observations did not fit the Pythagoreans' and Plato's model. Rather than discard earlier models, Eudoxus refined them within the constraints of his philosophical beliefs, for example, uniform circular motion.

Aristotle

Aristotle (384–322 B.C.E.) had the greatest influence on scientific thought of all ancient philosophers. Aristotle's writings and teaching established a unified set of principles that could be used to explain natural phenomenon. Although today we would consider his science primitive, it was based on sound logical arguments that incorporated much of the philosophical thought of his time. The teleological nature of Aristotle's explanations led to the subsequent acceptance of his ideas by the Catholic Church in Western Europe. Aristotle's philosophy guided European ideas and thought for 2,000 years.

Aristotle's *Meterologica* synthesized his ideas on matter. Aristotle believed that four qualities could be used to explain natural processes: hot, cold, dry, and moist. Hot and cold were active qualities. Hence, the addition or subtraction of heat leads to the transformation of things. Moist and dry were the result of the action of the active qualities. Only four possible combinations of these qualities could exist in substances: hot and moist, hot and dry, cold and moist, and cold and dry. Opposite qualities could not coexist. The combinations of hot and cold or moist and dry were impossible in Aristotle's system. The four allowable combinations determined the four basic elements. Air possessed the qualities hot and moist, fire possessed hot and dry, earth possessed dry and cold, and water possessed wet and cold. Aristotle's system is summarized in Figure 1.1 and is found in a number of ancient writings. Using the four qualities of matter and four elements as a starting point, Aristotle developed logical explanations for numerous

natural observations. Both the properties of matter and changes in matter could be explained using Aristotle's theory.

Aristotle explained the process of boiling as a combination of moisture and heat. If heat is added to a substance that contains moisture, the heat will draw the moisture out. The process results in the substance dividing into two parts. When the moisture leaves, the substance becomes thicker, while the separated moisture is lighter and rises. Aristotle used this type of reasoning to explain numerous physical and chemical processes including evaporation, decay, condensation, melting, drying, and putrefaction. While Aristotle's philosophy explained much, it did have its shortcomings. A simple example will help to illustrate how problems arise with any scientific theory. According to Aristotle, hard substances were hard because they had an abundance of earth and possessed the qualities of dry and cold. Aristotle claimed earth was heavy and so it moved down. Soft substances would not contain as much earth, but contained more water and air. These substances would not be attracted downward as strongly. Now consider ice and liquid water. Ice is dry and cold compared to liquid water, which is moist and cold. We would expect ice to have more of the element earth, and have a stronger downward attraction than liquid water. Yet, hard solid ice floats in liquid water. Even though many other problems existed with Aristotle's theory, it provided reasonable explanations for many observations.

Aristotle's system could also be used to explain the motion of objects. He explained his views on motion in his work *Physics*. Natural motion was the result of an object seeking its "natural" place in the universe. Solid objects, such as rocks, fell because they contained an abundance of Earth and desired to return to the Earth. Conversely, smoke rose because it had an abundance of

Figure 1.1
Aristotle's diagram on matter and change

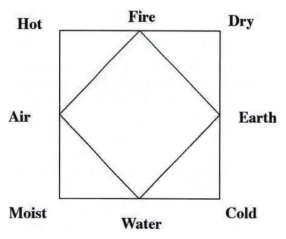

air. Violent motion was motion that opposed natural motion. For example, a rock can be lifted or thrown upward. According to Aristotle, violent motion required an agent in contact with the object in order to occur. This was readily observed when a cart was pushed uphill or an object lifted, but presented problems in explaining many other types of motion. For instance, when a rock is thrown upward, it leaves the thrower's hand and loses contact with the agent causing the motion. Aristotle invented various explanations to account for discrepancies in his ideas on motion. In the case of a projectile, he reasoned that air in front of the projectile, would be pushed behind it and propel it along. Aristotle also classified motion as linear, circular, or mixed. Terrestrial motion was linear and motion in the heavens was circular. In the atmosphere, between heaven and Earth, the two types could combine to describe the motion of objects.

With respect to the universe, Aristotle adopted and expanded upon Eudoxus' system. In contrast to Eudoxus and previous philosophers, Aristotle viewed the spherical motion as corresponding to physical reality and not merely a geometrical representation. Accordingly, Aristotle proposed Eudoxian spheres as three-dimensional shells. In his system, to counteract the affect the outer shells would have on the inner shells, Aristotle proposed additional spheres rotating in the opposite direction. These counter-rotating spheres negated the influence of one planet's set of shells on another. Aristotle's mechanical system increased the number of spheres needed to describe observed planetary motion to 50.

Ptolemy

Aristotle's natural philosophy was part of an integrated, comprehensive philosophy that provided a worldview that influenced thinking for the next 2,000 years. Other natural philosophers conformed to Aristotelian physics and proposed models for the universe that integrated Aristotelian principles with astronomical observations. The most successful of these models was that of Ptolemy (100–170 C.E.). Claudius Ptolemy (Figure 1.2) was a Greek astronomer and geographer who lived in Alexandria, Egypt. He made extensive observations over his life. Ptolemy integrated the work of his predecessors with that of his own to construct a geometric model that accurately predicted the motion of the Sun, Moon, and planets. One of his primary sources was Hipparchus (190–120 B.C.E.). Hipparchus had made discoveries such as the precision of the equinoxes and the different motions of the Moon.

The Ptolemaic system of the universe relied on clever geometric constructions to account for the movement of the Sun, Moon, and planets. Ptolemy's planetary motion consisted of a system of epicycles, deferents, and equants as displayed in Figure 1.3. Each planet moved in a circular orbit called an epicycle. The center of the epicycle itself revolved around a deferent. The deferent was positioned between the Earth on one side and an equant on the other. The epicycle moved at a constant speed with respect to the equant. By introducing more epicycles and circles, Ptolemy was able to construct a good approximation for the movement of the planets. His planetary system contained approximately 80 different circular motions. Ptolemy's Earth-centered universe was surrounded, in order, by the Moon, Mercury, Venus, the Sun, Mars, Jupiter, and Saturn.

While Ptolemy is best remembered for building a model of the universe that lasted 1,400 years, he also made other contributions to astronomy, physics, and geography. His principal work in science was entitled the *Almagest,* a 13-volume compilation of his work in astronomy and mathematics,

Figure 1.2
Ptolemy

and his observations. The *Almagest* built upon Aristotle's circular motions and presented the mathematics Ptolemy used to derive his system. His work addresses a wide range of astronomical topics including the seasons, eclipses, and the precession of the equinoxes. It also includes a star catalog with 48 constellations and hundreds of stars. Ptolemy also wrote a five-volume work entitled *Optics* in which he presented findings on color, reflection, mirrors, refraction, and vision.

Physics in the Dark and Middle Ages

The rise of the Roman Empire sharply curtailed advances in natural philosophy. The Romans did not emphasized science, and the disintegration of the Roman Empire

Figure 1.3
Ptolemy's system had a planet, P, moving in epicycles. The larger epicycle moved along a deferent, while its center moved at a constant velocity with respect to the equant. The equant was located on the opposite side of the center of the deferent from the Earth, E. The deferent and epicycles were composed of perfect circles. In this manner, Ptolemy was able to account for the complicated motion of the planets.

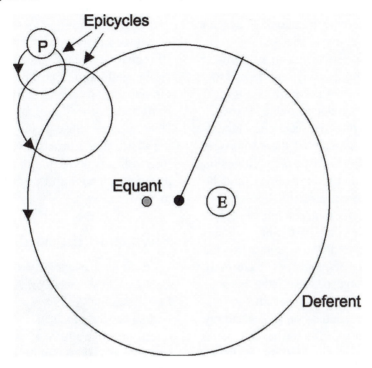

during the fifth century C.E commenced the period known as the Dark Ages in Europe. Ancient knowledge preserved in learning centers such as Alexandria and Athens and ignored for several centuries was revitalized with the rise of the Arab Empire in the middle of the first millennium. At its peak, the Arab Empire spanned from its eastern borders with China and India, across northern Africa into southern Spain. The Arabs translated and dispersed the works of the Greeks, and for a 500-year period from 600 to 1100 C.E. the Arabs made discoveries and advances in science, medicine, astronomy, and mathematics. The period

between 700 and 1100 C.E. was the peak of the Arab Empire. Initially, as Arabians conquered new areas they assimilated the ideas and knowledge of the conquered region. The vast geographic expanse of the Arab influence meant that they were exposed to a vast array of knowledge, and there was ample opportunity for the cross-fertilization of the work from many cultures. By 1000 C.E., nearly all of the ancient works had been translated into Arabic.

In addition to translating ancient works, Arab scholars advanced and refined this material. Baghdad (Iraq) and Damascus (Syria) developed as leading centers of

Arab scholarship. Several individuals made significant contributions to physics during the height of the Arab Empire. Thabit ibn Quarra (826–901) did work in the area of statics, using levers, and developed concepts related to the center of mass. Al-Battani (850–929) cataloged approximately 500 stars and critiqued the astronomical system proposed by Ptolemy in the *Almagest*. He showed that the Sun varied in its furthest distance from the Earth; in modern terms this would mean that the Earth's elliptical orbit varied. In conjunction with this work Al-Battani also provided accurate calculations for the precession of the equinoxes. Alhazen (965–1038) wrote 92 volumes on astronomy, geometry, and optics, and he is especially remembered for his work in the latter area. His experiments in optics led him to reject many of Ptolemy's ideas on light. He refuted Ptolemy's idea that the eyes were sources of light rays and correctly determined that objects were visible because they reflected light rays. Through his experiments, he developed ideas on lenses and ray tracing. Alhazen also found that the speed of light varies as it passes through different media. Using the rarefaction of light, Alhazen estimated that the atmosphere was 15 kilometers thick. Al-Biruni (973–1048) also conducted studies on light, as well as on hydrostatics and density. Avicenna (980–1037), a Persian physician best known for his studies in medicine, contributed to physical science by classifying machines as levers, wheels, screws, and so forth. Much of Avicenna's teachings questioned the status quo and teachings of Aristotle.

By Avicenna's time, around 1000 C.E., the Arab Empire was in decline from both internal and external forces. Factions of the Islamic faith battled one another. A general intolerance of science pervaded Arab culture, and scientists were not free to publish their ideas. Christian Crusaders from the West and Mongol invaders from the East exerted pressure on the Arabic world. As Europeans recaptured Arab regions, the classical knowledge that had been preserved and advanced by the Arabs influenced European thinking. Major Arab learning centers, such as Toledo, in Spain, and Sicily contained extensive libraries as sources to rekindle European science. Bilingual Christians and Muslims fluent in Arabic and Latin reintroduced the works of the ancient Greeks to European scholars. Early translations into Latin and some into Hebrew were subsequently translated into the vernacular European languages. From the twelfth century, major advances in the physical sciences shifted from Arab lands to Western Europe.

Physics in the Middle Ages

Scholars, especially clergy from the Catholic Church, were primarily responsible for translating Arabic, Syrian, and Persian writings into Latin. The translators not only reintroduced traditional Aristotelian philosophy, but also explained technological advances made during the period in which the Arabs flourished. Translated works brought to light advances in astronomy, mathematics, medicine, and natural philosophy. From these translations, it was apparent that the Arabs and Persians had advanced scientific knowledge significantly, especially during the period from 700 to 1100 C.E. Contributions from the Arab and Persian regions focused on the practical applications of using the method of experimentation, but much of the knowledge was based on Arabic translations of Greek philosophy.

Several Europeans were noteworthy in setting the stage for the development of modern physics. Albertus Magnus (1200–1280), also known as Albert the Great, played an important role in introducing Greek and

Arabic science and philosophy in the Middle Ages. Albert produced numerous commentaries on the works of Aristotle. These commentaries helped establish the value of studying natural philosophy along with the philosophical logic of Aristotle. Albert was canonized in 1931 and is considered the patron saint of all who study science. Albert had a major influence on his most famous student, Thomas Aquinas (1225–1274).

Major problems arose in trying to reconcile the Aristotelian worldview with the teachings of the Catholic Church, the dominant institution in Europe. Aristotle believed in an immutable universe of bodies in constant motion, but Church teaching claimed such views limited the power of God to change the universe. Aristotelian thought claimed the existence of natural laws that guided the behavior of objects apart from God, while Aristotle's view that matter was eternal contradicted the teaching of Creation. Thomas Aquinas reconciled Aristotelian logic and Christian teaching in his writings, especially *Summa Theologica*. It was in his *Summa* that Aquinas presented his philosophy that knowledge gained through reason and logic and through divine inspiration were compatible. In fact, Aquinas used observation and Aristotelian logic as proof of the existence of God. Since motion required a source in contact with a moving object, Aquinas reasoned that God was the Prime Mover of the stellar sphere, aided by His angelic agents who moved the planets. By establishing a coherent, consistent system that fit both Church dogma and Aristotelian philosophy, Aquinas' work established boundaries that guided orthodox thinking for the next 400 years.

Roger Bacon (1214–1294) was an English contemporary of Albertus Magnus. Bacon was a Franciscan clergyman who taught at Oxford and conducted studies in alchemy, physics, and astronomy. He conducted major studies on gunpowder. His major work in physics was in the area of optics. Bacon did pioneering work on the magnifying ability of lenses that predated the development of spectacles in Italy at the end of the thirteenth century. Bacon's major contribution to scientific thought was his development of modern scientific methods to guide the discovery of knowledge. Bacon purchased, modified, and built instruments and used assistants in conducting his experiments. In his *Opus Majus* (Major Work), *Opus Minus* (Minor Work) and *Opus Tertium* Bacon argued that the study of the natural world should be based on observation, measurement, and experimentation. Bacon proposed that the university curriculum should be revised to include mathematics, language, alchemy, and experimental science. Because of his teachings, Bacon often had difficulties with his superiors and spent nearly 15 years of his life in confinement. Although today we accept many of Bacon's ideas as the foundation of modern science, his methods were revolutionary in the thirteenth century.

Another primary figure of the Middle Ages was William of Ockham (1284–1347). William of Ockham, like Bacon, believed knowledge could be gathered through the senses, apart from scriptural teaching, and was persecuted by the Church for these beliefs. In physics, William of Ockham resurrected the impetus theory first proposed by John Philoponus (490–570) in the sixth century. According to Philoponus, the motions of heavenly bodies were due to an initial force, or impetus, supplied by God rather than constant pushing by angelic spirits, as proposed by Plato and Aristotle. The initial impetus imparted to an astronomical body caused its motion to continue because the vacuum present in space did not retard this motion. Aristotle believed a vacuum could not exist because if it did he would be at a loss to explain how projectiles moved through the air, as discussed previously. William of Ockham revived the impetus theory, and several other Middle Ages philosophers advanced this theory of motion, including Jean Buridan (1300–1358), Nicholas Oresme (1323–1382), and Nicholas of Cusa (1401–1464). Oresme rejected the idea of a stationary Earth. He proposed that the Earth rotated daily on its axis, and this motion was imparted by an initial impetus. He also did studies on light and introduced the idea of graphing two variables on a coordinate system. Nicholas of Cusa further advanced the idea of a mobile Earth, claiming that it rotated and moved around the Sun. He believed the heavens were made of the same materials present on Earth and that stars were other suns with their own planetary systems. The impetus theory never received universal support during the Middle Ages, and the ideas of Aristotle continued to dominate Western thought. William of Ockham's lasting contribution to science was his philosophical method. A general principle he developed is known as Ockham's Razor. This general principle states that when more than one theory is proposed to explain something, and both give similar results, the simplest theory is preferred. Physics and the other sciences continually apply Ockham's Razor. Scientists desire frugality; they attempt to explain as many facts as possible in the simplest terms.

The late Middle Ages, known as the Renaissance, marked the rebirth of classical learning in Europe. Initially centered in Italy, learning migrated north to Germany, England, France, Spain, and the Netherlands. In these areas, ideas were transformed into distinct schools of thought that laid the groundwork for the Protestant reformation.

The teachings of the Catholic Church were firmly rooted in Aristotelian logic as interpreted by Aquinas. Challenges to Catholic teachings invariably challenged the natural philosophy embodied within these teachings. While ancient knowledge was reintroduced during the Renaissance, critical analysis of these teachings raised questions about their validity. More importantly, the methods used to acquire knowledge about the natural world changed. Renaissance thinkers interpreting nature using a modern scientific approach through observation and experimentation challenged the scriptural interpretations proclaimed by clerical scholars. Against this backdrop, the stage was set for a complete revision of human thinking. This revision commenced with the work of Copernicus but took the next 300 years to come to fruition through the work of Newton. This will be the subject of the next chapter.

Summary

Physics can be considered the most basic science. While humans have constructed arbitrary divisions of knowledge into areas such as biology, chemistry, and geology, ultimately everything reduces to physics. Physics itself can be divided into areas, for convenience, that define various aspects of this field. This chapter has provided an overview of the main areas of physics. Additionally, some of the historical roots of physics have been presented. In the coming chapters, the main areas of physics will be examined in detail. The next chapter will continue to examine the development of classical physics, and starting with chapter 3, the basic concepts will be presented.

2

The Development of Classical Physics

Introduction

Physics for almost 2,000 years of human history had been dominated by the work of Aristotle. Aristotle's treatises such as *Physics* and *On the Heavens* provided a philosophical foundation for the properties of matter and its behavior. The motions of objects in the universe were explained to a high degree of precision by Ptolemy. For 2,000 years, the Aristotelian-Ptolemaic system dominated the human view of the universe. At the center of this universe an immovable fixed Earth presided over the heavens. By the beginning of the sixteenth century, the stage had been set for the birth of physics as a distinct discipline. Until this time, physics played a secondary role in natural philosophy and had made only modest advances from the days of Aristotle. This started to change as individuals adopted a modern scientific approach and subjected ancient ideas to experimental testing. Individuals within and outside the Catholic Church challenged its unquestioned authority concerning nature. While conflict between the Church and science intensified at the beginning of the seventeenth century,

the seeds of dissension had been planted 150 years earlier by Nicolaus Copernicus (1473–1543).

The Copernican System

Nicolaus Copernicus was born in Torún, Poland, to German parents. His father was a wealthy merchant but died when Copernicus was 10 years old, and a wealthy maternal uncle, Lucas Waczenrode, assumed Copernicus' upbringing. Waczenrode, who eventually was appointed a bishop, provided Copernicus an excellent education at several of the leading universities in Europe. Copernicus studied mathematics and astronomy at the University of Krakow (Poland), medicine at the University of Padua (Italy), and canon law at the University of Bologna (Italy). Upon receiving his doctorate he returned to Frauenberg, Poland, where he obtained an administrative position in his uncle's diocese. He was never ordained as a priest.

Throughout his life Copernicus studied the ancient philosophical works and mathematics and made astronomical observations. These observations were associated with his practice of medicine and his work with the Church. Medicine was closely associated with astrology, and medicinal practices involved taking into account planetary alignments and astrological signs. Another practical motivation for Copernicus' observations was to fix calendar dates for holy days associated with religious practice. Copernicus was consulted by Church officials in Rome regarding their use of the outdated Julian calendar; the current Gregorian calendar was introduced in 1582. Another reason for accurate astronomical observations dealt with navigation and the use of those observations to fix positions at sea.

Copernicus introduced his heliocentric century in a small, handwritten pamphlet called the *Commentariolus (Small Commentary)* around the year 1514. It should be noted that Aristarchus of Samos (310–230 B.C.E.) had proposed a heliocentric system, but it was never considered a valid model of the universe until after Copernicus reintroduced this concept 2,000 years later. Copernicus' model of the universe put the Sun near the center of the solar system, with the Earth and the other planets revolving about the Sun. He imposed several motions on the Earth: a daily rotation, an annual revolution around the Sun, and the precession of the axis. After introducing his ideas in the *Commentariolus,* Copernicus continued to work on his ideas for the remainder of his life. When he was 66, with the help of Georg Joachim Rheticus (1514–1574), a German scholar from Wittenberg, Copernicus assembled his life's work on astronomy into a treatise that was eventually published in Germany. The work entitled *De Revolutionibus Orbium Coelestium (On the Revolution of the Celestial Orbs).* The printing of the book was supervised under Andreas Osiander (1498–1552). Osiander, a converted Lutheran theologian, took the liberty of inserting a foreword that introduced Copernicus' work as a hypothetical model and stated that it that should

not be interpreted as physical reality. *De Revolutionibus* could thus be used as a model for mathematical calculations concerning the motion of the planets but not be read as a physical description of the universe. Osiander's foreword was unsigned, and it wasn't known until the 1800s that Copernicus did not write it. The publication of *De Revolutionibus* coincided with the death of Copernicus in 1543.

Copernicus' heliocentric system anchored the immobile Sun at the center of the universe and attributed its motion in the sky to the motion of the Earth itself. The Earth, its moon, and the other planets moved around the Sun. Stars were embedded in an immobile sphere that encased the entire universe. Copernicus also reasoned that since the stars held fixed positions in the sky that they must be located at a tremendous distance beyond Saturn, the last planet.

Copernicus' radical view of the universe was an attempt at mathematical simplification. It reduced Ptolemy's 80-some spheres to 34. Although Copernicus repositioned the order of the heavenly bodies, he preserved several characteristics of the ancient system. Foremost was his adherence to circular motion. According to the Copernican system, planets still moved in circular orbits around the Sun and also in epicycles in their orbits (Figure 2.1). While Copernicus' system reduced the number of spheres, it was still a fairly complex system that was no better in making astronomical predictions than Ptolemy's system.

The publication of *De Revolutionibus* produced little controversy during the immediate 75 years following its publication. Copernicus had died the same year it was published, and its preface presented the work as a hypothetical model. There were strong scientific and philosophical arguments that precluded its acceptance. One serious flaw was that parallax of the stars was not observed. If the Earth revolved around

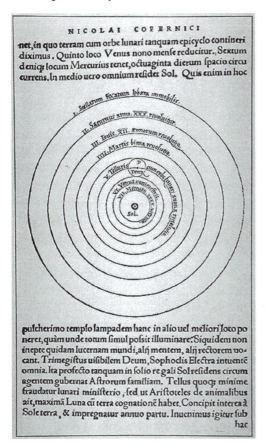

Figure 2.1
The Copernican system

land behind the spot. Additionally, an Earth rotating through the atmosphere would produce a tremendous wind. In order to resolve these problems, Copernicans proposed that the atmosphere was connected to the Earth and rotated along with it. The philosophical problems with the Copernican system were even more problematic. If the Earth was just another planet, then its place in the universe was not special, and other planets were made of the same earthly material. Linear motion could no longer be reserved as a special property of the Earth, and circular motion could not be solely associated with heavenly bodies.

The Copernican model was used freely by astronomers to predict astronomical events and fix dates. Reinhold's Prussian astronomical tables were published in 1551 and based on the Copernican system. These tables provided an alternative to the Alphonsine Tables published in the thirteenth century based on the Ptolemaic system. While many scholars could use a heliocentric framework as a mathematical tool without regard to whether it presented a realistic picture of the universe, there were others that were convinced that it represented physical reality. Better observations were required to resolve the utility of the Ptolemaic and Copernican systems, and these took place in the latter part of the sixteenth century.

Brahe and Kepler

The foremost observational astronomer in the years following *De Revolutionibus* was Tycho Brahe (1546–1601). Brahe was born into Danish nobility. For no apparent reason, Tycho was adopted at the age of two by his childless uncle and aunt. As part of the Danish nobility, Tycho was well educated at the local cathedral school and enrolled at the University of Copenhagen when he was 12. While attending the university, Tycho

the Sun, then its position would change by approximately 200 million miles (twice the mean distance between the Earth and the Sun of 93 million miles) over six months. This change in position should then produce a change in position of the stars. Since a change could not be observed, it was argued that the Earth could not be revolving around the Sun. For this reason, proponents of Copernicus had to argue that the stars were much farther away than commonly believed. The Copernican system was also inconsistent with Aristotelian physics. A rotating Earth would cause objects thrown directly upward from a spot on the Earth's surface to developed a keen interest for astronomy and devoted his life to its study. Tycho's reputation as an astronomer grew primarily as a result of his accurate observations made with large and precise instruments he had built. The King of Denmark, Frederick II, built Brahe an observatory on an island between Denmark and Sweden, which became known as Uraniborg. With ample support provided by Frederick II for assistants and equipment, Brahe continually charted the heavens from 1576–1597. Frederick died in 1588, but Brahe continued to receive support from the state. When Frederick's son Christian was old enough to assume the role of king, conflict developed between him and Brahe. Brahe lost control of Uraniborg, and after a brief stay in Copenhagen moved to Prague in 1599. Here the Holy Roman Emperor Rudolph II employed him for the last two years of his life. Brahe's observations were inconclusive with respect to the merits of the Ptolemaic and Copernican systems. The fact that Brahe did not observe a parallax shift in the stars convinced him that the Earth must be stationary. This led him to proposed yet another system that was a Ptolemaic-Copernican hybrid. Brahe's system placed the Earth at the center of the universe, with the Moon and Sun orbiting the Earth, and the planets orbiting the Sun (Figure 2.2).

While in Prague, Brahe employed Johannes Kepler (1571–1630) as his chief assistant. Upon Brahe's death, Kepler assumed the position as royal mathematician and astronomer for King Rudolph. As Brahe's successor, Kepler inherited the voluminous collection of observations that Brahe had made over the years, and from these advanced his own astronomical theories. Kepler was born in Germany, the son of a mercenary soldier and an innkeeper's daughter. His father died when he was five, and Kepler's early years were spent with his mother and grandparents running the family inn. Kepler's

brilliance enabled him to obtain state support to attend the University of Türbingen, where he was groomed for the Lutheran clergy. It was at Türbingen, the center for Lutheran astronomical studies, that Kepler was introduced to mathematics and astronomy, studying under Michael Maestlin (1550–1631). Although the Ptolemaic system was the predominant model used for astronomical studies at universities, students would also be exposed to Copernicus' and Brahe's work. Hence, the Ptolemaic system was still accepted as the primary system, but adherents to Copernicus' and Brahe's systems could also be found at the close of the sixteenth century. Kepler adopted Copernicus' heliocentric model and used it as the basis of his own astronomical studies.

The ancient Pythagorean School (see chapter 1) had a substantial influence on Kepler's astronomical ideas. He believed the planets' position and movement around the Sun followed a precise mathematical arrangement, and throughout his life sought to discover this arrangement. His first major work presenting his ideas was *Mysterium cosmographicum (Mystery of the Cosmos),* published in 1596. In this work Kepler presented a geometrical basis for the Copernican system. According to Kepler the six planets were located on spheres surrounding the Sun, separated by Euclid's five regular solids. The sphere of Saturn, the outermost planet, circumscribed a cube. This cube contained the sphere of Jupiter, and Jupiter's sphere circumscribes a tetrahedron that contained the sphere of Mars. Kepler's planetary system thus consists of six concentric spheres separated by the five perfect solids in the following order: Saturn-cube-Jupiter-terahedron-Mars-dodacahedron-Earth-icosahedron-Venus-octahedron-Mercury (Figure 2.3).

Kepler's book and the recommendation of Maestlin helped Kepler procure the position as Brahe's chief assistant. One of Kepler's primary tasks upon his arrival in

Figure 2.2
Brahe's system

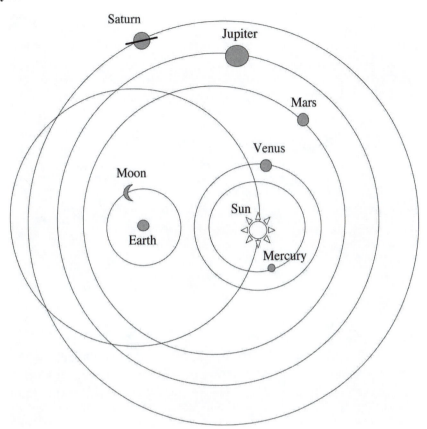

Prague was to assist Brahe in compiling a new set of astronomical tables based on Brahe's voluminous observations collected over the years. This task, which wasn't completed until publication of the Rudolphine Tables in 1627, required Kepler to analyze Brahe's observations and develop a coherent system for making accurate astronomical predictions. The motion of Mars presented the greatest challenge for Kepler and other astronomers in the early seventeenth century. Using various combinations of epicycles, eccentrics, and equants, Kepler sought to fit the motion of Mars using different combinations of circular motion. Kepler worked on

Brahe's Mars data for years. He employed ideas on magnetism developed by William Gilbert (1544–1603) to postulate that the Sun and planets acted as magnets. Kepler believed that the magnetic force between the Sun and planets was responsible for the orbital motion, and therefore, the farther the planet was from the Sun the slower its motion. After laboriously trying to fit the Mars data with circles, Kepler discovered that an elliptical orbit gave much better results (Figure 2.4). By eventually rejecting circular motion and adopting elliptical orbits, Kepler was able to propose a radical view that tipped the balance in favor of the Copernican system.

Figure 2.3
Kepler's model containing five platonic solids

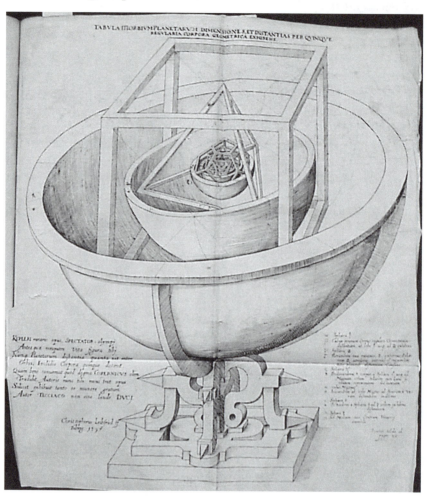

Figure 2.4
Diagram of an ellipse

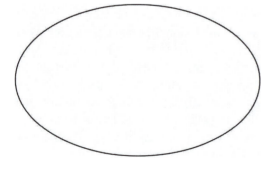

Kepler introduced his first two laws of planetary motion in 1609 in the work *Astronomia nova (New Astronomy)*. Kepler's first law states that the planets move about the Sun in elliptical orbits with the Sun positioned at one focus of the ellipse. Kepler's second law dealt with the speed of the planets as they orbit the Sun. Generalizing from Gilbert's theory of magnetism, Kepler proposed that a planet moved slower the farther it was from the Sun. Formally, Kepler's

Figure 2.5
Kepler's second law. Area 1 equals area 2. The time it takes for the planet to move between points 1 and 2 equals the time it takes to move between points 3 and 4

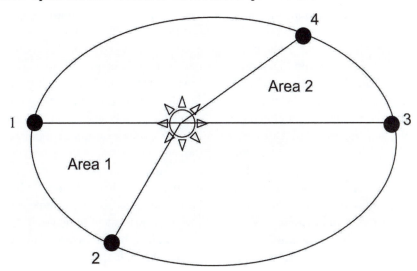

second law states that a line from the planet to the Sun will sweep out equal areas in equal time periods (Figure 2.5).

It took another decade for Kepler to introduce his third law. The decade following his publication of *Astronomia nova* was exceedingly difficult for Kepler. His first wife and seven-year-old son died during this period. Although he had simplified the Copernican model with his first two laws, other leading astronomers didn't readily accept Kepler's work. Contemporaries, such as Galileo, were reluctant to reject the idea of uniform circular motion that had been associated with the heavens for more than 2,000 years. Kepler's views led to his excommunication from the Lutheran Church in 1612. In spite of his difficulties, Kepler continued to search for the mathematical coherence in the heavens. In 1619 Kepler's work *Harmonices mundi (Harmony of the World)* appeared, in which he reviewed his ideas on positioning of the planets introduced in *Mysterium cosmographicum*. A major portion of this work is devoted to presenting a model of the universe in musical terms. For example, the planets were thought to move in a manner that resonated along musical scales. Planets moved through the heavens as musical notes moved along a scale. Each planet produced a tune that literally moved "in concert." Included in Kepler's *Harmony* is his third law. This law states that the square of the period of revolution of a planet divided by the cube of its mean distance from the Sun is a constant for each planet. Mathematically, it can be stated as

$$\frac{T^2}{D^3} = K$$

In this equation T is the period of revolution, D is the mean distance between the planet and the Sun, and K is the constant.

Galileo

The birth of classical physics is often associated with the life Galileo Galilei. Galileo was born near Pisa in 1564. His father, Vincenzo Galilei (1520–1591), was a musician who played the lute and studied and performed experiments on the mathematics of music. Galileo lived in Pisa during his early years and was privately tutored as a child. He moved to Florence at the age of 10 and continued his education in a monastery run by the Camaldolese order. Galileo was committed to joining this order and living a scholarly life in solitude, but his father opposed the idea and wanted Galileo to study medicine. When he was 17, Vencenzo sent Galileo back to Pisa to begin medical studies at the University there. The field of medicine never appealed to Galileo, but he had a keen interest in mathematics. While at Pisa, Galileo studied the classical works of Euclid and other ancient mathematicians and began to apply this knowledge to critically examine Aristotelian physics. Some of his earliest work involved an examination of the motion of a pendulum. Some authorities believe Galileo's interest in the pendulum began when he observed a lamp suspended from the Pisa Cathedral swinging back and forth in 1583, although his first notes on the subject do not appear until 1588. Galileo's study of the pendulum, like many subjects he explored, was examined intermittently over his life. In fact, he proposed his ideas on the use of a pendulum as clock mechanism in the last years of his life.

Galileo drifted away from medical studies, and more of his time was spent with mathematics during his years at Pisa. He left the University in 1585 and returned to Florence, where he obtained work as a teacher and private tutor in mathematics. While teaching mathematics, Galileo did studies on Archimedes' principles of density, specific gravity, and center of mass. Through this work, shared in correspondence with other mathematicians in Italy, Galileo built his own reputation as a mathematician and in 1589 was appointed professor of mathematics at his former school, the University of Pisa. While teaching at Pisa, he produced a work entitled *De motu (On Motion)*. This collection of essays was never published but explained some of Galileo's early ideas on motion and his use of the experimental method to test his theories. In *De motu* Galileo critically examined Aristotle's views and proposed alternative reasoning to explain motion. He clarified what it meant for an object to be heavier or lighter in terms of weight, volume, and density, rather than using individual arbitrary properties such as size or volume. Galileo then went on to expand upon Archimedes' ideas of floating bodies to explain how objects move through different media, for example, how a solid moves through water. Galileo also reinterpreted Aristotle's concept of natural motion by showing how heavy bodies of "earth" may move upward (such as a piece of wood floating to the surface of a liquid) depending on the interaction of a body and the media in which it exists. Galileo rejected motion occurring as the result of a natural ordering of fire, air, earth, and water and assumed all bodies move according to either an internal property, due to its own weight or due to an external force imposed by the media in which it exists. Galileo also rejected Aristotle's idea that motion cannot exist in a vacuum and reasoned that all objects would fall with constant speed through a vacuum (only later did he conclude that objects would fall with uniform acceleration). Galileo argued that falling bodies of different weight dropped from the same height would strike the earth at approximately the same time. While the common notion is that Galileo arrived at this idea by dropping objects from Pisa's leaning tower, he actually came to this conclusion through logical arguments. In fact, his notes on the experiment seem to indicate that the lighter object moved slightly ahead of the heavier object during the initial stage of free fall, but that the heavier object caught up to and eventually passed the lighter object, so that the heavier object struck the ground slightly ahead of the lighter one. Modern researchers have interpreted this discrepancy as caused by an unconscious physiological factor. A person holding two unequal weights in each hand must exert a firmer grip on the heavier object, and consequently, even though the person believes the objects are being released simultaneously, the lighter object is actually released an instant before the heavier object. The fact that the heavier object passes the lighter object in flight can be attributed to a difference in wind resistance. Galileo was aware of this resistance, and this is why he proposed equal rates of fall in a vacuum. *De motu* presents Galileo's seminal ideas on motion. While Aristotelian physics had been under attack for several centuries, it was Galileo's systematic reinterpretation and presentation of both logical and experimental arguments against it that resulted in a new physics.

Galileo's tenure at Pisa lasted only three years, from 1589 to 1592. In 1592 Galileo accepted the professor of mathematics position at the University of Padua, in Venice. Galileo remained there for 18 years, and this seemed to be the most fruitful and satisfying period of his life. At Padua Galileo continued to refine his ideas on motion and mechanics. Galileo's great impact on science goes beyond the scientific principles he introduced. His methods established a model for modern science in which theories are formulated by a continual process of subjecting ideas to experimental testing. Galileo used the pendulum and inclined planed to test his ideas on motion. He realized that in order to slow down the vertical speed of an object, an inclined plane could be employed. In essence, the inclined plane was Galileo's version of time-lapse photography. Galileo's use of the inclined plane required that he interpret motion as consisting of two independent components: a constant horizontal component and an accelerated vertical component. These two components, added together, would describe the motion down the plane. Using this interpretation, Galileo described the path of a projectile as a parabola. An important consequence for this was that Galileo could now explain the motion of objects on a rotating Earth. One of the principal arguments against a rotating Earth involved the Aristotelian argument that an object, such as a cannon ball, launched vertically into the air would land behind the point at which it was launched. Galileo argued that the horizontal motion imparted by a rotating Earth would be imparted to the vertically launched object and carried by it during flight, and therefore, it would land at the same spot from which it was launched. Galileo used the analogy of dropping an object from a tall mast to the deck of a ship to explain his reasoning. An object dropped straight down from a stationary ship would hit the deck directly below where it was dropped. According to Aristotle, when the object was dropped from a moving ship, the ship's forward motion caused the object to strike the deck behind the base of the mast. Galileo's interpretation was that when the object was released it possessed the same horizontal component of motion as the moving ship. Since the horizontal component of motion of the dropped object was identical to the ship's forward motion, it would strike the deck directly below the mast. An observer outside the frame of reference of the moving ship would describe the path as parabolic. According to Galileo's physics, the object hit the deck directly below the mast when the ship was stationary and when it was moving. A person on the ship would observe just the vertical motion of the object and, if the ship was the only frame of reference, could not tell whether the ship was stationary or moving. Galileo reasoned that the atmosphere was carried along with the rotating Earth, and that observed motions would be the same on a stationary and a rotating Earth. Interpreting motion from different frames of reference was a critical aspect as Einstein developed his theory of relativity. This subject is discussed in chapter 15.

Early in 1609, Galileo heard reports of telescopes (the telescope was invented in Holland, not by Galileo) being used in northern Europe. From descriptions in his correspondence, and possibly observing the new device on ships sailing into Venice, Galileo built his first telescope by the fall of 1609. Galileo saw the telescope as a practical tool for use in commerce, military operations, and scientific investigation. As he developed the skill to grind lenses and use the principles of refraction, Galileo was able to build more powerful telescopes. His original scopes had a magnification power of 3 to 4, but soon he was able to produce scopes with a magnification power greater than 20. He used his scopes to advance his reputation and gain favor with the Tuscan nobility, as well as with Church officials. He would periodically hold viewing sessions, where he invited influential Venetians to peer through his telescopes at the heavens and at distant objects. He was successful in persuading the Venetian Senate to grant him a large salary increase in exchange for a license to produce telescopes based on his design. Using his telescopes, Galileo observed that the Moon was not a smooth crystalline sphere but contained mountains and valleys; he observed the phases of Venus, saw sunspots, and discovered four moons revolving around Jupiter. His astronomical findings using the telescope were

published in May 1610 in a work entitled *Sidereus nuncios (Starry Messenger)*. With publication of this work, Galileo solidified his status as one of Europe's leading natural philosophers and was an honored guest not only in Venice, but also in other centers of learning, such as Florence and Rome.

To gain favor with the Grand Duke Cosimo de Medici, Galileo named the moons of Jupiter the Medicean stars. He also presented the Duke with one of his telescopes as a gift. Through his accomplishments and self-promotion, Galileo was offered the position of mathematician and philosopher to the Grand Duke. The position offered Galileo a hefty salary increase and freed him from any teaching duties. This appointment allowed Galileo to devote his full energy to the pursuit of knowledge, unencumbered by university duties.

Galileo's work with the telescope confirmed his belief in the Copernican system. Galileo was a contemporary of and corresponded with Kepler. Although he was not a staunch supporter of Copernicanism early in his career, his work using the telescope provided Galileo with strong support for the heliocentric model. His discovery of the moons of Jupiter and the observation of a supernova demonstrated that the heavens were not fixed and all heavenly bodies did not revolve around the Earth. More importantly, Galileo's study of Venus and it's phases over the course of several months corresponded with the Copernican system and not the Ptolemaic system. As Galileo's evidence mounted in favor of a Copernican interpretation of the universe, so did his boldness and confidence in promoting this worldview. Unfortunately for Galileo, this was in direct opposition to Catholic theology. Galileo was a devoted Catholic with two daughters serving as sisters in a convent. Galileo believed he could reconcile the Copernican system with Holy Scripture if it was not interpreted literally.

Galileo's own observations convinced him that a literal interpretation of the Bible was inappropriate, and he believed he could convince others. In this respect, he was undoubtedly overconfident.

While Galileo had many supporters in the Catholic Church, he also had enemies, who were no doubt envious of his fame and believed he presented a danger to the faithful. These individuals conspired to have Galileo silenced by lobbying the Inquisition to take action against him. As a result of their action and the Church's proclamation against the teaching of Copernicanism, Galileo was issued a warning in 1616 by Cardinal Bellarmine (1542–1621), one of the Church's leading theologians, to refrain from teaching or defending the heliocentric system. It could still be presented as a hypothetical construct but was not to be presented and taught as physical reality. This essentially was a warning to Galileo to "cool it" with respect to his arguments supporting the Copernican theory. The time was not yet ripe to unleash a full defense of the Copernican system.

In the years following the 1616 "gag order," Galileo continued his studies on motion, charted the movement of Jupiter's moon in an attempt to determine longitude at sea, and did a study on comets. His work on comets and his scientific method were presented in *Il saggiatore (The Assayer)*. In *The Assayer*, Galileo stressed the importance of mathematical reasoning rather than philosophical interpretation in studying nature. His work can best be summarized in the following quote:

Philosophy is written in this grand book—the universe—which stands continuously open to our gaze. But the book cannot be understood unless one first learns to comprehend the language and interpret the characters in which it is written. It is written in the language of mathematics, and its characters are triangles, circles, and other geometrical figures, without which it is humanly impossible to understand a single word of it; without these one is wandering about in a dark labyrinth. (As quoted by Machamer in *The Cambridge Companion to Galileo*, pp. 64f.)

Galileo published *Il saggiatore* in 1623 and dedicated the work to the newly elected pope, Mafeo Barberini, who took the name Urban VIII. Barberini was an admirer of Galileo and upon presentation of *Il saggiatore* to Urban VIII, Galileo sought permission to publish a book on the merits of the Copernican and Ptolemaic systems. Since Urban VIII was an old ally of Galileo and almost a decade had elapsed since he had been silenced, Galileo believed he could craft a work that presented the merits of the Copernican system without violating the 1616 order.

Due to his poor health and other problems Galileo book entitled *Dialogo (Dialogue Concerning the Two Chief World Systems)* was not published until 1632. Galileo wrote in a popular style in Italian rather than Latin to appeal to a wider audience. The book consisted of a conversation that took place over four days between three gentlemen: Simplicio, who reasons using Aristotelian thought; Salivati, who speaks for Galileo; and Sagrado, a Venetian nobleman who serves as a neutral interlocutor. *Dialogo* presents Galileo's arguments for the Copernican system based on his life's work in physics and astronomy. On the first day the conversation presents Galileo's argument against Aristotelian philosophy and dispels the notion that the heavens and Earth are composed of different materials. On the second day the conversation turns to Galileo's idea of inertia, in which he uses his ship analogy to argue that motion on a stationary and rotating Earth would be identical. The third day is based on Galileo's findings in astronomy and argues against the Aristotelian idea that the universe is unchanging and permanent. The last part of the book deals with the movement of the planets in circular orbits and Galileo's view of the solar system. In this final portion of the book Galileo proposes the Earth's rotation as the cause of the tides. In fact, Galileo's originally proposed title was *Dialogue on the Tides.*

Although *Dialogo* had been reviewed by Church censors before its publication, Galileo's enemies were furious when the work was published. They successfully appealed to Rome that Galileo violated the 1616 decree forbidding him to teach and defend the Copernican theory. In truth, there was confusion as to what actually had transpired between Galileo and Cardinal Bellarmine in 1616. An unsigned report had been located in the Office of the Inquisition concerning the meeting between Galileo and Bellarmine. This report forbade Galileo from any further discussion of the heliocentric theory. Galileo, unaware that such a report existed, had evidently been blindsided by his detractors, who were now out for blood. Galileo's publisher in Florence was ordered to cease publication of *Dialogo,* although the first thousand copies had already been sold. Galileo, who was now almost 70 years old, was ordered to Rome to answer to charges of heresy. The trial of Galileo has often been viewed in retrospect as science on trial. Pope Urban was led to believe Galileo had duped him when he granted him permission to write a balanced evaluation of the Ptolemaic and Copernican systems. Galileo's supporters could not sway Church officials to be lenient, while his enemies wanted to make an example of Galileo to others who might consider challenging Church authority. Galileo was forced to recant his belief in the heliocentric system and remained under house arrest for the remainder of his life.

Galileo lived the last decade of his life in his Florentine countryside villa, continuing his study of nature. Although feeble and blind, Galileo nevertheless still conducted experiments assisted by assistants and colleagues who held him in high esteem. During these years Galileo wrote his last major work, which summarized his findings in physics over his life. The book, entitled *Discourses and Mathematical Demonstrations Concerning Two New Sciences,* was published in 1638. *Discourses* presented a new physics based on Galileo's accumulated work, starting from his earliest studies in Pisa. The book was sent in parts via friends and mail to Holland, where it was published. Galileo discussed the strength of materials and properties of fluids in the first part of *Discourses.* He then presented his ideas on motion and finished with a discussion of projectile motion. *Discourses* established the science of mechanics and laid the foundation for the study of physics as a science requiring mathematics and experimentation. Galileo provided a new description of motion (the description of motion is known as kinematics) that continues to form the foundation for the study of physics. Important kinematic ideas Galileo derived are that bodies of different weights fall at almost identical speeds, bodies will eventually reach a terminal velocity, projectile motion will be parabolic, a pendulum's period is independent of weight and displacement for small amplitudes and is proportional to the length of the pendulum, and motion can be resolved into a uniform horizontal motion and an accelerated vertical motion. With respect to uniform accelerated motion under the influence of gravity, Galileo demonstrated that the successive distances traveled in equal periods of time increase in a pattern corresponding to the odd numbers: 1, 3, 5, and so forth. For example, if after 1 s if the distance traveled was 1 ft, then in the next second it would travel 4 ft (1 + 3 = 4), and in the third second it would travel 9 ft (1 + 3 + 5). This is equivalent to stating that the vertical distance traveled is proportional to time squared; the formula for acceleration under the influence of gravity is

$$\text{distance} = \frac{1}{2}at^2$$

In this equation a represents acceleration and t time (see chapter 3). Galileo also proposed fundamental ideas on the concept of inertia. He showed that an external agent was not required for motion to take place, but that motion could continue after a force causing the motion was removed. Although Galileo wrongly believed that inertia would cause a body to continue in a circular path around the Earth in the absence of an outside force, he was correct in the basic idea that motion would continue unimpeded unless other forces caused the motion to change.

After publication of *Discourses,* Galileo spent his remaining years continuing his scientific work. His attempts to build a pendulum clock by describing designs to his son Vincenzo were never realized. He died a condemned man at his villa in 1642 and could not even be buried in the family tomb in the church at Santa Croce. His body was eventually transferred to the church in 1737, but Church authorities a century later stilled viewed Galileo as a heretic. It wasn't until 1992 that Pope John Paul II admitted the Church had erred in Galileo's trial, and he ruled the case closed without overturning Galileo's conviction.

Newton

One result of the Vatican's opposition to free scientific inquiry was that leading centers of science moved to northern Europe. Protestant areas, which originally adopted a strict antiheliocentric stance, became receptive to Copernican ideas in the mid seventeenth century. In the post-Galileo years, advances in natural philosophy took place in England, France, and Holland. Natural philosophers in these countries studied the classical texts, but two centuries had elapsed since the publication of *De Revolutionibus,* and a modern scientific approach was taking hold. René Descartes (1596–1650), a contemporary of Galileo, was a Copernican who proposed alternative views to Galileo's mechanics and argued that the laws of physics had a rationalist basis. Francis Bacon (1561–1626) believed an inductive empirical approach was required to unlock the secrets of nature and that this method would lead to a general improvement in civilization.

The new approach to science and the accumulated knowledge of Renaissance philosophers paved the way for Isaac Newton to provide a coherent model to unify physical thought at the start of the eighteenth century. Newton (1642–1727) was born on Christmas day. (Newton's birth in 1642 is dated according to the old Julian calendar, which England was still using at his birth; in terms of the modern Gregorian calendar his birth was on January 4, 1643.) He was born in Lincolnshire, England, into a farming family. Newton's father died before his birth, and he was raised primarily by his grandmother. By all accounts he was a sickly child. His mother planned for him to leave school and operate the family farm, but his uncle noticed his mechanical and academic ability and supported his preparation for a university education. Newton entered Trinity College, Cambridge, in 1661 and obtained his bachelor of arts degree in 1665. He planned to continue his formal studies at Trinity but was forced to return to the Lincolnshire countryside because the college was closed due to the plague. While on this forced sabbatical, Newton worked on many of the seminal ideas in physics and math that distinguished him as a genius. He studied classical mathematics and developed his own mathematical methods including the binomial theorem, infinite series, and method of fluxions. His method of fluxions was the beginning of calculus. The method of fluxions is what is known today as differentiation. The inverse method of fluxions is integration. Newton recognized these as complementary methods. Newton used his method of fluxions to calculate tangents, maxima, minima, and radii of curvature along different points of a curve. Newton returned to Cambridge after it reopened in 1666 and continued his scientific work. He obtained his master of arts in 1668 and was appointed Lucasian Professor of Mathematics at Trinity the following year.

Newton remained at Trinity for 25 years, working independently, but he shared his work through correspondences with other European scientists and was elected to the Royal Society in 1672. Newton studied a wide range of topics while at Trinity. His experiments with light demonstrated that natural light could be separated into distinct colors with a prism and then recombined into its original form. Many of Newton's studies dealt with alchemistry, chemistry, theology, and mysticism. Most of his work remained as unpublished notes and writings. He was protective of his work and regarded rival scientists who challenged his ideas as inferior. He was especially critical of Gottfried Wilhelm Leibniz (1646–1716), who independently derived a method of calculus that was subsequently adopted by other European mathematicians and philosophers.

Newton's most prominent rival was Robert Hooke (1635–1702), who along with Newton and others wrestled with the problem of celestial mechanics and the influence of gravity on motion. Newton had examined these problems while at Lincolnshire when Trinity was closed. He returned to the problem later and by his own account had derived a general theory to explain the motion of the planets and motion on Earth by 1680. Edmund Halley (1656–1742), one of Newton's trusted colleagues and an eminent scientist whose name is associated with the comet he predicted would return every 76 years, approached Newton in 1684 to ask him for help in calculating the motion of the planets under the inverse square law. Newton informed Halley that he had calculated these motions and found them to conform to Kepler's laws. Newton was familiar with the work of Halley, Hooke, and Christiaan Huygens (1629–1695). These were three prominent scientists who all wrestled with the problem of planetary motion but were unable to provide an acceptable mathematical proof using the inverse square law. Huygens proposed the concept of centrifugal force and said this force was inversely proportional to the square of the radius of curvature. Hooke argued that gravity also followed an inverse square law. Encouraged by Halley, Newton promised to provide proof of the inverse square law and, with Halley's assistance, wrote a comprehensive treatise on motion that synthesized the work of his contemporaries as well as that of predecessors such as Kepler and Galileo. The final result, published in 1687 with financial support from Halley (the Royal Society lacked sufficient funds to publish the work), was entitled *Philsophaie Naturalis Principia Mathematica (Mathematical Principles of Natural Philosophy)*. The work is presented as three books. In the opening book of *Principia,* Newton presented the mathematics required for his presentation. He did not use his law of fluxions but depended mostly on classical Euclidean geometry for his proofs. Newton then set forth his three laws of motion, the concept of centripetal force, and the dynamics of motion. In the second section of *Principia,* Newton continued to develop his ideas on motion and applied these to fluids. The third book deals with Newton's "system of the world." He applied his laws of motion and mathematics to show how an inverse square of gravitation can be used to explain the Moon's orbit around the Earth. He also accounted for the planets' orbital motions around the Sun in conformance with Kepler's laws. Newton's theory of gravitation, like any good theory, accounted not only for planetary motion, but other celestial motion (movement of comets, precession of the Earth's axis) as well as motion on Earth, for example, projectile motion and tides. Newton's *Principia* gave precise meaning to the fundamental concepts of motion such as mass, force, acceleration, and inertia, and he used these concepts to present a coherent theory to resolve problems that had existed for 2,000 years.

Newton's *Principia* established Newton as England's leading scientist. Shortly after its publication, Newton reduced his scientific activities and assumed several roles in government. He was elected a member of Parliament in 1689. In 1692, he suffered a mental breakdown, but eventually recovered to assume the position of Warden of the Mint in 1696 and Master of the Mint in 1699. In 1703 he was elected president of the Royal Society, a position he held till his death. His lasting scientific work during the latter part of his life consisted of revising parts of the *Principia.* In 1704, his life's work on light was published as the *Optiks.* Much of his last years was spent on alchemical, mystical, and theological subjects.

Newton's work was the crowning achievement of the Scientific Revolution. Almost 300 years after Copernicus' *De Revolutionibus,* Newton provided a framework that would last 200 years until other observations required Einstein's theory of relativity. Still, the vast majority of common observations can be explained with Newtonian mechanics. Newton's theory was used to discover Neptune and Pluto when unexplained perturbations in the known planets' orbits could not be explained. It was used to put men on the Moon and is continually applied to everyday problems. It can be used to determine how fast a car was going before a crash, the tidal range in coastal cities, and how to design a roller coaster. The next chapter will present Newtonian mechanics in greater depth.

Summary

Classical physics started to replace the ancient Aristotelian-Copernican worldview around 1500. The replacement of ancient ideas with a classical view to describe nature was so dramatic it is often referred to as a revolution, more specifically the Copernican Revolution. Copernicus laid the foundation for looking at the universe from an entirely different perspective. This not only had ramifications for science, but society as a whole. While Copernicus planted the seeds for the scientific revolution to follow, it was left to others to cultivate these seeds into a coherent system for describing nature. The individuals responsible for this work are considered by modern society as some of the greatest scientists to have lived. Leading this group were Galileo and Newton. The active scientific lives of Galileo and Newton spanned 125 years. At the start of Galileo's career, it was heretical to consider the planets orbiting the Sun and the Earth rotating on its axis. By the end of Newton's life, these and a host of other natural motions were not only accepted, but cogent theories had been put forth to explain motion. It was realized that the Earth was not exempt from the laws of the universe. The laws that explained an apple falling to Earth also explained how the planets moved through the sky.

3

Translational Motion

Introduction

Chapter 2 presented the historical development of ideas about motion. These ideas were synthesized by Newton at the end of the seventeenth century in his *Principia* and have shaped our thinking of motion ever since. Concepts, once questioned and debated, are now commonly accepted, and each of us has a common understanding of the basic principles of motion. Terms describing motion, such as velocity and acceleration, are used colloquially to describe motion, often with little thought of their precise physical meaning. In this chapter the principles of mechanics introduced in the previous chapter will be examined in greater detail. The focus in this chapter will be translational, or linear, motion. Translational motion is characterized by movement along a linear path between two points.

Scalars and Vectors

Before translational motion can be examined, it is important to distinguish between scalar and vector quantities. A **scalar quantity,** or just scalar, is defined by a magnitude and appropriate units. Ten dollars, 3 meters,

and 32 °F are examples of scalar quantities. A quantity that is defined by a magnitude and direction with appropriate units is a **vector quantity,** or vector. Wind velocity is a good example of a vector. When wind velocity is reported, both the magnitude and direction are given, for example, 10 miles per hour out of the north. Vectors require a specified or assumed direction. A vector quantity is incomplete without a direction. The importance of including the direction for a vector can be illustrated by considering a person standing at the end of a narrow dock. If the person were told to move three steps forward versus three steps directly backward, it would probably make the difference between falling in the water and staying dry. Three steps is a scalar, whereas three steps backward is a vector. Throughout this book, physical quantities will be defined as scalars or vectors.

Speed and Velocity

Speed and **velocity** are terms that are often used interchangeably but are not equivalent. Speed is a scalar and velocity is a vector. In order to understand the difference between speed and velocity, a distinction has to be made between distance and displacement. Before making this distinction, it is important to assume a frame of reference to describe motion. Our most common frame of reference is the Earth. Even though the Earth itself is in motion, it is commonly treated as stationary, and movement is measured with respect to this "stationary" surface. For example, when we are told to measure the distance from a point, it is assumed that the point is fixed at a specific location and the distance is measured from this fixed point. In our treatment of motion, a fixed Cartesian coordinate system attached to the Earth will be assumed, with horizontal directions referenced to compass directions (Figure 3.1). Vertical motion will simply be referred to as up or down.

Distance is scalar measure of the amount of linear space covered. **Displacement** is a vector that points from an object's initial position to its final position. The magnitude of the displacement is simply the straight-line distance between the initial and final positions. If a person jogs around a 400 meter track, then the distance covered is 400 meters, but the displacement is zero since the jogger ends up right back at the starting line. Using the Cartesian coordinate system of Figure 3.1, the distinction between distance and displacement can be demonstrated by considering a bicyclist traveling 4 miles to the east and then 3 miles north. The distance covered by the bicyclist is 7 miles. The displacement calculated using the Pythagorean theorem is 5 miles in a direction 30° north of east (Figure 3.2).

Distance and displacement can be used to define speed and velocity. Speed is a scalar quantity defined by the distance an object travels divided by the time of travel, whereas velocity is a vector defined by displacement divided by the time of travel. To determine time, the final time is subtracted from the initial time. For example, if the bicyclist

started a trip at 1:00 P.M. and ended an hour later at 2:00 P.M., the time of travel would be 1 hour. Since time is always determine by subtracting the start time from the end time, it can be represented mathematically as $t_{final} - t_{initital}$ or Δt. The Δ is called the delta sign and means "change in"; therefore, Δt means the change in time. The average speed of the bicyclist during her trip is found by dividing the distance covered of 7 miles by Δt, which is 1 hour, to give 7 miles per hour (mph). The units for speed will be length divided by time, for example, feet per second (ft/s), or kilometers per hour (km/h). The most familiar unit for speed in the United States is miles per hour. The magnitude of the average velocity is found by dividing the displacement of 5 miles by 1 hour to give 5 mph. This is only the magnitude of the velocity. Since velocity is a vector, a direction must also be given. The average velocity should be reported as 5 mph in a direction 37° to the north of east.

The speed and velocity calculated were specified as the average speed and average velocity. The formulas for the average speed and velocity are

$$\text{average speed} = \frac{\text{distance}}{\Delta t}$$

$$\text{average velocity} = \frac{\text{displacement}}{\Delta t}$$

The equations and examples in this section illustrate the importance of using the correct terminology in physics. When reporting a velocity, a direction should always be given along with the magnitude, but at times the direction can be assumed. For instance, for a jet taking off down a runway with an average velocity reported as 200 mph, the direction is assumed down the runway.

The average speed and velocity were calculated for the bicyclist in the previous

Figure 3.1
Motion is referred to a Cartesian coordinate system, with the compass directions indicating horizontal motion

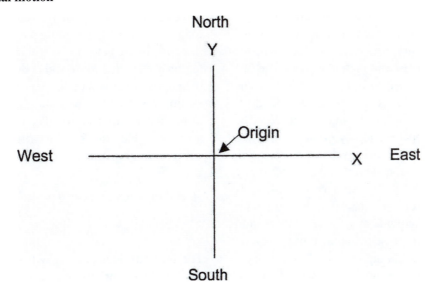

Figure 3.2
Moving a distance 4 miles east and then 3 miles north results in a displacement of 5 miles at an angle of approximately 37° north of east

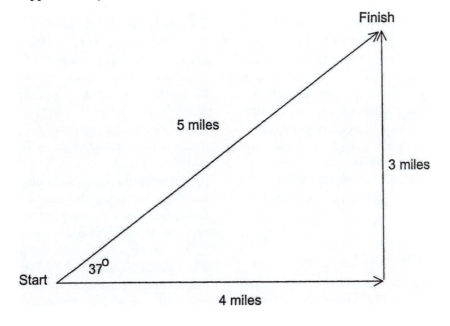

example. Average speed and velocity are useful for reporting the general progress of a trip but don't represent what may be happening at any particular time during the trip. The bicyclist could have pedaled quickly and covered the first 4 miles in 20 minutes, rested for 20 minutes, and then completed the last 3 miles in the remaining 20 minutes. The speed and velocity can be characterized for each of these intervals. In order to describe what is happening at any point in time, the instantaneous speed and velocity can be used. The instantaneous speed is the speed at any instant, while the instantaneous velocity is the velocity at any instant. The latter includes both the magnitude and direction of motion at any instant. Instantaneous motion can be found by calculating the speeds and velocities over smaller and smaller time intervals. A 20 minute interval is much too large, but the question arises as to how small a time interval must be to determine instantaneous motion. Just how long is an instant? In the bicycle example, instantaneous motion can be approximated over short time intervals such as a few seconds. For example, the 1 hour bicycle trip could be broken up into 1,200 3 second intervals and a speed and velocity calculated for each of these 3 second intervals. The 3 second intervals could be broken down further. To determine the true instantaneous velocity requires the use of calculus and taking the limit as Δt approaches zero in the equations for average speed and velocity. A formal derivation using calculus for instantaneous motion will not be presented here; the approximate definition for instantaneous speed and velocity as speed and velocity measured over a short time interval will be adequate.

Acceleration

Acceleration is another important variable used to describe motion and refers to the change in velocity over a period of time. Since acceleration results from a change in velocity, which is a vector, it also is a vector. Acceleration may result from a change in magnitude, direction, or both magnitude and direction of the velocity vector. In this chapter, only a change in speed is considered; in the next chapter it will be seen that circular motion results from a change in direction of the acceleration vector. The equation for average acceleration can be written as

$$\text{average acceleration} = \frac{\Delta \text{velolcity}}{\Delta \text{time}}$$

The units of acceleration will be a length per time per time, for example, kilometers per hour per second. Acceleration will typically be expressed as m/s/s in metric units. In this case, where time units are the same, the acceleration would be expressed as m/s^2. The basic unit for acceleration in the English system is ft/s^2.

One of the most familiar examples of acceleration is during acceleration or braking in traffic. When accelerating from a red light, the initial speed of a vehicle is 0 mph. Several seconds later the vehicle has obtained some speed, and using the time it takes to obtain this speed, the magnitude of the acceleration vector can be calculated. For example, if it takes 10 s to reach 40 mph, the acceleration is 4 mph/s. In this case the direction is assumed to be in the direction of the road. Acceleration is a measure of performance for cars and is often reported indirectly by stating the time it takes for a vehicle to reach 60 mph (or 100 km/h in countries using the metric system). Reported values can be used to calculate the average acceleration for a car. A 2003 Ford Mustang has a reported time of 5.6 s, so its acceleration is 10.7 mph/s. This compares to a Honda

Civic that takes 10.2 s to reach 60 mph, giving an acceleration of 5.9 mph/s. A dragster reaches 300 mph in approximately 4.5 s down a quarter mile straightaway, giving an acceleration of almost 67 mph/s. This is about the same as the maximum acceleration of the Space Shuttle during take-off. Greater yet is the acceleration of bullets fired from a gun, which is several hundred miles per hour per second. It must be remembered that these acceleration values are only the magnitude of the acceleration, and that it is assumed that the acceleration vector is in the direction of the motion.

A deceleration is the same as a negative acceleration. This can be seen when a car that is moving 40 mph comes to a stop in 10 s. In this case the change in velocity is –40 mph, so the acceleration is –4 mph/s. Each second the car moves 4 mph slower. The acceleration vector points in the opposite direction of the motion. Just as with velocity, a distinction is made between average acceleration and instantaneous acceleration. If an object accelerates at a uniform rate, then the acceleration is constant over the course of motion. On the other hand, if acceleration is not constant, short time intervals can be used to approximate the instantaneous acceleration. The instantaneous acceleration at any point would equal the limit of the change in velocity divided by the change in time as the time interval approaches zero. The instantaneous acceleration can be approximated by dividing the change in velocity by a short time interval.

General Formulas for Kinematics

Displacement, velocity, and acceleration can be used to describe motion. The description of motion is known as **kinematics.** When the acceleration is constant, several equations can be used to describe the kinematics of motion in one dimension. These are summarized in Table 3.1.

The first equation in the table states that the final velocity (throughout this section velocity will be assumed to mean only the magnitude of the velocity vector or speed) equals the initial velocity plus the gain in velocity due to any acceleration. If a car is traveling at 35 mph and accelerates at 2 mph/s, then after 5 s it will be going 45 mph (35 mph + 2 mph/s × 5 s = 45 mph). The second equation states that displacement is equal to the average velocity times time. Multiplying the initial velocity times the time and adding the displacement due to acceleration can also be used to find displacement, as expressed in Equation 3. Some numbers can serve to illustrate the use of Equations 2 and 3. If the initial velocity of a motorcycle is 40 mph and its velocity 20 s later is 60 mph, then its acceleration is 1 mph/s. Using the second equation, the displacement would be the motorcycle's average velocity of 50 mph times 20 s. In order to perform this calculation, units must be made consistent. Since velocity is in miles per hour and time is

Table 3.1

1.	$v_f = v_i + at$
2.	$\Delta x = \left(\dfrac{v_f + v_i}{2} \right) t$
3.	$\Delta x = v_i t + \dfrac{1}{2} a t^2$
4.	$2a\Delta x = v_f^2 - v_i^2$

v_f = final velocity, v_i = initial velocity, x = position, Δx = displacement, a = acceleration, t = time.

in seconds, 20 s is converted to hours by using the conversion factor of 3,600 s/h and then multiplying by the average velocity as follows:

$$(20 \text{ s})\left(\frac{1 \text{ h}}{3,600 \text{ s}}\right)\left(50\frac{\text{mi}}{\text{h}}\right) = 0.28 \text{ mi}$$

Using the third equation the calculation is

$$\left(40\frac{\text{mi}}{\text{h}}\right)(20 \text{ s})\left(\frac{1\text{h}}{3,600 \text{ s}}\right)$$

$$+\frac{1}{2}\left(\frac{1\text{mi}}{\text{h-s}}\right)\left(\frac{1\text{h}}{3,600 \text{ s}}\right)(20 \text{ s})^2 = 0.28 \text{ mi}$$

The fourth equation can be applied to the motorcycle example:

$$2\left(\frac{1 \text{ mi}}{\text{h-s}}\right)\left(\frac{3,600 \text{ s}}{\text{h}}\right)(0.28 \text{ mi})$$

$$=\left(\frac{60 \text{ mi}}{\text{h}}\right)^2 - \left(\frac{40 \text{ mi}}{\text{h}}\right)^2$$

$$2,016\frac{\text{mi}^2}{\text{h}^2} \approx 2,000\frac{\text{mi}^2}{\text{h}^2}$$

The last equation is only approximate because of rounding error. An equality is found if the actual displacement of 0.277 . . . mi is used rather than 0.28 mi.

Acceleration Due to Gravity

One of the most common examples of uniformly accelerated motion in one dimension is that of the case of free fall under the influence of gravity. Galileo discovered that the displacement due to gravity was proportional to time squared (t^2). This is exactly what is seen in Equation 3 in Table 3.1. A dropped object has an initial velocity of zero, and its displacement from its starting position will be equal to $1/2(at^2)$. According to Newton's law of universal gravitation (see next the section and chapter 4), the acceleration due to gravity at the Earth's surface is 9.8 m/s², or 32 ft/s². The symbol g is used for the gravitational acceleration. The value of g is not constant but varies slightly over the Earth's surface due to changes in elevation and geology. Variations in g are only important when very accurate measurements need to be made; therefore, g can be assumed constant for most calculations. Due to wind resistance, an object will accelerate at a rate of less than g. If wind resistance is ignored, an object will accelerate toward the center of the Earth at 9.8 m/s². This means that each second the object's speed increases by 9.8 m/s. A marble dropped from a ledge overhanging the Grand Canyon will have a speed of 98 m/s after 10 s, according to Equation 1. The average velocity of the marble during its descent is 49 m/s (toward the center of the Earth), so after 10 s its displacement is 490 m downward. The displacement can also be determined using Equation 3:

$$d = \frac{1}{2}at^2 = \frac{1}{2}\left(9.8 \text{ m/s}^2\right)(10 \text{ s})^2 = 490 \text{ m}$$

This example can also be considered in reverse. If the marble were fired upward at a velocity of 98 m/s from the canyon floor, it would reach a maximum height of 490 m.

The acceleration due to gravity affects motion in the vertical direction and is independent of motion in the horizontal direction. A classic scene in movies and TV shows is a car plummeting over a cliff. The car's horizontal motion suddenly turns into projectile motion as it leaves the road and becomes airborne. Projectile motion is a combination of horizontal motion with no acceleration and vertical motion under the influence of gravity. There are many familiar examples of projectile motion: throwing or hitting a baseball, firing a rifle, performing a long jump, or a ball rolling off a flat surface. If we consider the car example and assume the villain is fleeing from the law at a constant speed of 80 mph, then the 80 mph represents the horizontal component of motion. We'll assume the villain is not accelerating and driving along a straight stretch of mountain road when he suddenly discovers he has no brakes just as the road curves. According to Newton's first law, which will be addressed shortly, the car plunges over the cliff. As soon as the car leaves the road, it acquires vertical motion and begins to accelerate downward at 32 ft/s^2. Neglecting wind resistance, a constant horizontal velocity of 80 mph continues as the car plummets downward. The projectile motion of the car can be analyzed by considering the combined motion in the horizontal (x) and the vertical (y) directions. Table 3.2 summarizes the initial conditions for the motion of the car.

If we assume the car is in the air 3 s before it hits bottom, the displacement and velocity components during the car's projectile motion after 1s, 2 s, and 3 s can be calculated and are represented in Figure 3.3. In Figure 3.3, it is seen that the horizontal velocity is constant at 117 ft/s (80 mph) throughout the fall. The vertical velocity component increases according to the equation $v = gt$. The horizontal displacement increases by 117 ft each second, but due to the accelerated motion in the vertical direction, the vertical displacements are 16 ft, 64 ft, and 144 ft for 1 s, 2 s, and 3 s, respectively. These can be calculated using the equation $d = 1/2(gt^2) \cdot (1/2)gt^2$. Figure 3.3 shows that the car traveled a horizontal distance of 351 ft. The horizontal distance an object covers during flight is termed the range. Since the horizontal velocity is constant, the range is found by simply multiplying the time of flight by the horizontal velocity. It can be shown that the maximum range occurs when a projectile is fired at an angle of 45°. One way to demonstrate this is by using a garden hose and seeing how far away water can be squirted. An angle close to 45° should give the maximum range (see activity 4, chapter 16). Furthermore, equivalent ranges are found for angles that sum to 90°, for example, 30° and 60°.

The independence of motion in the horizontal and vertical directions means that an object will accelerate toward Earth at the same rate whether it is dropped or moving. A bullet fired horizontally from a height of 1 meter and one simultaneously dropped from 1 meter should both reach the ground at

Table 3.2
Initial Conditions of Car Driving over Cliff

	Horizontal Dimension	Vertical Dimension
Component	x	y
Initial Speed	80 mph	0 mph
Direction	Down road	Toward center of Earth
Acceleration	0	g

the same time. A classical problem in physics deals with shooting an elevated target. Assume a sharpshooter aims at the bull's-eye of the target illustrated in Figure 3.4. The sharpshooter knows that the instant the rifle is fired the target will be released. The question is "Knowing that the target will be released, where should the sharpshooter aim to hit the bull's-eye?" Many people will answer that the sharpshooter should aim below the target to take into account that the target is falling, but the correct answer is to take dead aim at the bull's-eye. The reason is that both the bullet and target will accelerate toward the Earth at a rate equal to g. In fact, if the target is stationary and a shot is fired directly at the bull's-eye, it will hit below dead center because the bullet has

experienced a downward acceleration due to the Earth's gravity. Of course, the error may only be an imperceptible fraction of a millimeter, but the basic fact remains that the fired bullet moves as a projectile that continually experiences a downward acceleration over the course of its flight.

Galileo studied motion on an inclined plane. This technique allowed Galileo to resolve the acceleration due to the gravity vector, g, into components perpendicular and parallel to the inclined plane's surface. He could then use water clocks and other timing devices available at the time to measure the displacement of balls rolling down the incline. Figure 3.5 illustrates how trigonometry is used to resolve g into x and y components parallel and perpendicular to

Figure 3.3
Projectile motion of a car over a cliff. The original speed of the car is 80 mph, which is equal to 117 ft/s. The double-headed arrows are displacements and the single headed arrows are velocity vectors. Motion is shown at one, two, and three seconds.

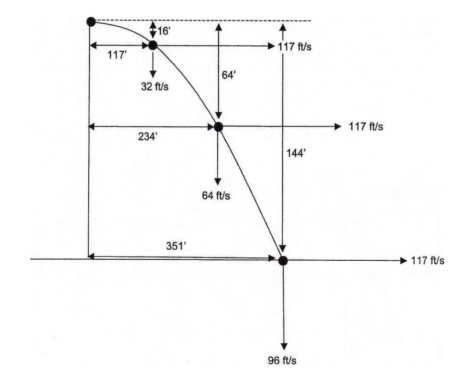

Figure 3.4
Aiming at a target. If a shooter fires at a target the instant it is released, the bullet will strike the bullseye after it travels along a parabolic path. Motion is greatly exaggerated to illustrate the concept.

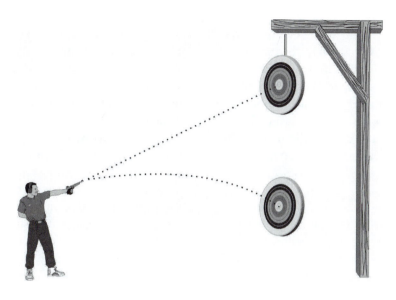

Figure 3.5
A ball rolling down an incline plane elevated an angle of θ. The acceleration due to the gravity vector g can be resolved into components down the plane a_x equal to $g(\sin \theta)$ and perpendicular to the plane a_y equal to $g(\cos \theta)$.

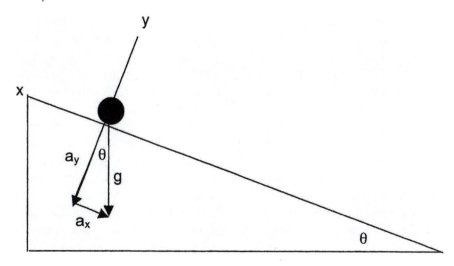

the plane, respectively. The acceleration down the plane is equal to *g* times the sin θ. In the extreme cases, when θ assumes values of 0° and 90°, it is found that the accelerations would be zero and *g*, respectively. One technique employed by Galileo was to space bells along the length of the incline plane such that a ball rolling down the plane would cause the bells to chime. By spacing the bells to get a constant interval between chimes, Galileo was able to measure the distance covered and deduce the relationship between distance and acceleration.

Newton's Laws of Motion

Galileo's work in kinematics dispelled many of the Aristotelian ideas on motion but was still incomplete because it didn't fully explain the reasons causing motion. A full understanding of motion requires knowledge of **dynamics.** Dynamics deals with the cause of motion and the relationship between force and motion. Force is vector quantity that we all have an intuitive feel for but is difficult to define. A force can be thought of as an interaction between two objects. Forces can be thought of as a simple push or pull. Forces can be classified as contact or noncontact forces. Hitting a baseball with a bat, pushing a piece of furniture across the floor, and towing a car are examples of contact forces. Contact forces result when the objects interacting touch each other. Aristotle believed all motion was the result of contact forces, and that if there was not contact between two objects then motion ceased. Noncontact forces result in action at a distance. The gravitational interaction between the Moon and the Earth, the attraction of a piece of iron to a magnet, and the attraction between the nucleus and electrons in an atom are examples of noncontact forces. Forces can also be categorized according to their nature by using a descriptive term. Common categories include gravitational force, magnetic force, buoyant force, frictional force, normal force, and electrical force.

Isaac Newton provided a coherent explanation of motion that can be summarized in his three laws of motion. Newton's first law of motion states that an object's velocity will remain constant unless acted upon by a net force. Another way of stating this is that an object will remain at rest or in motion in a straight line with constant speed unless acted upon by a net force. When an object is at rest, its velocity is zero. Its velocity remains zero unless a net force acts on the object. It is important to specify that it is the net force that changes the state of motion. Several forces may act on an object simultaneously, but if these cancel each other, there is no change in motion. When an object is in motion, its velocity stays constant unless a net force acts on it. If an object's velocity is constant, this means that its speed and direction must remain constant. The ability of an object to retain in a state of rest or a constant velocity (it should be remembered that a constant velocity of zero is equivalent to a state of rest) is termed **inertia. Mass** is a measure of inertia. In the metric system, mass is measured in kilograms. An object with a greater mass is obviously more difficult to move than one with less mass because it possesses greater inertia. Conversely, a massive object that is moving is much more difficult to stop or to change direction. For instance, an aircraft carrier moving at 20 mph takes several miles to stop.

We have all experienced Newton's first law many times. Every time an object is placed down it is natural to assume it will remain at rest until it is either retrieved or acted upon by an unbalanced force. When riding in a car and the brakes are suddenly applied, a person will continue to move in a forward direction. Luckily, seat belts restrain the forward motion. The same sensation occurs when riding a skateboard that hits gravel and suddenly stops. The rider continues forward and flies off the skateboard. Another example of experiencing Newton's first law is when trying to round an icy corner. Often, instead of following a curved path, motion continues in a straight line consistent with the first law.

Newton's second law of motion explains what happens when a net force acts on an object. A net force acting on an object causes it to accelerate, and the acceleration produced is directly proportional to the net force applied to the object and inversely proportional to its mass. Newton's second law is apparent whenever the wind blows. A mild wind is sufficient to blow napkins and empty paper plates off a picnic table but is not capable of moving more massive objects. If a gust occurs, this greater force scatters the light paper projects and may even cause heavier objects to move. Mathematically, Newton's second law of motion can be written as

$$a = \frac{F}{m} \text{ or } F = ma$$

In these equations a is the acceleration of the object, F the net force applied to the object, and m the mass of the object. Using SI units (Système International is the modern metric system), mass is measured in kilograms and acceleration in meters per second^2. For a 1 kg object to obtain an acceleration of 1 m/s^2 would require a force of 1 kg/m-s^2. The metric unit for force equal to 1 kg/m-s^2 is defined as the newton and is abbreviated N.

$F = ma$ is one of the most fundamental equations in physics. Knowing any two of the three variables allows calculation of the third unknown variable. For example, using the car data from the previous section, the acceleration

of a Honda Civic is 5.9 mph/s. This equals 2.6 m/s^2. The mass of a Honda Civic is approximately 1,100 kg. This means that the force required to accelerate the Honda Civic is roughly 3,000 N. A single-engine plane requires about 10 times this force to accelerate to liftoff, while a large loaded jet may require as much as 100 times this force.

Gravity and Weight

A special case of Newton's second law involves the **weight** of an object on Earth. Weight is the Earth's gravitational pull on an object. Weight can be found by multiplying an object's mass by g. Letting W equal weight and m equal mass, the weight of an object on Earth is given by the equation $W = mg$. The value of the acceleration due to gravity, g, is approximately 9.8 m/s^2 or 32 ft/s^2. This value is a result of Newton's law of universal gravitation, which states that all matter in the universe is attracted by a universal gravitational force. If two objects are considered, the magnitudes of the gravitational forces exerted on each object are equal and directly proportional to the product of the two masses and inversely proportional to the distance separating the masses. The two forces, one acting on each object, are opposite in direction. The force on each object points toward the center of the other object. Mathematically, Newton's law of universal gravitation for two objects with masses m_1 and m_2 separated by a distance r is given by

$$F_g = \frac{Gm_1m_2}{r^2}$$

where F_g is the magnitude of the gravitational force and G is the universal gravitational constant. The value of G is found experimentally and is equal to 6.67259×10^{-11} N-m^2/kg^2. The value of G was first determined by Henry Cavendish (1731–1810) in 1798. (Cavendish did not actually determine G directly, but rather the density of the Earth from which the mass of the Earth and value of G could be determined.) Cavendish used an apparatus called a torsion balance, designed by John Michell (1724–1793). Michell's design was based on an apparatus Charles-Augustin de Coulomb (1736–1806) had used for determining the force between charged objects. The torsion balance consists of a rod with two weights at each end suspended from a thin wire filament (Figure 3.6). When large masses are brought close to the smaller masses attached to the rod, the gravitational attraction between the spheres causes the wire to twist. By measuring the small angle of deflection, the gravitational constant, G, can be determined. Cavendish's torsion balance consisted of a 6 foot wooden rod with 2 inch lead spheres. He used 350 pound lead weights to cause the rotation and measured the angle of rotation with a telescope and special scale. Cavendish enclosed his torsion balance in a glass case to protect it from air currents and other external influences. Cavendish determined that the density of the Earth was 5.48 times that of water, which translates into an experimental value of G of approximately 6.75×10^{-11} N-m^2/kg^2.

For objects near the Earth's surface, the values of m_1 and m_2 in the equation for universal gravitation can be replaced by m_e and m_o, where m_e and m_o are the masses of the Earth and object, respectively. It is also assumed that the mass of the Earth is concentrated at a single point at its center. The value of r can be assumed to be the mean radius of the Earth, r_e. Substituting the actual values into the equation gives

$$F_g = \frac{Gm_e m_o}{r_e^2}$$
$$= \frac{\left(6.67 \times 10^{-11}\,\text{N} - \text{m}^2/\text{kg}^2\right)\left(5.98 \times 10^{24}\,\text{kg}\right)}{\left(6.38 \times 10^6\,\text{m}\right)^2} m_o$$
$$= \left(9.8 \frac{\text{m}}{\text{s}^2}\right) m_o$$

This equation demonstrates why the acceleration due to gravity is approximately 9.8 m/s^2 near the Earth's surface. It is only approximate because it varies slightly due to the changes in the distance between an

Figure 3.6
In the torsion balance the angle of rotation of the small balls toward the larger balls can be used to determine G

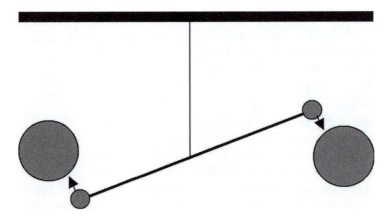

object and the center of the Earth. Because the Earth is not a perfect sphere but an oblate spheroid with its radius decreasing toward the poles, the value of g increases toward the poles. The value at the equator is closer to 9.78 m/s^2, while at the poles it is about 9.83 m/s^2. Besides latitude, other factors that affect g at the Earth's surface include altitude, variations in geology, and the tides. The value of g is lower on mountaintops, for instance it is about 9.77 m/s^2 at the top of Mt. Everest, and higher in valleys and on the seafloor.

It is important to note that weight is a force, and therefore, it is a vector quantity. We have been conditioned to consider mass and weight as synonymous, but they are two different properties. Mass is a measure of the quantity of matter and is a numerical measure of the amount of inertia of an object. Mass is an intrinsic scalar property that does not vary with location. A book with a mass of 2 kg will have a mass of 2 kg everywhere on Earth. Weight measures the gravitational force on an object. On Earth, weight depends on the acceleration due to gravity and varies slightly depending on the local value of g. A 2 kg mass's weight is found by using the equation $F = mg$ and is equal to 19.6 N. The distinction between mass and weight extends outside the Earth. The 2 kg book will still have a mass of 2 kg on the Moon, but because the Moon's surface gravitational attraction is only about 1/6 that of Earth, this means the book will have a weight of 1/6 that of its Earth weight, or about 3.3 N. The appropriate units for weight are newtons in the metric system and pounds in the British system. When we weigh ourselves on a scale and report our weight in pounds, we are using a unit of force. If we use metric units of kilograms to report our weight, we are not using a correct unit. Kilograms are used for mass, and we should be using newtons for weight. Because the value of g

is assumed to be constant over the Earth, a direct relationship between mass and weight exists, and grams and kilograms are often used as weight units for common measurements. For example, food packages will report the net weight in grams or kilograms. It would be correct for the package to read "net mass" or give the value in newtons. The British unit for mass is the slug, but it is not commonly used in the United States.

When an object rests on a level surface, such as a table, the force of gravity acts downward on the object. It is obvious that the object does not accelerate downward, so another force equal and opposite to gravity must be acting upward. This force is the **normal force.** The normal force is the perpendicular (normal means perpendicular) force exerted by a surface in contact with an object. For an object resting on a level surface, the gravitational and normal forces give rise to a situation known as equilibrium. An object is in a state of equilibrium when its acceleration is zero. In the case of an object resting on a level surface, it is apparent that the acceleration is zero, but equilibrium also exists when an object is moving with a constant velocity.

Newton's Third Law

Newton's third law states that for every action there is an equal and opposite reaction. This means that forces always exist in pairs that are equal in magnitude but directly opposed to each other. It is important to realize that in Newton's third law the action and reaction forces act on different objects. If you push on the wall with a force equal to 10 N, then the wall pushes back on you with a force equal to 10 N. As a person walks across the ground, he exerts a force on the Earth, while simultaneously the Earth exerts a force on the person. The Earth's force causes the person to accelerate forward according to

Newton's second law, $a = F/m$. At the same time the Earth is accelerated in a direction opposite of the person walking. Because of the Earth's tremendous mass compared to a person, its acceleration can be considered to be zero.

There are many familiar examples of Newton's third law. Jet and rocket engines work by expelling hot gases and propelling the vehicle forward. The Space Shuttle uses this same concept to make corrections in flight with short bursts from horizontal and vertical thrusters. Anyone who has fired a rifle experiences a certain degree of recoil with each shot. The recoil is the rifle's reaction from the action of firing the bullet. As a boat is rowed, the oars exert a force on the water and the water exerts an equal and opposite force on the boat, moving it forward.

Momentum and Impulse

According to Newton's first law, an object in motion will remain in motion in a straight line unless acted upon by a net force. An object's inertia depends on its mass whether it is at rest or moving, but in the latter case it also depends on the object's velocity. The product of mass and velocity is a vector quantity called **momentum.** Momentum, often symbolized with the letter p, is a measure of how hard it is to stop or change the velocity of a moving object. The equation for momentum can be written as $p = mv$. In order to change the velocity of a moving object, a net force must be applied to the object. If the velocity changes, then the object has undergone an acceleration. If the original velocity of the object is v_i and the final velocity is v_f, then according to Newton's second law the net force, F, acting on the object is $F = ma = m(v_f - v_i)/\Delta t$. Rearranging this equation gives $F\Delta t = mv_f - mv_i$. The first part of this equation, $F\Delta t$, is the net force multiplied by the time over which

the force acts. Force times Δt is a physical quantity called **impulse.** The second part of the equation is the change in momentum of the object; this equation is known as the impulse-momentum equation and states that the impulse is equal to the change in momentum. According to the impulse-momentum equation, the change in momentum of an object depends on both the net force acting on the object and the duration of the net force.

Consequences of the impulse-momentum equation are readily apparent in collisions and other situations when an object is brought to a stop. For example, consider the situation of jumping the same vertical distance onto pavement versus jumping into a net. In both cases the change in momentum will be the same, but in the first case the force of the pavement causes an abrupt change over a relatively short period of time. When jumping into a net, the net force acts over an extended period, cushioning the jump. The force that acts to change the momentum is not constant but changes during contact. In the use of the impulse-momentum equation, the force considered is the average force during contact. According to the impulse-momentum equation, the average force exerted by the pavement will be much greater than that exerted by the net. The cushioning effect of the net is a consequence of extending the time and decreasing the average force. Car bumpers, bungee ropes, cushioned running shoes, and padded guardrails are examples where the change in momentum is spread over a longer time period in order to decrease the average force on the object.

The Conservation of Momentum

A physical property characterized by a quantity that remains constant in the universe is said to be a **conservative property.**

Figure 3.7

Two cue balls approach each other and collide on a pool table. The total momentum of the system is the same before and after collision.

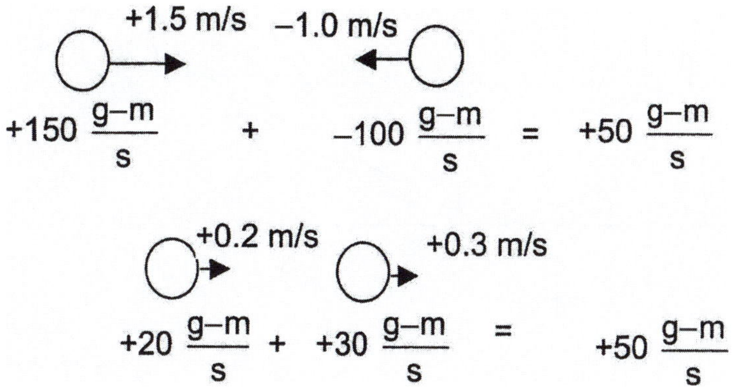

A number of fundamental laws of physics are based on conservative properties, for example, energy and mass. Another conservative property in nature is momentum. According to the conservation of linear momentum principle, the quantity of linear momentum in the universe is constant. It is impossible to measure the total linear momentum possessed by all objects in the universe, but by examining isolated parts of the universe it is possible to demonstrate that linear momentum is conserved. An isolated part of the universe is termed a **system.** A system may consist of a test tube or the entire planet and is defined according to the subject of interest. A ballistics expert may define a system as a gun, bullet, and wooden target into which the bullet will be fired, whereas a meteorologist's system may consist of the Earth's atmosphere. For a physicist studying momentum changes involved during the collision of two objects, the system can be defined as that small region of the universe where the collision takes place. During the collision of the two objects, they exert equal and opposite forces on each other according to Newton's third law. These forces are termed internal forces because they originate from within the system. In addition to **internal forces,** external forces may exist.

External forces originate outside the system's boundaries. In the case of two objects colliding on a surface (such as pool balls on a table), the normal force and gravity would be examples of external forces.

A system in which all the external forces sum to zero is termed an isolated system. It can be demonstrated that for an isolated system that linear momentum is conserved. In the case of two objects interacting with masses m_1 and m_2 and velocities v_1 and v_2, respectively, and using the subscripts i for initial and f for final, the conservation of linear momentum can be written as

$$(m_1)(v_{1i}) + (m_2)(v_{2i}) = (m_1)(v_{1f}) + m_2(v_{2f})$$

This equation states that the total linear momentum before and after interaction will be the same. To illustrate the concept of conservation of momentum, consider two cue balls rolling along a pool table toward each other (Figure 3.7). The masses of the two balls can be assumed to be equal at 100 g. The velocity of ball 1 is +1.5 m/s and ball 2 is −1.0 m/s. If the velocity of ball 1 is +0.2 m/s after striking ball 2, then ball 2's velocity must be +0.3 m/s (Figure 3.7). Conservation of momentum applies to the pool

Figure 3.8
Surfaces, even though they may appear and feel smooth, are highly irregular, giving rise to frictional forces

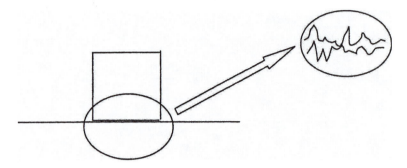

table example because the external normal and gravity forces cancel, making the colliding balls an isolated system.

Friction

The discussion of motion in this chapter has been simplified to ignore the effects of friction. Frictional forces impede motion and must be considered in a thorough analysis of motion. Two main types of friction that exist are static friction and kinetic friction. Static friction keeps an object from moving when a force is applied. A person pushing on a heavy object is unable to budge it because of static friction. Static friction depends on the normal force exerted on the object. In the case of an object resting on a level surface, the normal force on the object is equal to its weight. As the surface is elevated, the normal force progressively decreases toward zero as the angle of elevation approaches 90°. Static friction also depends on the relative roughness of the surface. Surfaces when examined microscopically show irregular features that consist of numerous bumps and depressions (Figure 3.8). Even seemingly smooth surfaces display these features. This contouring causes surfaces to interlock and impedes movement.

The coefficient of static friction, symbolized using μ_s, is a measure of the relative roughness between surfaces. Table 3.3 gives values of μ_s for different surfaces. For many cases the maximum static frictional force can be approximated using the equation $F_s = \mu_s F_n$, where F_n is the normal force on the object. Using values from Table 3.3, we can estimate the maximum force required to start an object moving. For example, to slide a 30 kg wooden dresser over a wooden floor requires a maximum force of about 90 N:

$$(30 \text{ kg})(9.8 \text{ m/s}^2)(0.3) \approx 90 \text{ N}$$

When an object is moving over a surface, the frictional force on the object is termed kinetic friction. Similar to static friction, kinetic friction can be estimated using a coefficient of kinetic friction, μ_k. Values of μ_k are included in Table 3.3. Values of μ_k are typically smaller than μ_s; this means than when pushing an object, less force is required to keep it moving than to start it moving. The magnitude of the kinetic frictional force acting on a moving object is calculated using the equation: $F_k = \mu_k F_n$.

Table 3.3
Approximate Coefficients of Static and Kinetic Friction

Surfaces	μ_s	μ_k
Rubber on dry concrete	1.0	0.75
Rubber on wet concrete	0.30	0.25
Rubber on ice		0.15
Waxed snowboard on snow	0.14	0.05
Wood on wood	0.2–0.4	0.2
Ice on ice	0.1	0.03

Summary

Linear motion characterizes much of the movement observed every day. Linear motion can be describe with a few basic equations. The most basic of these are Newton's laws of motion. Newton's laws allow us to determine such things as how long it will take an object to drop a specific distance or how much force is required to move an object. Placing these type of calculations in a practical context allows engineers to specify the size of engines to fly planes, what the speed limits on roads should be, the amount of friction required to stop a truck in a runaway ramp, and countless other applications. This chapter limited the discussion to linear motion. The concepts in this chapter will be expanded to include rotational motion in the next chapter.

4

Rotational Motion

Introduction

In translational motion, an object moves from one point to another. Another type of motion occurs when an object moves in a curved path. This type of motion is called rotational motion. Examples of rotational motion are all around us. A spinning compact disc, the wheels of a car, and a blowing fan are a few common examples of rotational motion. As you sit reading this book, you are in constant rotational motion as the Earth spins on its axis. Rotational motion may accompany translational motion, as when a ball rolls down a ramp. It may also occur in the absence of translational motion, as when a top spins on its axis. Rotation refers to the circular movement of an object about an axis that passes through the object. Another type of curved motion occurs when an object moves about an axis outside of it. This is called a revolution. Both rotational and revolutionary motion will be considered in this chapter, and the term rotational motion will be used as a generic description for curved motion. Many of the concepts for rotational motion parallel those for translational motion. The

simplest type of rotational motion is uniform circular motion.

Uniform Circular Motion

Circular is used as a general term to describe motion along a curved path and should not be interpreted to mean motion following a perfect circle; for instance, elliptical motion is classified as circular motion. Common examples of circular motion include the orbital motion of satellites, amusement park rides, rounding a curve in a car, and twirling an object on the end of a string. Many of the concepts describing translational motion have analogous counterparts for circular motion. These will be developed in this chapter. The concepts in this chapter apply to both rotational and revolutionary motion. Rotation involves circular motion around an axis that passes through the object itself, for example, a rotating top. Revolution involves circular motion around an axis outside of the object. The Earth rotates on its axis and revolves around the Sun.

In describing translational motion, a simple Cartesian coordinate system was

Figure 4.1

Coordinate system used to describe circular motion

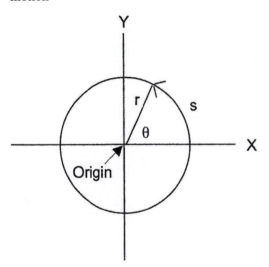

Figure 4.2

At every point on the circumference the tangential speed is constant for uniform circular motion. The tangential velocity vector is tangent at each point along the circumference.

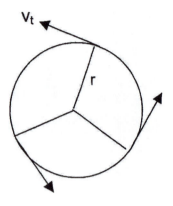

used to reference position. This system can also be used for circular motion. When using the Cartesian coordinate system for circular motion, the position of an object can be referenced to the origin and the positive x axis (Figure 4.1). Angles are measured counterclockwise starting from the positive x axis. The angular displacement is the angle swept out during a specific time period. For any two points along the circumference, the angular displacement is equal to $\Delta\theta$. When starting from the positive x axis, where θ is defined as zero, the angular displacement is just θ. The distance an object moves in one revolution around the circle is the circumference of the circle, $2\pi r$. The time it takes for an object to make one revolution around the circle is called the period, symbolized by T. The **tangential** or linear speed of an object moving in a circular path of radius r around the origin is equal to the change in distance along the circumference divided by the change in time, $\Delta s/\Delta t$ (if referenced from the positive x axis, the distance is just s). Since an object will travel a distance

equal to the circumference in one period, the tangential speed can be found by dividing the circumference by the period:

$$\text{tangential speed} = \frac{2\pi r}{T}$$

When the tangential speed is constant, the motion is described as **uniform circular motion.** The tangential velocity, v_t, is the vector analogy to tangential speed. Its magnitude is constant and equal to the tangential speed for uniform circular motion, and its direction is tangent to the circle at any instance (Figure 4.2).

In addition to its tangential speed and velocity, an object moving in uniform circular motion also has an **angular velocity,** symbolized by ω. The angular velocity is found by dividing the angular displacement by the time over which this displacement occurs. If the angular displacement, $\Delta\theta$, is the angle swept out by the object during a specific period of time, Δt, then the angular velocity is equal to $\Delta\theta/\Delta t$.

Referring to Figure 4.1, if it takes an object 5 s to move through an angle of 60°, then the magnitude of its angular velocity is 12°/s. The standard unit for expressing angular velocity is radian(s) per second or just s^{-1}. A radian is a unit of angle measurement and is defined as the arc length divided by the radius. Again referring to Figure 4.1, $\Delta\theta$ in radians would be equal to s/rad. Since there are 360° in a circle and the circumference of a circle is $2\pi r$, 360° is equal to 2π radians. The arc distance can be found by multiplying the angular displacement in radians times the radius: $s = r\Delta\theta$. If both sides of this equation are divided by time (Δt), the equation relating tangential speed to angular velocity is found: tangential speed $= r\omega$.

The tangential speed of an object on the Earth's surface depends on its latitude. The tangential speed is a maximum at the equator and decreases to zero at each pole. At the equator the Earth's radius is 6.38×10^6 m and its circumference is 4.0×10^7 m. The period of rotation is 24 h or 86,400 s; the tangential speed at the equator is therefore

$$\frac{4.0 \times 10^7 \text{ m}}{86,400 \text{ s}} = 460 \text{ m/s} = 1,030 \text{ mph}$$

The magnitude of the angular velocity at the equator is

$$\frac{2\pi \text{ rad}}{86,400 \text{ s}} = 7.3 \times 10^{-5} \text{ rad/s}$$

The tangential speed can also be found using the formula $r\omega$:

$$(6.38 \times 10^6 \text{ m})(7.3 \times 10^{-5} \text{ rad/s}) = 460 \text{ m/s}$$

The direction of the angular velocity vector points in a direction perpendicular to the plane of rotation. The direction

Figure 4.3
Right-hand rule for determining direction of angular velocity vector

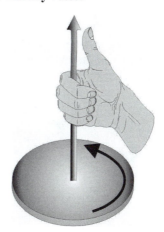

can be determined by using a convention called the **right-hand rule.** To apply the right-hand rule, the four fingers of the right hand are curled in the direction of motion. When the thumb is extended, it points in the direction of the angular velocity vector (Figure 4.3).

Centripetal Acceleration and Force

According to Newton's first law of motion, a moving object will travel in a straight line unless acted upon by a force. This is demonstrated when an object whirled in a circle overhead is released and flies off in a direction tangent to the circle. In uniform circular motion, the magnitude of the velocity vector is constant, but its direction is constantly changing. A constant change in velocity means that the object experiences a constant acceleration (or deceleration). This constant acceleration directed toward the center of the circle is called the **centripetal acceleration** (Figure 4.4). Centripetal means "center-seeking." The centripetal

Figure 4.4

In uniform circular motion the tangential velocity vector undergoes a constant change in direction. From V_1 to V_2 an instant later the velocity changes by ΔV. This change in velocity is due to a constant acceleration toward the center of the circle and is termed centripetal acceleration, a_c.

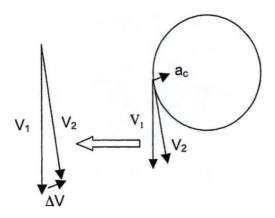

acceleration, a_c, is directed everywhere perpendicular to the tangential velocity vector and toward the center of circular motion.

The magnitude of the centripetal acceleration is equal to the tangential speed (v) squared divided by the radius:

$$a_c = \frac{v^2}{r}$$

Since $r\omega$ is equal to v, it can be substituted into the above equation to give $a_c = r\omega^2$. The centripetal acceleration is produced by a force directed toward the center of circular motion termed the **centripetal force.** If F_c represents centripetal force and m the mass of an object in uniform circular motion, F_c is found by applying Newton's second law:

$$F_c = ma = m\frac{v^2}{r}$$

Centripetal is a general term that identifies the force as directed toward the center of a circle. The actual source of the centripetal force can be gravity, friction, electromagnetic, tension, or various other forces. When a car rounds a corner, the centripetal force is supplied by friction between the tires and the road. The equation for centripetal force shows that the centripetal force is directly proportional to the square of the tangential speed and inversely proportional to the radius. This means that as the car travels faster and the turn gets tighter, a greater centripetal force is required to make the turn. There is a limit to how much frictional force can be supplied by the tires, dictated by the coefficient of friction between the road and the rubber tire. If the road is wet or icy, the frictional force decreases, and driving too fast or trying to make a tight turn results in lost of control and sliding off the road. One way to supply a greater centripetal force and allow turns at higher speeds is to bank or elevate the roadway toward the outside. In this manner, both friction and the normal force of the roadway on the car supply the centripetal force. Speedways, such as Daytona, use this principle to enable cars to maintain high speeds around the turns. The racetrack essentially pushes the car around the corner.

Gravity is the centripetal force that keeps the planets in orbit around the Sun and the Moon around the Earth. Newton was able to demonstrate mathematically that the centripetal acceleration of the Moon toward the Earth conformed to the inverse square law. The Moon's period around the Earth with respect to the distant stars is 27.3 days (2.36×10^6 s). This period is known as the **sidereal** month. It is shorter than the lunar month of 29.5 days due to the fact that the lunar month is measured with respect to the Earth, which itself is moving around the Sun. Assuming the Moon orbits the Earth in uniform circular motion at a radius (distance between centers of Earth and Moon) of approximately 3.90×10^8 m, the magnitude of the angular velocity of the Moon around the Earth is

$$\omega_{moon} = \frac{2\pi \text{ rad}}{2.36 \times 10^6 \text{ s}} = 2.66 \times 10^{-6} \text{s}^{-1}$$

and the angular acceleration of the Moon toward the Earth is

$$\begin{aligned} a_{cmoon} &= \omega_{moon}{}^2 r = (2.66 \times 10^{-6} \text{ s}^{-1})^2 \\ &\quad (3.90 \times 10^8 \text{ m}) \\ &= 0.00276 \text{ m/s}^2 \end{aligned}$$

This value can be compared to the value of the Earth's gravitation at the Moon's location calculated using the inverse square law. The distance between the Earth and Moon is approximately 60 times the radius of the Earth. The value of g at the Earth's surface is 9.8 m/s². According to the inverse square law, the gravitational force is inversely proportion to the square of the distance: $G \propto 1/d^2$. This means the value of g at a distance of 60 times the Earth radius should be about

$$\left(9.8 \text{ m/s}^2\right)\left(\frac{1}{60^2}\right) = 0.00272 \text{ m/s}^2$$

This value is identical (accounting for rounding error and using approximate values in the calculation) to the calculated centripetal acceleration. Therefore, the centripetal acceleration of the Moon toward the Earth is provided by gravity. Newton demonstrated that the same force that causes an object to fall to the ground on Earth is responsible for keeping the Moon in orbit around the Earth.

The motion of the Moon around the Earth means the Moon is constantly falling toward the Earth. It may seem strange that an object could be constantly falling toward Earth but never hit the ground, until the Earth's curvature is considered. Consider for example standing at the equator

Figure 4.5
When throwing a baseball, the ball follows a parabolic path and lands some distance away, depending on how hard it is thrown. If the ball is thrown hard enough its parabolic path would follow the curvature of the Earth.

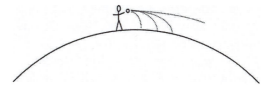

and throwing a baseball due east. The ball travels in a parabolic path and strikes the Earth some distance from where it was thrown. As the ball is thrown harder, it covers more horizontal distance before striking the ground (Figure 4.5). When throwing a baseball, the Earth's curvature is of no consequence; in essence the Earth can be considered flat. Now consider what would happen if Superman throws the baseball. Superman can throw a ball so far that the Earth can no longer be considered flat. In fact, if Superman throws the ball hard enough, the parabolic path it makes as it falls to Earth will match the curvature of the Earth and it will never strike the ground. The speed Superman needs to throw the ball can be estimated by assuming that the Earth's gravitational force is the only force acting on the ball. This force supplies the centripetal force need to keep the ball in orbit. Equating gravitational force and centripetal force gives

$$\frac{Gm_b m_e}{r_e^2} = \frac{m_b v_b^2}{r_e}$$

gravitational force = centripetal force

where the subscripts b and e are used for the ball and Earth, respectively.

Rearranging and solving for v_b gives

$$v_b = \sqrt{\frac{Gm_e}{r_e}}$$

Substituting the value for the gravitational constant, Earth's mass, and Earth's radius, the ball's speed is 7.91×10^3 m/s or almost 17,700 mph. This speed represents the **orbital velocity** of an object near the surface of the Earth. The real-world equivalent of throwing a baseball takes place when satellites are fired into orbit. A satellite's orbital velocity has to be approximately 18,000 mph to maintain an orbit near the surface of the Earth. The first manned space flights in the early 1960s and the Space Shuttle have orbits within a few hundred miles of the Earth's surface. The orbital velocity of near-Earth flights, 18,000 mph, equates to 5 miles every second. At this rate it takes about 90 minutes to orbit the Earth. Certain satellites, known as **geosynchronous,** orbit the Earth such that their position is fixed with respect to the earth's surface. Geosynchronous satellites orbit the Earth at an altitude of approximately 22,000 miles. At this altitude the satellite's period is approximately 24 hours, and its orbit keeps it at a stationary point with respect to the Earth's surface. Geosynchronous satellites are widely used for communications, national defense, and weather forecasting.

The use of equations for uniform circular motion to predict orbital periods, altitudes, and velocities are close approximations. The orbits of satellites, moons, and planets are not perfect circles but, as Kepler stated, elliptical. For elliptical orbits, the period of a satellite, T, is equal to $2\pi\sqrt{a^3/\mu}$ where a is the length of the semimajor axis in kilometers, and μ is the gravitational parameter equal to 398,600 km²/s². This equation is just another version of Kepler's third law. Another point concerning orbital velocity is that the velocity to maintain an orbit is less than the actual velocity needed to escape the Earth's gravitational pull. The **escape velocity** is the velocity needed to escape the Earth gravitational influence and is about 25,000 mph or 7 mi/s. A rocket fired at a velocity less than 25,000 mph will eventually fall back to Earth, but falling back to Earth in space results in an orbital motion when the fall matches the curvature of the Earth. Another consideration for near-Earth orbits is that there is a very slight atmospheric drag on space objects that retards their motion. This slight drag continually slows down the object, and when no compensation is made by firing engines to overcome drag, the object loses altitude and spirals inward toward the Earth. This process may take from days to decades, depending on the original altitude. The process accelerates as an object loses altitude and experiences an increasingly greater amount of atmospheric drag. Small objects disintegrate in the atmosphere, but many large parts or pieces can survive reentry and strike the Earth. Periodically, international news focuses on a large piece of space junk that will tumble back and impact the Earth.

Orbital Motion, Gravity, and Apparent Weight

One of the most characteristic scenes of humans traveling in space is that of astronauts and objects floating through a spacecraft. Such scenes are often attributed to space being a weightless environment. In reality, the location of near-Earth space flights is within a few hundred miles of the Earth's surface, and the force of gravity on a spacecraft and its contents is only slightly less than g. Since weight equals mg, the weight of an individual in a spacecraft orbiting near the Earth is only slightly reduced. For example, an astronaut in a space shuttle orbiting at 300 miles would weigh about 15% less than on Earth. The space surrounding the Earth does not possess unique properties that cancel the effect of Earth's gravity, but the motion of the orbiting craft produces an apparent weightless environment. This is due to the fact that the craft and its contents are all accelerating toward the Earth in a continuous free fall.

Apparent weightlessness can be understood by examining how we think of weight and how the situation on Earth and orbiting in space differ. When standing on a scale to measure your weight, two forces can be considered. The force of gravity pulls you toward the center of the Earth, and an equal and opposite normal force of the Earth pushes on the scale. This force is transmitted to the bottoms of your feet. Equilibrium exists between the gravitational and normal; the normal force causes a spring in the scale to be compressed, resulting in the scale's reading, for example, 150 pounds. Now consider what happens when you weigh yourself in an elevator. As long as the elevator is not moving, your weight in the elevator is the same as that in your bathroom. If you weigh yourself on an ascending elevator, the normal force exerted by the floor of the elevator must be greater than your weight. It must be greater in order for you to accelerate upward. Conversely, when you descend, the normal force must be less than your weight. The normal force becomes progressively smaller the faster you descend. If the elevator cable should snap causing the elevator to go into a state of free fall, the elevator and you would both accelerate toward Earth at the rate of g. Your weight would still be given by mg, but the normal force would be zero, and the scale would record a weight of zero. You would

float through the elevator just as astronauts float through their craft in a state of apparent weightlessness. The situation that exists in an orbiting spacecraft is analogous to an elevator in free fall. An orbiting spacecraft is in a state of continuous free fall. During takeoff the situation is analogous to the ascending elevator and astronauts experience a force several times g.

In order to approach a zero-gravity (microgravity) environment without traveling into space, NASA employs a modified plane aptly named the Vomit Comet (VC). Astronauts can experience apparent weightlessness for brief periods riding the VC. To produce apparent weightlessness, the VC ascends to 22,000 feet and then undergoes a series of parabolic roller-coaster dives. During the dive portion of the flight, occupants experience apparent weightlessness for a approximately half a minute.

Artificial Gravity

Although the apparent weightlessness present in an orbiting spacecraft allows astronauts to perform a variety of stunts, it presents a formidable practical and physiological challenge for extended space flights. Any object not anchored down will float through the capsule, and relative directions such as up and down lose their meaning. A more serious problem is the physiological effects of a weightless environment on the human body. These effects include weakening of bone structure due to bone demineralization, muscle atrophy including weakening of the heart, disorientation, lethargy, nausea, decrease of red blood cells, and fluid redistribution in the body. Endurance records of more than a year in space have established the practical limits of space travel. Extended trips in space, such as a mission to Mars, would require several years and the need to mitigate the health effects of an extended

Figure 4.6
A rotating space station several hundred meters in diameter would produce an artificial gravity environment

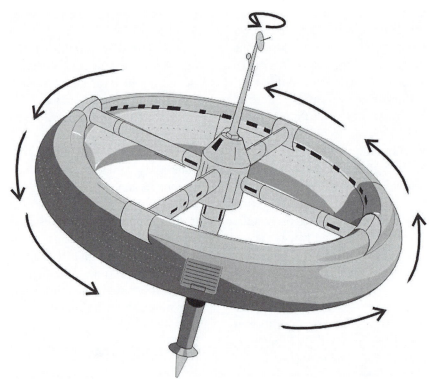

space flight. One way to do this is by creating an artificial gravity environment.

Artificial gravity environments are created using rotation. Gravity is simulated by the **centrifugal force** created from the rotation. Centrifugal means "center fleeing" and is often misinterpreted when discussing rotational motion. The misinterpretation of centrifugal force arises from failing to realize that motion is often referenced from an accelerating or **noninertial** frame of reference. For example, turning a sharp corner in a car causes the car's occupants to be thrown outward by a centrifugal force. In actuality, when the car turns, the occupants will momentarily continue in a straight line according to Newton's first law. The outward motion will be arrested by seatbelts, friction between the seat and passengers, and the

outside of the car; these all serve to provide the centripetal force to allow the passengers to corner with the car. A person sitting on a slippery seat, unrestrained, next to an open door would fly out of the car. For that person it would seem that an outward force threw them out of the car. This force is very real for the person referencing motion with respect to the noninertial accelerating car but does not exist for someone outside this frame of reference. A person viewing the situation from a helicopter hovering above the car would state that the person continued in a straight line, while the car turned the corner.

A spinning noninertial frame of reference can be employed to simulate gravity. Consider a cylindrical space station rotating around its central axis (Figure 4.6). In order to keep the astronaut moving in a circular motion, a centripetal force must be applied to the astronaut. The force depends on the radius of the space station and the angular velocity according to the equation ma_c. To simulate the Earth's environment, a centripetal acceleration (a_c) of 9.8 m/s² would be desirable. The angular velocity of the station can be calculated using the formula for centripetal acceleration: $a_c = r\omega^2$. A space station with a radius of 5 m would need an angular acceleration of 1.4 s⁻¹, which is equivalent to about 13 rotations every minute. For an outside observer, the centripetal acceleration acts on the occupants to keep them in rotation. For the astronauts within the station, the constant centrifugal force acting on them has the same effect as gravity does on Earth. A close approximation to the situation in the space station exists during certain amusement park rides. On one particular ride, known as the Gravitron, people stand facing the center against a vertical wall encircling a horizontal round platform. The platform is set in spinning motion, and after obtaining a certain speed, the floor falls out. The riders' experience is that of being pressed up against the wall by a centrifugal force.

Many obstacles need to be overcome to employ artificial gravity. One problem is that of motion sickness produced by a rotating environment. Experiments indicate that rates must be limited to a few rotations per minute. At rotations this slow, a space station would have to have a radius of 100 meters. While this may be adequate for a stationary colony, it would be impractical for current space travel. One plan for space travel is to equip spacecraft with centrifuges where astronauts could undergo periodic gravity sessions to retard the effects of prolonged travel in a microgravity environment.

Angular Acceleration

In uniform circular motion, the acceleration is due to a constant centripetal force causing a continual change in direction of the object in circular motion. The tangential speed remains constant, and the angular acceleration is also zero. When circular motion is nonuniform, the tangential speed changes, and there is a tangential and angular acceleration. The average tangential acceleration is the change in tangential velocity divided by the change in time. The equation $\Delta\theta/\Delta t$ defines the magnitude of the average angular velocity, ω. The instantaneous angular velocity is found by calculating the angular velocity over a very small time interval, similar to instantaneous speed defined in chapter 3. A good way to distinguish between average and instantaneous angular velocity is to consider a car's tachometer. A car's tachometer gives the engine's instantaneous angular velocity at any time. If a car initially at rest accelerates uniformly until the tachometer reads 200 rpm, then the average angular velocity is 100 rpm, while the instantaneous angular velocity is 200 rpm. Angular velocity is related to tangential velocity according to the equation $v_t = r\omega$.

The change in magnitude of the angular velocity divided by the change in time is the average angular acceleration, α:

$$\alpha = \frac{\Delta\omega}{\Delta t}$$

The units for angular acceleration are typically radians per second², or rad-s⁻². The instantaneous angular acceleration is calculated by using a small time interval. Using the uniformly accelerating car example, the average and instantaneous acceleration of the car are equal (due to the fact that acceleration is constant). If the car took 20 s to accelerate to 200 rpm (equivalent to approximately 21 rad/s), then the angular acceleration would be

$$\alpha = \frac{\Delta\omega}{\Delta t} = \frac{21 \text{ rad/s} - 0 \text{ radians/s}}{20 \text{ s}}$$
$$= 1.05 \text{ s}^{-2}$$

This means that each second the engine's crankshaft is turning 1.05 rad/s faster. The relationship between tangential acceleration and angular acceleration is $a_t = r\alpha$.

Rotational Dynamics

In translational motion, an equilibrium condition exists when an object is not accelerating and there is no net force acting on an object. **Rotational equilibrium** exists when an object is not rotating. In order for an object to be in rotational equilibrium, forces acting on an object that cause rotation must sum to zero. A force that acts to produce rotation is known as a **torque.** A torque depends on both the force and where the force acts. For example, consider a book resting on a smooth table. If you use your finger to push the book in the middle along its axis of symmetry, the book will slide across the table. (Figure 4.7). Now consider what happens when you push the book at one of its edges. Rather than sliding along the table, the book rotates around an axis perpendicular to the center of the book. The difference in the two situations in Figure 4.7 arises due to the fact that an unbalanced torque exists when pushing the book along the edge. In both cases the force acting is the same, but the position where the force is applied can create an unbalanced torque. Another example where torque's dependence on both the force and position where it is applied is illustrated when opening a door with a handle mechanism that extends horizontally across the door. The door opens by rotating along a vertical line through the hinges. It is much easier to open the door by pushing on the handle on the end opposite the hinges. It is very difficult, if not impossible, to open the door when pushing on the handle close to the hinges. The difference when opening the door is due to the fact that the torque on the door increases as the distance that the force is exerted from the hinges increases.

The distance the force is applied from the axis of rotation is called the **lever arm.** The lever arm is defined as the perpendicular distance from the line of action of the force and the axis of rotation. In the case where the force is applied perpendicular to the object, the lever arm is the distance between where the force is applied and the axis of rotation (Figure 4.8). When the force is applied at an angle other than 90°, then the lever arm is reduced. In the extreme case, when the force passes through the axis of rotation, such as pushing the book in the center, the lever arm (and torque) is zero, resulting in no rotation.

Torque is often represented by the Greek letter τ (tau) and is defined mathematically as the product of the magnitude of the force and lever arm (l): $\tau = Fl$. A positive torque results in a counterclockwise rotation, and a negative torque produces clockwise rotation. The unit for torque is newton-meter (N-m). The equation for torque shows why it is difficult to open a door by applying a force near the hinges. In this case, the lever arm is small, resulting in a small torque. The equation also explains many other familiar situations. For example, a light person on a seesaw can balance a much heavier person if the lighter person sits further away from the seesaw's fulcrum. Although the lighter person exerts less force, the greater lever arm produces a torque that balances the torque produced by the heavier person. Another common example applying torque is when a pipe is used to extend the handle length of a wrench to remove a stubborn bolt. The pipe lengthens the lever arm, thereby producing a greater torque to allow rotation of the bolt.

Figure 4.7
Pushing a book along its centerline causes a translational displacement, while pushing along the edge causes the book to rotate

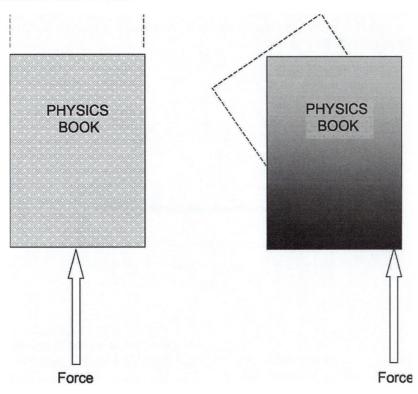

Torque is the rotational equivalent to force. According to Newton's second law, a net force produces a linear acceleration given by the formula $F = ma$. For rotation, a net torque produces rotational acceleration. In order to develop an equation using torque, consider the example of an object of mass m positioned at the end of a rod rotating around an axis passing through the opposite end (Figure 4.9). If a force acts in the tangential direction, Newton's second law for the object is $F = ma$. The acceleration in this example is the tangential acceleration, a_t. Replacing a with a_t and multiplying both sides of the equation $F = ma$ by r gives $Fr = mra_t$. Since $a_t = r\alpha$, this equation can be written using angular acceleration by substituting for a_t: $Fr = mr^2\alpha$. The left side of this equation is the torque, τ, acting on the object. The right-hand side contains the terms mr^2 and α. The term mr^2 is referred to as the **moment of inertia** of the object and is symbolized by I. The moment of inertia is a measure of the difficulty to start or stop rotation. The greater the moment of inertia, the more difficult it is to commence rotation or, if an object is rotating, to change or stop rotation. The moment of inertia is analogous to mass in translational motion. Just as mass is a measure of inertia for translational moment, moment of inertia is a measure of inertia for rotational motion. Newton's second law for rotational motion can be written as $\tau = I\alpha$, which states that the net torque

Figure 4.8
The lever arm depends on where the force is applied, the direction of the force, and the distance between the line of action of the force and axis of rotation. The axis of rotation runs through the point of the fulcrum perpendicular to the page. The lever arms for forces applied at the same position, but different directions are shown. The lever arm for 3 is zero because the line of action extends through the axis of rotation.

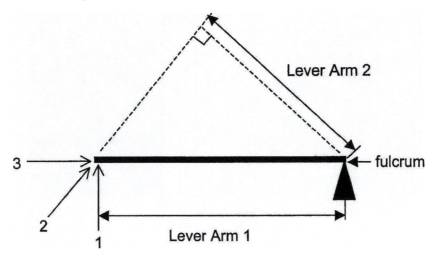

Figure 4.9
An object of mass *m* rotates at a distance *r* around an axis. The rod can be considered massless.

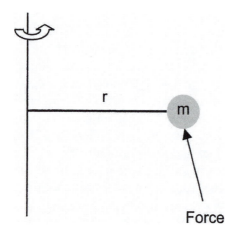

equals the moment of inertia times angular acceleration.

The case of an object rotating at the end of a rod is a simplified example. The mass of the object was considered to be concentrated at one point and the rod massless. For solid rigid objects, the moment of inertia can be approximated by dividing the object into small parts, finding *I* for each part, and summing. The exact procedure for finding the moment of inertia involves using calculus and will not be dealt with here, but the results of this procedure will be presented for several rigid shapes. The moment of inertia depends on how the mass is distributed around the axis of rotation. For example, consider two cylinders of equal radius: one hollow and one solid (Figure 4.10). If the masses of the two cylinders are equal, the moment of inertia of the hollow cylinder is greater than that of the solid cylinder. This can be seen if *I* is approximated by dividing each cylinder into small parts, and mr^2 is calculated for each of these parts. The moment of inertia of each cylinder would be the sum of the moments of inertia of all the parts, or $\Sigma m_i r_i^2$, where the subscript i is used to designate

Figure 4.10
Two cylinders of equal radii and mass but different mass distributions have different moments of inertia

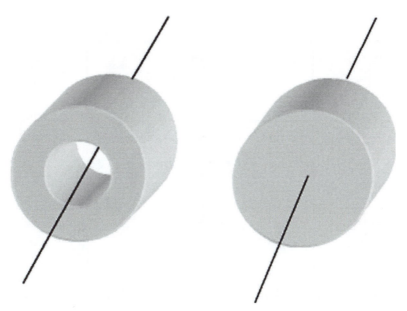

each part. Because the mass for the hollow cylinder is concentrated further from the axis of rotation, its $\Sigma m_i r_i^2$ and, therefore, its moment of inertia will be greater. The actual formula for the moment of inertia of a solid cylinder of radius r is $I = 1/2(mr^2)$, while the formula for a hollow cylinder with an outside radius of r_o and inside radius of r_i is $I = 1/2m(r_o^2 + r_i^2)$. If a solid cylinder and hollow cylinder were released down an incline, the solid cylinder would reach the bottom faster than the hollow cylinder. This is because the solid cylinder's relatively smaller moment of inertia means it has less resistance to rotational motion and will thus accelerate faster down the incline compared to a hollow cylinder. The solid cylinder's relatively greater moment of inertia means it has more resistance to rotation and accelerates down the incline slower than the solid cylinder. The moments of inertia of several common rigid objects are given in Table 4.1. Table 4.1 demonstrates that the axis of rotation is important when determining the moment of inertia. For example, a rod's moment of inertia is less if it rotates around an axis through its center as opposed to an axis passing through its end.

Balancing and Center of Mass

Torque and rotational motion are related to the common practice of balancing. When an object is balanced, it is in a state of equilibrium. No net force or torque acts on the object. To balance a broomstick on the end of your finger, you must constantly adjust your finger to keep the **center of mass** of the broomstick over the support point of your finger. The center of mass can be considered a point where all the mass of the object is concentrated as a result of averaging all the masses of particles making up the object. The center of mass of a uniform sphere is the center of the sphere, and the center of mass of a meter stick is at the

Table 4.1
Moment of Inertia for Solid Objects of Mass m about Indicated Axes

Object	Shape	Moment of Inertia
Solid cylinder of radius r		$I = mr^2$
Thin rod of length l		$I = \dfrac{1}{12} m\, l^2$
Thin rod of length l		$I = \dfrac{1}{3} m\, l^2$
Solid sphere of radius r		$I = \dfrac{2}{5} m\, r^2$
Thin cylinder or hoop of radius r		$I = mr^2$
Thin rod of length l		$I = \dfrac{1}{3} m\, l^2$

50 centimeter mark. Center of gravity is often used interchangeably with center of mass, but there is slight difference between the two. The center of gravity depends on the local gravitational field. For locations on Earth the slight changes in the gravitational field at different locations mean that the relationship between mass and weight (which is determined by gravity) is not an exact constant. The difference in the gravitational field over the surface of the Earth is so small that for all practical purposes the center of mass and center of gravity can be considered identical.

Any object that is balanced has its center of gravity located over a support point such as its base. As long as the center of mass of an object stays above a point of support, a state of equilibrium exists and the object can stay balanced. In the example of balancing a broomstick on your fingertip, you constantly shift your finger so it remains under the center of gravity. If a pushbroom is balanced, it is much easier to balance the broomstick when the heavier sweeping part is attached and at the upward end opposite your finger. This is because in this configuration the broom has a much larger moment of inertia around the axis of rotation that runs along your finger as compared to the broomstick without the sweeper part attached. The larger moment of inertia means there will be a greater resistance to rotation around your finger. Fewer rapid corrections will need to be made to maintain rotational equilibrium with the sweeper part attached. A similar result can be experienced when balancing a baseball bat on the handle end as compared to the barrel end. Increasing our moment of inertia by extending our arms is a common reaction to maintain our balance when we walk across a narrow path, for example, a log over a creek. Likewise, a tightrope walker moving along a high wire carries a long pole to increase the moment of inertia around the wire, thereby allowing her ample time to make corrections to maintain balance.

Angular Momentum

In chapter 3, linear momentum was defined as the product of mass and velocity, and the conservation of linear momentum was introduced. Just as a moving object tends to retain its linear momentum, a rotating object tends to continue its rotation. The **angular momentum** of an object is the rotational analogy to linear momentum. Whereas linear momentum is equal to mv, angular momentum is equal to the product of the moment of inertia and angular velocity, $I\omega$. Just as linear momentum is conserved in the absence of a net external force, the angular momentum of a system is conserved in the absence of a net external torque acting on the system (see chapter 5 for a discussion of the system). The conservation of angular momentum can be stated mathematically as $I\omega$ = constant.

The conservation of angular momentum explains why it is more difficult to remain upright on a slow-moving bike. Pedaling slowly on a bike means that the tire-wheel system has less angular momentum compared to when the bike is pedaled faster. A system with a greater angular momentum will require relatively more torque to change. When pedaling at a fast rate, a bicyclist has no difficulty maintaining balance and following a straight-line direction of travel because of the angular momentum of the tire-wheel system. Conversely, it is difficult to maintain balance and follow a straight line on a slowly pedaled bike and impossible to balance for more than a second or two on a stationary bike. The stability that the spinning tires impart to a bicycle is also evident in other spinning objects such as tops, gyroscopes, Frisbees, and yo-yos. In each case, rotation in a specific direction is maintained as angular momentum is conserved.

A familiar example of conservation of momentum is observed when a figure skater goes into a spin. Initially, the skater's arm and perhaps a leg will be extended, and the rotation will be relatively slow (Figure 4.11). As the skater's appendages are pulled in, the skater's moment of inertia, I, decreases. If I decreases, then the angular velocity, ω, must increase in order for $I\omega$ to remain constant. This is seen as the skater's rotation increases rapidly. A similar result occurs when a diver or gymnast goes into a tuck in order to perform a spinning move.

Figure 4.11
Conservation of momentum. The skaters angular velocity increases as she brings her arms in, changing her moment of inertia.

The conservation of angular momentum can also be applied to an object revolving around a point, for example, uniform circular motion. In this case the angular momentum can be written in the form $mv_t r$. This equation shows that the angular momentum of a revolving object is the product of the object's linear momentum and radius. Conservation of angular momentum is the basis of Kepler's third law. As a planet revolves around the Sun in an elliptical orbit, its radius changes. When the radius decreases, the tangential velocity must increase for angular momentum to be conserved.

Summary of Translational and Circular Motion Equations

Throughout chapters 3 and 4 equations were presented to describe both translational and rotational motion. Equations developed for translational motion have analogous equations that apply to rotational motion. Table 4.2 lists the main

Table 4.2
Kinematic Equations for Translational and Rotational Motion

	Translational Motion	Rotational Motion
Final velocity	$v_f = v_i + at$	$\omega_f = \omega_i + \alpha t$
Displacement	$\Delta x = \dfrac{v_f + v_i}{2}t$	$\Delta \theta = \dfrac{(\omega_f + \omega_i)}{2}$
Displacement	$\Delta x = v_i t + \dfrac{1}{2}at^2$	$\Delta \theta = \omega_i t + \dfrac{1}{2}\alpha t^2$
Change in velocity	$2a\Delta x = v_f^2 - v_i^2$	$2\alpha\Delta \theta = \omega_f^2 - \omega_i^2$
Acceleration	$a = \dfrac{\Delta v}{\Delta t}$	$\alpha = \dfrac{\Delta \omega}{\Delta t}$
Momentum	mv	$L\omega$
Newton's second law	$F = ma$	$\tau = L\alpha$

equations for translational and rotational motion. The first four equations are the kinematic equations presented in Table 3.1. Each of the equations for translational motion has a corresponding equation for rotational motion. The four equations for rotational motion apply when the angular acceleration is constant, for example, uniform circular motion. The time interval, t, for the first four equations (both translational and rotational) is the time over which the change in motion takes place.

5

Work, Energy, and Heat

Introduction

Life requires a continual input of energy into the Earth. This energy is distributed over the Earth by wind and water currents. Our ancestors learned how to harness these forms of energy to perform work. The rigging of a sail on a boat or the use of a watermill to grind wheat are simple examples of how energy is transformed into work. Modern society primarily depends on the chemical energy locked inside fossils fuels for its energy. Energy is required to support a high standard of living, and the development of an industrial society depends on an adequate supply of energy. Because of this, energy is often examined as it relates to politics, the environment, and technology. We use the terms **work, energy, and heat** every day, but they have very specific meanings in the context of physics. When someone says they are *going to work* or to *turn up the heat,* we know exactly what is meant. Problems can arise, though, when we try to relate the colloquial meanings of these terms to their physical meaning. This chapter will focus on the scientific meaning of work and energy and introduce how heat

relates to these two concepts. Heat and how it affects matter will be more fully explored in the next chapter.

System and Surroundings

In order to examine work, energy, and heat it is helpful to define a system to which these concepts apply. A **system** is that part of the universe a person is interested in and wants to examine. The concept of a system often seems abstract and arbitrary. At times it is very easy to define a system, while in other cases it might not be obvious what the system includes. For example, when you take an ice cube out of the freezer and place it in a glass at room temperature, the system can be defined as the glass and its contents. In addition to the glass, the system includes the ice cube, liquid water as the ice cube melts, and the air in the glass. Another person may look at the same situation and define the system as just the contents of the glass and not include the glass itself. When two cars collide, one person may define the system as the two cars, while another person may define it as the cars and the road between the two cars. A climatologist may define a system as the entire

Earth and its surrounding atmosphere or just one portion of the atmosphere. The ultimate system is the entire universe. It is important to realize that how a system is defined is subjective and depends on an individual's particular point of view. The important point is that once a system is defined for an analysis, that it not be altered.

Everything outside the system is considered the **surroundings.** This means the entire universe excluding the system comprises the surroundings. Although everything outside the system comprises the surroundings, generally speaking only the immediate surroundings are considered when analyzing a situation. A useful conceptual formula is

system + surroundings = universe

Systems may be defined as open, closed, or isolated. In an **open system** both matter and energy can move freely across the system's boundary. An open jar containing water would be an open system, since the water can evaporate from the jar and energy can pass freely across the system's boundary. In a **closed system** matter is confined to the system, but energy can move across the system's boundary. If a lid were placed on a jar containing water, it would then be a closed system. In an isolated system neither matter nor energy is exchanged between the system and its surroundings. A closed thermos bottle approximates an **isolated system.** A thermos is well insulated to minimize heat transfer between it and its surroundings and when sealed matter cannot enter or leave the system.

Potential, Kinetic, and Internal Energy

Energy is one of the most basic scientific concepts and is a unifying principle that has applications in all scientific disciplines. Energy is defined as having the ability to do work or produce heat (or both). Many different forms of energy exist, and these are often classified according to their source, including electrical, nuclear, chemical, solar, and geothermal. While these categories may seem distinct, in many respects they actually represent different forms coming from a common source. For example, electrical energy obtained from a power outlet often results from the chemical energy released from the burning of fossil fuels. In turn, fossil fuels were once living matter that captured solar energy and converted it to biomass. Most forms of energy on Earth can be traced back to the Sun. The energy contained in fossil fuels was originally captured through photosynthesis. Hydropower results from the solar energy used to evaporate water and drive the water cycle. Likewise, differential solar heating of the Earth drives energy obtained from the wind.

Energy may be stored within a system in several forms. Three general forms by which energy is stored in a system include **potential energy, kinetic energy, and internal energy.** Potential energy can be considered stored energy that results from the position or configuration of the system. The gravitational potential energy of an object is due to its position with respect to the Earth's surface or another appropriate reference level. Gravitational potential energy is defined as the product of mass, the gravitational constant, and the height above the reference level:

gravitational potential energy = mgh

It is important to specify the reference level when reporting the gravitational potential energy. While the Earth's surface is a natural reference level, it often makes sense to use another reference level. For instance, a person in a room on the third floor might choose the floor of the room rather than the Earth's surface as a more logical reference level for calculating gravitational potential energy. Another form of potential energy is chemical potential energy. Chemical potential energy results from the position of electrons with respect to nuclei and is stored in the bonds of chemical compounds. During a chemical reaction atoms are rearranged, and it is this reconfiguration of the atoms that causes the energy changes accompanying the reaction. The arrangement and reconfiguration of atomic nuclei is the source of nuclear potential energy. Elastic potential energy results from the deformation on an object. For example, when a spring is compressed or stretched, energy is stored in the spring as elastic potential energy.

Kinetic energy is the energy of motion. Anything that moves possesses kinetic energy. Translational kinetic energy is equal to one-half times the product of mass and the magnitude of velocity squared:

$$\text{translational kinetic energy} = 1/2(mv^2)$$

Rotational kinetic energy is equal to one-half times the product of the moment of inertial and the magnitude of rotational velocity squared:

$$\text{rotational kinetic energy} = 1/2(I\omega^2)$$

The fact that translational kinetic energy is related to the square of the velocity has consequences during storms when wind speeds increase significantly above normal conditions. A 50 mph wind compared to a 10 mph wind has 25 times the kinetic energy, while hurricane winds of 100 mph would have 100 times the kinetic energy of a 10 mph wind.

An object resting on the ground has zero kinetic energy and zero gravitational potential energy with respect to the Earth's surface, but it would be wrong to assume it possesses no energy. The object is made of atoms, and each atom contains electrons revolving around an atomic nucleus. By virtue of their translational, rotational, and vibrational motion, atoms possess kinetic energy. Matter also possesses chemical potential energy stored within chemical bonds, as mentioned previously. The internal energy is the sum total of the kinetic and potential energy possessed by the atoms and molecules comprising matter.

Work

Inherent in the definition of energy are the terms work and heat. Work and heat are transfer properties and can be considered processes that transfer energy across a system's boundary. Work can be transferred to the system across the system boundary from the surroundings or from the system to the surroundings. Work is done when a force is applied to an object, causing it to move. You can push on a wall all day but no work is done on the wall because it doesn't move. Work, W, is the product of force, F, and the magnitude of displacement or distance over which the force acts, $W = Fd$. In the simple case where a constant force is applied to an object and the force and motion are parallel and in the same direction, work is calculated by multiplying the resultant force applied to the object by how far it moves. Often the force applied to the object is applied at an angle. In this case only the component of force in the direction of motion is used to determine work, and so the equation for work becomes $W = Fd \cos \theta$, where θ is the angle between the force and direction of motion (Figure 5.1). When the force and motion point in the same direction, work is defined as positive, while work is negative when the force and displacement point in

Figure 5.1
Work is defined as the component of force in the direction of motion times the magnitude of displacement or distance the object moves

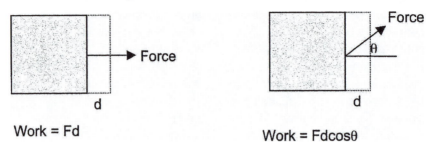

Work = Fd

Work = Fdcosθ

opposite directions. An example of the latter takes place with friction. As a person pushes an object across the floor, the person does positive work on the object, but the force of friction does negative work. In the previous example, the force applied to the object was considered constant. When the force is variable, it is required to consider the force exerted over small increments and use the methods of integral calculus to determine the work done on the object.

Heat and Temperature

Heat is another means by which energy is transferred between a system and its surroundings. Heat is defined as a transfer of **thermal energy** from a body of higher temperature to a body of lower temperature. Energy and heat are often considered as synonymous terms, but heat involves the transfer of thermal energy. A system possesses a specific amount of energy (kinetic, potential, internal), but not heat. Heat only has meaning in the context of the movement or transfer of energy. When saying that heat is added to or subtracted from a substance, what is meant is that heat flows into or from the substance.

Until as late as the middle of the nineteenth century many scientists believed that heat was an actual substance. The caloric theory held that heat was a fluid substance

that was released when substances interacted. Caloric was believed to be similar to the proposed substance phlogiston that was used to explain combustion, until dispelled at the end of the eighteenth century. Several scientists dispelled the theory of caloric, but major credit is given to Benjamin Thompson (1753–1814). Thompson was raised in Massachusetts but fled to Europe after he was discovered spying for the British during the Revolutionary War. In Europe he settled in Bavaria and climbed the social ladder until he was appointed as a Count in the Holy Roman Empire. He took the name Rumford from the town Rumford, Massachusetts, where he lived his adult life while in the colonies. Count Rumford's ideas on caloric grew out of his work at the Munich munitions works where cannons were bored. Rumford used the boring process to design several experiments that dispelled the idea of heat consisting of the fluid caloric. Rumford showed that the metal chips could not account for the heat generated in the boring process. By using a dull boring cylinder, immersing the cannon in water, and observing the temperatures during boring, Rumford demonstrated that the heat produced did not come from the metal chips and ruled out several other potential sources of caloric. Rumford observed that the amount of caloric transferred to the cannon from the boring bit seemed to be inexhaustible. In his 1798 paper on heat to the Royal Society, Rumford concluded: "It is in hardly necessary to add that anything which any insulated body, or system of bodies, can continue to furnish without limitation cannot possibly be a material substance: and it appears to me to be extremely difficult, if not quite impossible, to form any distinct idea of anything, capable of being excited and communicated, in the manner the heat was excited and communication in these, except it be motion." Ironically, Thompson married Antoine Lavoisier's (1743–1794) widow Marie Anne Lavoisier (1758–1836), but the marriage lasted only several months. Lavoisier had dispelled the phlogiston theory in chemistry and believed that caloric was a basic substance similar to chemical elements, but Count Rumford dispelled the caloric theory.

Since heat is related to temperature, it is important to have a clear understanding of the meaning of temperature. Temperature is one of those terms that's continually used but rarely given much thought. As long as we can remember, we've had our temperature taken, observed daily temperatures in weather reports, and baked foods at various temperatures. Intuitively we think of temperature in terms of hot and cold, but what does temperature actually measure? Temperature is a measure of the random motion of the particles (atoms, molecules, ions) making up a substance. Specifically, temperature is a measure of the average kinetic energy of the particles in a substance. As defined previously, kinetic energy is energy associated with motion. To illustrate the concept of temperature, consider the atmosphere surrounding us. The atmosphere is comprised almost entirely of nitrogen and oxygen molecules. Each of these molecules contains a certain amount of kinetic energy by virtue of the fact that each has mass and is moving. If we could somehow measure the speed of gaseous molecules in the atmosphere, we would discover that they move at various speeds. While most move near some average speed, some are moving much slower, and others much faster than average. A graph of the distribution of molecules at certain speeds is shown in Figure 5.2. A velocity of 400 m/s is roughly 800 mph and represents the typical speed gas molecules would have at room temperature. The shape of graph 5.2 is dependent on temperature. At a higher temperature a greater percentage of the molecules would be moving with greater velocities, and at a lower temperature a greater percentage with slower velocities (Figure 5.3).

Higher and lower velocities translate into higher and lower kinetic energies of the molecules, respectively. A change in temperature can be interpreted in terms of the kinetic energy at the molecular level. An increase in temperature implies more energetic particles, and a decrease in temperature less energetic particles. When a mercury thermometer is placed in a hot oven, it is exposed to more collisions with greater kinetic energies. The kinetic energy of the gas particles is transferred to the kinetic energy of the mercury, causing it to expand and register a higher temperature. The expansion of the mercury can be compared to a rack of billiard balls being struck by the cue ball. At the lower temperature, the column of mercury in the thermometer is like the racked balls. An energetic cue ball is like an energetic gas molecule in the oven. The cue ball transfers its kinetic energy to the racked balls, causing them to spread out.

Similarly, the energetic gas molecules in the oven impart their energy to the mercury atoms, causing them to expand up the thermometer column. Cooling would cause the mercury's volume to decrease. If the mercury were cooled to its freezing point of $-39°C$, the formation of solid mercury might be

Figure 5.2
Distribution of molecular speeds at a given temperature. The distribution indicates that molecules move at a variety of speeds.

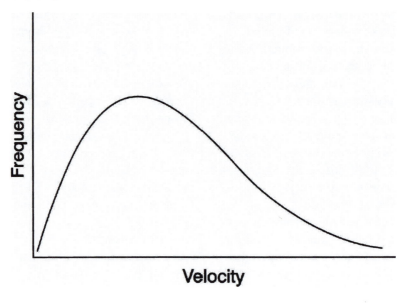

Figure 5.3
As the temperature increases, a greater percentage of molecules move at higher speeds

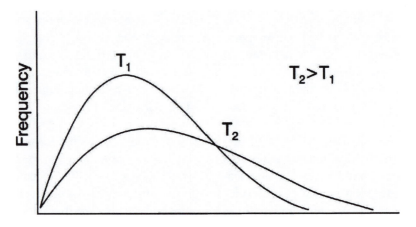

considered analogous to racking up the billiard balls.

The two most common scales used to measure temperature are the Fahrenheit and Celsius scales. Daniel Fahrenheit (1686–1736), a German, proposed his temperature scale in 1714 in Holland. Fahrenheit invented the modern mercury thermometer and calibrated his thermometer using three different temperature standards.

He assigned a value of zero to the lowest temperature he could obtain using a mixture of ice, salt, and liquid water. A value of 30 was used for a mixture of ice and freshwater. The third point was set at 96, based on the oral temperature of a healthy man. Using his scale, Fahrenheit determined the boiling point of water would be 212. He later changed the standard of the ice-freshwater mixture from 30 to 32 in order to have an even 180 temperature divisions between the freezing and boiling points of water.

Anders Celsius (1701–1744), from Sweden, devised his temperature scale in 1742. Celsius assigned a value of zero to the boiling point of water and 100 to the temperature of thawing ice. Instrument makers soon reversed the zero and 100 to give us the modern freezing and boiling points of water as 0°C and 100°C, respectively. The relationship between the Fahrenheit and Celsius (formerly called centigrade) temperature scales is given by the two equations

$$\text{degrees Celsius} = \frac{5}{9}\,(\text{degrees Fahrenheit} - 32)$$

$$\text{degrees Fahrenheit} = \left(\frac{9}{5} \times \text{degrees Celsius}\right) + 32$$

Both the Fahrenheit and Celsius temperature scales are based on the physical characteristics of water. The use of water to establish a zero point means that both 0°F and 0°C are arbitrary and not true zero points. Since temperature is a measure of the random motion of the particles making up a substance, true zero on either the Fahrenheit or the Celsius scale would imply that there is no motion at 0°F and 0°C. Substances at these temperatures are made of particles that possess an ample amount of motion and

kinetic energy. A temperature scale that has a true zero is known as an absolute temperature scale. One problem that arises when using the Fahrenheit and Celsius temperature scales, which are not are not absolute scales, is that mathematical comparisons do not portray a true quantitative relationship. For example, if we compare equal quantities of water at 10°C and 20°C, we might expect the water at 20°C to be twice as hot or have twice as much energy as the water at 10°C. This is not the case. As an analogy consider a child who is 3 feet tall and an adult who is 6 feet tall. Our absolute scale for measuring a person's height is the vertical distance from the bottom of the feet, and using this absolute scale the adult is twice the height of the child. But what if both the child and adult stood on top of 5 foot ladders, and we measured their heights from the ground? The child would now measure 8 feet and the adult 11 feet. The adult on this revised scale is no longer twice as tall as the child. The Fahrenheit and Celsius temperature scales are like measuring heights of people standing on ladders. It might sound ridiculous to measure people standing on ladders, but the heights would seem normal if that's how we've always measured height. Similarly, reporting temperatures with Fahrenheit and Celsius units seems perfectly normal because that's how its always done.

The most common absolute temperature scale used by scientists is the Kelvin temperature scale, proposed by William Thomson, Lord Kelvin (1824–1907). Kelvin's temperature scale has an **absolute zero.** At absolute zero, all molecular motion ceases, but because of the Heisenberg uncertainty principle (see chapter 10), atoms and molecules still possess **zero-point motion.** So it is incorrect to claim that all motion ceases, since vibrations and rotations still occur. True comparisons can be made using the Kelvin scale. A substance at a temperature of 400 kelvins contains particles with twice as much kinetic energy as a substance at 200 kelvins. Absolute zero is the temperature where the random motion of particles in a substance stops. It is the absence of temperature. Absolute zero is equivalent to –273.16°C. The relationship between the Kelvin and Celsius temperature scales is

kelvins = degrees Celsius + 273.16

For most work it is generally sufficient to round off 273.16 to 273.

Units for Energy, Work, and Heat

Work, heat, and energy are expressed in the same units, but these vary according to the situation. The SI unit for energy is the **joule,** abbreviated J. The joule was named in honor of the English physicist James Prescott Joule (1818–1889). The joule is a rather small unit of energy. It takes approximately 10,000 J to raise the temperature of a coffee cup of water by 10°C. Often kilojoules (kJ) are used, with 1 kJ equal to 1,000 J. Another popular unit is the **calorie,** abbreviated with a "c." A calorie is the amount of energy necessary to raise 1 g of water by 1°C. One calorie is equal to 4.18 J. The physical unit of a calorie should not be confused with a food calorie. A food calorie, symbolized with a capital C, is equivalent to 1,000 calories. So a piece of cake with 400 Calories actually contains 400,000 calories.

Conservation of Energy

The conservation of energy law states that the amount of energy contained in the universe is constant. Since the universe consists of a system and the surroundings, the conservation of energy means that the energy of a system and its surroundings will be constant. While it is impossible to measure the amount of energy the universe contains,

changes in the energy stored in a system may be examined. Energy can be stored in a system as potential, kinetic, and internal energy. Furthermore, energy can be transferred across the system's boundary as work. Another form in which energy can be transferred across a system's boundary is as heat. The transference of energy as work and heat results in changes in the energy stored within the system. While other forms of energy transfer can take place through mass transfer (such as evaporation out of an open jar), sound, nuclear radiation, and light, these will be minor in many situations and will not be considered. By examining the energy stored in a system and energy transference by work and heat, several fundamental principles relating energy, work, and heat can be derived as special cases of the conservation of energy law.

When a force is applied on an object (it will be assumed that the object itself comprises the system) causing movement, work has been done on the object. The force causes the object to accelerate according to Newton's second law: $F = ma$. The acceleration of the object can be written as $(v_f - v_i)/\Delta t$, so F equals $m(v_f - v_i)/\Delta t$. The work done on the object is Fd. The magnitude of the displacement, d, is equal to the average velocity times the time interval $(v_f + v_i)\Delta t/2$. Multiplying the equations for force and displacement gives an 2 expression relating work and kinetic energy:

$$\text{work} = Fd = m\frac{(v_f - v_i)}{\Delta t}\frac{(v_f - v_i)}{2}\Delta t$$

$$= \frac{1}{2}mv_f^2 - \frac{1}{2}mv_i^2 = \Delta \text{ kinetic energy}$$

This statement is one form of the work-energy theorem. It states that when a net force performs work on an object, and this work results only in changing the kinetic energy of the object, that the work equals the change in kinetic energy of the object. In this situation it is important to remember that the object itself is considered the system, with all its mass located at a point at its center of mass. The work-energy theorem does not apply for more complex systems and where forces such as friction are considered.

Another relationship between work and energy involves the change in the gravitational potential energy of an object. If an object (the system) is lifted off the ground, where its gravitational potential energy is zero, to a height, h, where its gravitational potential energy is mgh, the work done on the object by gravity is $-mgh$. The work is negative because the force of gravity acts in a direction opposite the direction that the object is lifted. Additionally, the work needed to raise an object is mgh. The height h is actually the change in height, and the work is calculated from the equation work $= Fd$, where F is equal to mg and d is the change in height, h. In this situation where the object is defined as the system, no net work is done on the object. The work done by lifting and by gravity sum to zero ($mgh + -mgh$). This result is also found by applying the work-energy theorem, since the change in kinetic energy of the object is zero. If the system were defined as the object and the Earth, the work done on the object would be mgh. This is because the gravitational force would be within the system's boundary, and work is defined as the transference of energy across the boundary. Likewise, if the system were defined as the object and the person lifting the object, the work done on the object by gravity would be $-mgh$. This illustrates the importance of defining the system, as discussed at the beginning of the chapter. Defining the system as the object and the Earth, the work done is mgh. Since mgh equals the change in potential energy of the object, work equals the change in potential energy: work $= \Delta$ potential energy. This is another version of the work-energy theorem

but actually represents another simplified special case of conservation of energy.

When an object (the system) falls from a particular height, h, where its gravitational potential energy is mgh with respect to the ground, gravitational potential energy is transformed into kinetic energy. Neglecting losses due to friction and other sources, the loss in gravitational potential energy results in a gain in kinetic energy as the object accelerates toward the ground. The object's gravitational potential energy will have been totally converted into kinetic energy the instant before the object strikes the ground. The kinetic and gravitational potential energy of the object are its mechanical energy. The relationship between gravitational potential energy and kinetic energy in the absence of other external forces is given by the conservation of mechanical energy. The conservation of mechanical energy states that the change in kinetic and gravitational potential energy for an isolated system is zero: Δ kinetic energy $+ \Delta$ potential energy $= 0$. In the case of an object dropped from height, h, the change in potential energy is given by $-mgh$. The change is found by subtracting the initial gravitational potential energy, mgh, from the final gravitational potential energy, which is zero. Letting v_f equal the speed of the object just before it strikes the ground, then the change in kinetic energy is $1/2(mv_f^2)$ (the initial kinetic energy is zero). Applying conservation of mechanical energy gives $mgh = 1/2(mv_f^2)$. Rearranging this equation and solving for v_f gives $v_f = \sqrt{2gh}$ or $4.43\sqrt{h}$, where h is in meters and v_f is in meters per second. An object dropped from 25 m should, therefore, have a speed of approximately 22.15 m/s just before it strikes the ground. The speed would actually be less than this due to frictional forces and the conversion of some of the initial gravitational potential energy to heat as the object falls.

Table 5.1
Sign Conventions for Q, W, and ΔE

Q	W	ΔE	Description
+	−	+ or −	Heat flows into the system, and the system does work on surroundings, so the internal energy could increase or decrease.
+	+	+	Heat flows into the system, and work is done on the system, so the internal energy increases.
−	−	−	Heat flows out of the system, and the system does work on its surroundings, so the internal energy decreases.

Work can change the kinetic and/or potential energy of a system but can also change the internal energy of a system. For example, when a tire pump is compressed, work is done on the gas within the cylinder of the pump. If energy is conserved, the internal energy of the system increases. Another way to increase the internal energy of the system is through the addition of thermal energy, in other words, by heat. It is important to remember that heat refers to the transference of thermal energy and is not the energy itself. Work done on a system and heat added to the system increase the system's internal energy. Conversely, the system may do work on its surroundings, and there may be a heat loss to the surroundings. This would lower the internal energy of the system. The change in internal energy due to transference across a system's boundary by work and heat is given by the first law of thermodynamics. This law states that the change in internal energy, ΔE, equals the work done on or by the system and heat transfer across the system's boundary: $\Delta E = Q + W$ (Q is heat and W is work). The sign conventions for heat, work, and the change in internal energy are summarized in Table 5.1.

Machines

A machine is a device that does work. Machines convert energy into useful work and are typically designed to multiply a force and/or alter the direction of the force. One of the simplest machines is the lever. Examples of the lever include using the claw of a hammer to pry a nail, a pick to lift a rock, or a bottle opener to remove a cap. Neglecting friction, conservation of energy dictates that the work input must equal the work output of the machine. For example, consider the lever shown in Figure 5.4. The lever allows a relatively small force exerted over a relatively large distance to be transformed into a relatively large force over a relatively small distance. All machines that multiply force do so at the expense of distance. Consider, for example, using a wrench to turn a bolt. In order to loosen the bolt, the force is multiplied by using a wrench. The bolt itself moves only a few

Figure 5.4
A force can be multiplied at the expense of the distance the force is applied using a machine. The work output equals the work input.

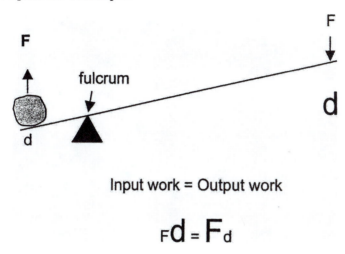

centimeters, while the wrench moves several times this distance.

The ratio of the force applied to the force delivered when using a machine is termed the **mechanical advantage** of the machine. If the distance between the fulcrum and force applied in Figure 5.4 is four times that of the distance between the fulcrum and the rock, then the mechanical advantage of the lever would be 4. An applied force of 50 newtons could theoretically be increased to 200 newtons, but to lift the rock 25 centimeters would require the force to be applied over a distance of 100 centimeters. The mechanical advantage gives the theoretical output based on the input. In reality, all the energy used to perform work cannot be realized as output work. The efficiency of a machine is the ratio of the work performed to the energy used to perform the work:

$$\text{efficiency} = \frac{\text{work output}}{\text{energy input}}$$

Simple machines have existed for thousands of years and all serve to make work easier. The lever is one simple machine. Archimedes (287–212 B.C.E.) is said to have discovered the law of the lever and exclaimed to King Hiero of Syracuse: "Give me a place to stand and I'll move the Earth." Five other simple machines include the wheel and axle, the pulley, the inclined plane, the wedge, and the screw (Figure 5.5). Although listed separately, the six simple machines are related and may be considered different configurations of a common machine. The wheel and axle transform linear motion along the circumference of a wheel to rotary motion. The force exerted at the circumference of the wheel is magnified at the axle. The force along the wheel is exerted over a longer distance compared to the axle. Conversely, a relatively large force exerted at the axle can produce a smaller force over a longer distance along the circumference of the wheel. Pulleys are modification of the wheel and axle in which a rope or cable is fed over the

Figure 5.5
Simple machines

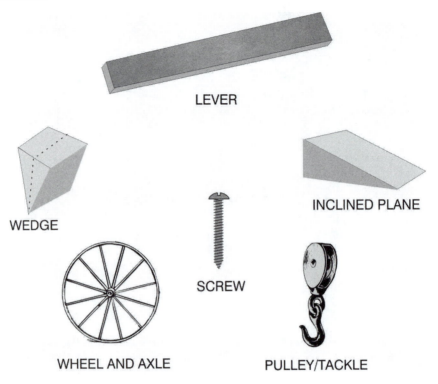

wheel or wheels in the pulley system. A single pulley serves to change the direction of a force. Two or more pulleys can be configured, such as in a block-and-tackle system, to magnify force. An inclined plane makes it easier to overcome the force of gravity. An inclined plane, such as a ramp, requires that a force be exerted along the hypotenuse of the ramp in order to lift an object a specific vertical distance. For example, in trying to lift a heavy object into the back of a pickup truck the force required can be significantly reduced by using a ramp with rollers to push the object along the ramp rather than trying to lift the object directly into the truck. A screw is an inclined plane that has been wrapped into a helical coil. Archimedes employed the screw over 2,000 years ago as a mechanical device to pump water. Archimedes' screw consisted of a screw encased in a solid cylinder that was tilted with one end immersed in water (Figure 5.6). By turning the screw at the opposite end, water was transported up the tube. Archimedes screws were used to pump bilge from boats and to irrigate fields. Versions of the Archimedes screw are still used today where primitive forms of agriculture are still employed.

Thermodynamics

The study of the relationship between work, heat, and energy comprises that branch of physics know as **thermodynamics.** The basic principles of thermodynamics are summarized in the laws of thermodynamics.

Figure 5.6
Archimedes screw

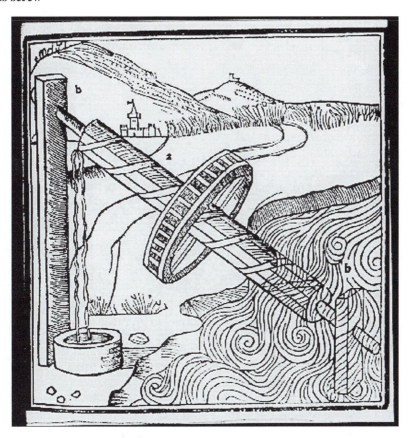

The **zeroth law of thermodynamics** defines thermal equilibrium. It states that if two systems are in thermal equilibrium with a third system, then they are in thermal equilibrium with each other. Thermal equilibrium means that systems are at same temperature, and therefore, no net thermal energy is transferred between the systems.

The **first law of thermodynamics** is the same as the conservation of energy principle discussed previously. It is stated mathematically as $\Delta E = Q + W$. It should be noted that many books will state the first law of thermodynamics as $\Delta E = Q - W$. In stating it in this form, the sign on the work term is negative because work is defined as positive when the system does work on the surroundings. The positive sign for W in the equation $\Delta E = Q + W$ is based on the fact that work is defined as positive when the surroundings do work on the system. Both conventions are acceptable, but it is important to remain consistent with positive and negative signs when using the first law.

While the first law of thermodynamics states that energy cannot be created or destroyed, the **second law of thermodynamics** deals with energy transformations and the direction of energy transfer. The second law can be stated in many different ways, but one form is that heat will not flow from a region of low temperature to a region of high temperature. Another version of the second law states that it is impossible for an engine to use 100% of the heat input to do work. The laws of thermodynamics can be illustrated by examining how these laws apply to common items such as engines, steam turbines, air conditioners, and refrigerators.

When energy transfer occurs and a change in volume takes place, work is done. The change in volume that occurs when solids and liquids are heated or cooled (and not converted to gases) is negligible, and therefore, the work represents only a small proportion of the change in internal energy. Liquids converted to gases (and gases themselves) can undergo appreciable volume changes when heated or cooled. Therefore, gases are typically employed when the goal is to produce mechanical work. Work obtained when a gas expands against a piston in a cylinder is illustrated in Figure 5.7. As a gas expands from an initial state to a final state, work is done on the surroundings by the system. The work done is the force exerted on the piston times the distance the piston moves. Since force/area = pressure, the force is equal to the pressure times the area. The distance the piston moves is equal to the change in volume, ΔV, divided by the area of the piston. Work in the case of expansion of the gas can thus be written as

$$\text{work} = \text{area of piston} \times \text{pressure}$$
$$\times \frac{\Delta v}{\text{area of piston}} = \text{pressure} \times \Delta v$$

Heat engines convert thermal energy into mechanical energy used to perform work. Common examples include gasoline and diesel engines used in transportation, and machinery and steam turbines used

Figure 5.7
Cylinder and work. As a gas expands against external pressure, work is done on the surroundings. This work equal the pressure times the change in volume, $P\Delta V$.

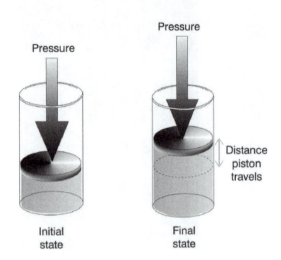

P-V WORK DIAGRAM

to produce electricity in power plants. A typical heat engine operates in a cycle in which heat is added to a working fluid such as water, causing it to expand and perform mechanical work. The expansion in a steam turbine involves the vaporization of liquid water into high-energy steam. This steam is forced through a turbine, where it loses energy to produce work. After leaving the turbine, the low-energy steam is condensed back to a liquid, and the cycle is repeated (Figure 5.8). When the low-energy steam condenses, unused or waste heat is expelled to the surroundings. In a gasoline engine, the combustion of fuel in the cylinder causes expansion in the cylinder, and waste heat is expelled to the surroundings with the exhaust.

The second law of thermodynamics limits the efficiency of heat engines. Heat engines cannot convert all the input energy

Figure 5.8
Steam turbine. Steam from a boiler expands through a turbine to produce work. The steam is condensed and pumped back to a boiler where the process is repeated.

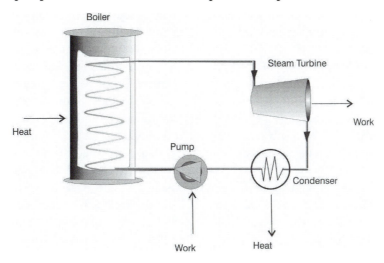

to useful mechanical energy. Some energy, in the form of heat, is always expelled to the surroundings. In a car engine this occurs when exhaust gases are forced out of the cylinders. The fraction of the original thermal energy entering the engine that is converted to net mechanical work is termed the efficiency of the heat engine. The maximum possible efficiency of a heat engine was derived by the French engineer Sadi Carnot (1796–1832) and is given by the equation

$$\text{maximum efficiency} = 1 - \frac{T_{\text{cold}}}{T_{\text{hot}}}$$

In this equation, T_{cold} is the absolute temperature of the cold reservoir, and T_{hot} is the absolute temperature of the hot reservoir. In a heat engine, the work is extracted when heat flows from the hot reservoir, or source, to the cold reservoir, or sink. In a power plant that uses superheated steam at 500 K and a sink at 300 K, the maximum efficiency would be 40%. Actual power plants run at efficiencies slightly less than this.

In a heat engine, work is obtained from the flow of heat from high-temperature source to low-temperature sink. Work can be used to force heat to flow from a region of low temperature to a region of high temperature. This is the reverse of the natural movement of heat from high temperature to low temperature. Heat pumps, refrigerators, and air conditioners are based on this principle. Energy is added to the working fluid in a heat pump by compressing the fluid. The process can be compared to using a bicycle tire pump for heating a space. If we pumped furiously on the tire pump, we would notice that the pump's cylinder becomes warm. Now if the pumping took place outside a building where the temperature was colder than the building, and there was a way to transfer the relatively warm air surrounding the pump cylinder into the building, then the building could theoretically be heated. Of course, this is another example where Superman would be needed to pump fast enough to produce a practical flow of heat to make a difference, but it illustrates, in part, how heat pumps, refrigerators, and air conditioners operate. Work is added to increase the energy content of the fluid. This is like using a pump to move water uphill against gravity.

The energy to perform work on the fluid in many devices is supplied by a compressor, and the fluid itself is often a substance that undergoes a phase change. For example, in a refrigerator the fluid is a substance like Freon, which vaporizes in the coils inside the freezer. The vaporization process requires heat, and this heat is supplied by the freezer compartment. The transfer of heat from the freezer to the fluid keeps the freezer cold. The heat absorbed by the fluid is expelled to the room as the vapor circulates and partially condenses in the coils on the exterior of the refrigerator. The fluid then returns to the compressor, where condensation is completed, and the cycle repeats itself. Whereas in a turbine or heat engine, a working fluid expands (vaporizes) to produce work, in refrigerators, air conditioners, and similar devices, work is used to compress and condense a fluid.

Power

The rate at which energy is transformed or transferred is defined as power. Included within this broader definition is the restricted definition of power as the rate at which work is performed. Power can be calculated by dividing the amount of energy transformed or work done by the time required to transform the energy or perform the work:

$$\text{power} = \frac{\text{energy}}{\text{time}} = \frac{\text{work}}{\text{time}}$$

The equation for power indicates that if work is performed twice as fast, then power

is doubled. The SI unit for power is the watt, which is equivalent to 1 J/s. The watt unit is name in honor of James Watt (1736–1819). Watt made numerous improvements to the steam engine and marketed his engines in the latter part of the eighteenth century. Major customers of Watt were the coal companies. These coal companies used horses to perform most of the labor such as hauling coal and driving pumps. In order to equate the work his steam engine could perform to that performed by horses, Watt defined the horsepower and rated his steam engines using this unit. One horsepower is equivalent to 33,000 ft-lb/min. This would mean that a horse capable of pulling 3,300 lb over a distance of 10 ft in 1 min would deliver 1 hp. One horsepower is equivalent to 746 W. Horsepower is still a commonly used unit to quantify the power of engines.

Summary

Energy, work, and heat are three related concepts that are constantly present in our daily lives. In fact, life and the entire universe can be considered a continual process of converting energy to useful work and, in the process, creating heat. The universe is the ultimate system for analyzing energy transformations. Within the universe, highly ordered forms of energy, such as fossil fuels, are burned to create work resulting in heat. Likewise, stars radiate energy to surrounding colder regions. If heat is thought of as random, disorganized energy, then the universe is continually moving toward a random disordered state. At some point in the distant future, once all highly ordered energy sources have been converted to heat, the universe will attain one constant temperature, and it will be impossible to perform useful work. This stage is called the heat death of the universe. While we don't have to concern ourselves with the heat death of the universe, more pressing energy concerns for humans involve using energy efficiently, developing new energy sources, and distributing energy to the world population.

This chapter has introduced several of the basic laws that describe the use of energy. The most fundamental of these is the first law of thermodynamics, which states that energy cannot be created or destroyed. This chapter has focused on energy as a property distinct from matter. In chapter 15, this idea will be placed in the context of relativity theory, and it will be seen that energy and matter are not distinct properties but are related.

6

Heat

Introduction

One of the earliest concepts we learn concerns the distinction between hot and cold. Each of us uses these words casually and in a relative sense. A temperature of 50° F may seem cold in July, but hot in December, depending on where you live. We quantitatively describe regions of hot and cold using temperature. Regions of hot and cold result in thermal energy transfer from a region of higher temperature to a colder region. The transfer of thermal energy between regions of different temperatures defines heat. When thermal energy is transferred between substances, a number of consequences can occur. These include temperature changes, phase changes such as melting and boiling, physical changes such as expansion and contraction, and chemical changes. In this chapter, we will examine several important physical changes that take place due to heat. Additionally, the three main modes of heat transfer—conduction, convection, and radiation—will be examined.

Heat and Temperature Change

The previous discussion of heat was limited to heat as it relates to work and internal energy. Heat was defined as a net transfer of thermal energy between regions at two different temperatures. When energy transfer in the form of heat occurs between an object and its surroundings, the most common result is a temperature change of the object. If a temperature change accompanies heat transfer, the heat transfer is termed **sensible heat** because it can be sensed or measured. While a temperature change is expected when heat transfer occurs, heat transfer does not always result in a temperature change. Heat transfer may cause a phase change, and no temperature change occurs as long as two phases are present. In this situation, heat is referred to as **latent heat.** The relationship between heat and phase changes will be examined in the next section.

The relationship between heat transfer, Q, and the change in temperature of a substance depend s on the **specific heat** capacity of the substance. The specific heat capacity of a substance is a measure of the amount of heat necessary to raise the temperature of 1 g of the substance by 1°C. The specific heats of several common substances are listed in Table 6.1. Table 6.1 demonstrates that the specific heat of a substance depends on its phase. The specific heat of liquid water is approximately twice that of ice and steam. Water has one of the highest specific heats compared to other liquids. The high specific heat capacity of liquid water is directly related to its chemical structure and the presence of hydrogen bonds. The high specific heat of water explains why coastal environments have more moderate weather than areas at similar latitudes located inland. Water's high specific heat capacity means coastal regions will not experience drastic temperature changes as compared to inland regions.

The relationship between heat, specific heat capacity, and temperature change of a substance is given by the equation $Q = mc\Delta T$. In this equation, Q is the amount of thermal energy (often Q is referred to simply as heat) transferred in joules, m is the mass in grams, c is the specific heat capacity of the substance, and ΔT is the change in temperature. The temperature change is equal to the final temperature minus the initial temperature. As an application of this equation, consider what happens when heating a pot of water on the stove. Assume that the pot contains about 1 L of water. Since the density of water is approximately 1 kg/L, the pot contains 1 kg or 1,000 g of water. In order to heat the water from room temperature, which is about 20°C, to 50°C would require

$$Q = mc\Delta T = (1,000 \text{ g})t$$
$$(4.2 \ \frac{J}{g = °C})(50 \, °C - 20 \, °C)$$
$$= 126,000 \text{ J} = 126 \text{ kJ}$$

By knowing three of the four variables in the equation $Q = mc\Delta T$, the fourth variable can be calculated. For example, if the quantity of heat transferred, mass, and temperature change were known, then the specific heat capacity could be calculated. The amount of heat required to raise the temperature of water in the previous example would actually be more than 126 kJ because not all of the heat is transferred to the water. Some is used to heat the metal pot itself, and some is lost to the surroundings. **Calorimetry** is an experimental method used to collect data to make heat transfer calculations. Calorimetry measurements are made with a **calorimeter.** Various calorimeters exist, but the most common type consists of an insulated container containing water. Figure 6.1 shows a simple calorimeter constructed out of Styrofoam cups. The specific heat of an object can be determined with calorimetry. To determine the specific heat of an object, for example a metal, a known mass of the metal is heated to a specific temperature. The metal is then

Table 6.1
Specific Heat Capacity of Some Common Substances

Substance	Specific Heat J/g=°C
Steel	0.45
Wood	1.7
Ice	2.1
Liquid water	4.2
Steam	2.0
Air	1.0
Alcohol	2.5

quickly transferred to the calorimeter that contains a known amount of water at a known temperature. After a short period of time, the metal and water reach thermal equilibrium, and the final temperature of the water, which is the same as the metal, is noted. The heat lost by the metal is transferred to the water. Since the calorimeter is insulated, very little of the heat is lost to the surroundings. The amount of heat gained by the water can be calculated using the equation $Q = mc\Delta T$, where m is the mass of the water, c is water's specific heat, and ΔT the change in temperature of the water. The amount of heat gained by the water is equal to the heat lost by the metal. Using the mass of the metal and its change in temperature, the specific heat capacity of the metal can be calculated.

Calorimetry is also used to determine the energy involved in chemical reactions including the metabolic rate of organisms. The energy content of food is determined using a bomb calorimeter (Figure 6.2). In bomb calorimetry, a quantity of food is compressed into a small pellet and ignited electronically within the bomb calorimeter's chamber. The amount of heat released is indicated by the change in water temperature within the calorimeter and converted to Calories. A food Calorie, indicated with a capital C, is equivalent to 1,000 calories; the English physical unit is spelled with a

Figure 6.1
Simple calorimeter

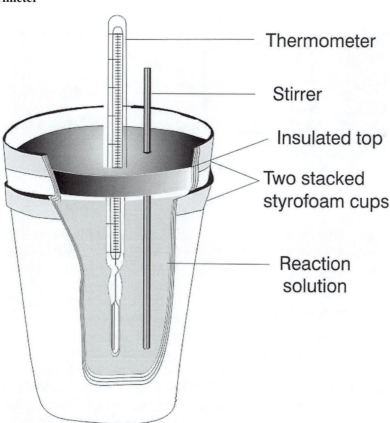

Thermometer

Stirrer

Insulated top

Two stacked
styrofoam cups

Reaction
solution

Figure 6.2
Bomb calorimeter

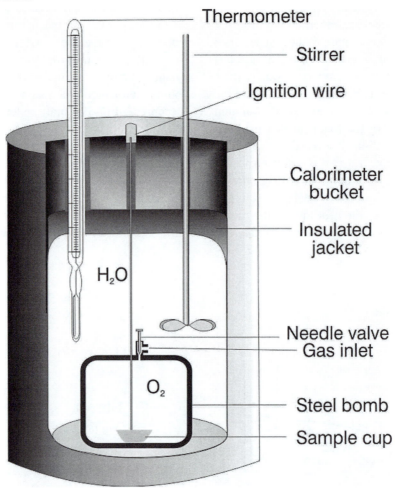

Thermometer

Stirrer

Ignition wire

Calorimeter bucket

Insulated jacket

H_2O

Needle valve
Gas inlet

O_2

Steel bomb

Sample cup

lowercase c. Therefore, a piece of cake containing 300 food Calories contains 300,000 physical calories.

Heat and Phase Changes

When heat is transferred to or from a substance, it is normal to think that its temperature will change, but this is not necessarily true. For example, a glass of ice water in a kitchen at room temperature will continually absorb heat from its surrounding but will remain at a constant temperature of approximately $0°C$ until all the ice is melted. Similarly, a pot of boiling water on a burner continually absorbs heat but remains at $100°C$. The common factor in these two examples is that when a system consists of multiple phases or states of matter, then its temperature remains constant regardless of the amount of heat added. The three most common states of matter are solid, liquid, and gas. Two less familiar states are **plasma** and **Bose-Einstein condensate.** Plasma is a gas in which most atoms or molecules have loss a proportion of their electrons. This state generally occurs when a gas is subjected to tremendous temperatures, is electrically excited, or is bombarded by radiation. Plasma is the principal component of interstellar space and comprises over 99% of the universe. The plasma state is also present in gas-emitting light sources such as neon lights. The state of matter known as the Bose-Einstein condensate was theoretically predicted by Satyendra Bose (1894–1974) and Albert Einstein (1879–1955) in 1924. A Bose-Einstein condensate forms when gaseous matter is cooled to temperatures a few millionths degree above absolute zero. At this temperature, atoms, which normally have electrons at many energy levels, coalesce. Different atoms are indistinguishable from one another and form what has been termed a "superatom." The first experimentally produced Bose-Einstein condensate was reported in 1995, and currently a number of research groups are investigating this newly discovered state of matter.

Focusing on the three common states of matter, six different phase changes are possible. These are summarized in Table 6.2. The most familiar phase change occurs between the solid and liquid states (freezing/melting) and the liquid and gaseous states (vaporization/condensation). Figure 6.3 illustrates the heating curve for water and depicts how temperature changes as heat is added to a substance. Initially, when ice is below its freezing point, heating will increase the ice's temperature just as with other solids. Once the ice reaches its melting point of $0°C$ (at 1 atm pressure), the ice starts to melt. At this stage both solid and liquid exist. The heating curve plateaus at the melting point even though heat is still being added to the ice-liquid mixture. It may seem strange that the temperature does not increase even though energy is added, but

Table 6.2
Phase Changes

Melting	Freezing
Solid→Liquid	Liquid→Solid
Sublimation	Deposition
Solid→Gas	Gas→Solid
Vaporization	Condensation
Liquid→Gas	Gas→Liquid

this can be explained by the fact that the added energy is used to convert ice to liquid water. The energy supplied at the melting point is not being used to increase the random kinetic energy of the water molecules, but to overcome intermolecular attractions between water molecules. The energy supplied at the melting point goes into breaking the water molecules free from the crystalline structure, resulting in the less-structured liquid state. As long as ice is present, any energy added goes into causing the phase change. The heat necessary to melt the ice is termed the **heat of fusion** for water, and its value is 6.0 kJ per mole of water. The heat of fusion of a substance is a measure of how much energy is required to convert a solid into a liquid. It would be expected that solids that are tightly held together would have high heats of fusion. Sodium chloride with its strong ionic bonds has a heat of fusion of 30.0 kJ per mole, aluminum has a heat of fusion 10.7 kJ per mole. Substances held together by weak forces have low heats of fusion. The heat of fusion of oxygen is 0.45 kJ per mole.

Once all the ice has melted, the energy added to the liquid water can now go to increasing the kinetic energy of the water

Figure 6.3
Heating curve for water

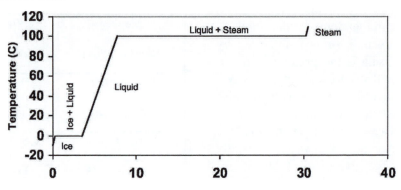

molecules, and the temperature begins to rise again. The rate at which the temperature rises is governed by the specific heat of liquid water, 4.2 J/g-°C. The specific heat of liquid water is twice that of ice, therefore, the rate at which the temperature increases for liquid water is only half of what it is for ice. The heating curve reaches a second plateau at water's boiling point of 100°C. At this temperature, any heat added to the water goes into breaking the hydrogen bonds between water molecules as liquid water is converted to steam. Just as with the first plateau at the melting point, the temperature remains constant as long as the liquid and gas phases coexist. The wide plateau at the boiling point indicates that much more energy is required for the vaporization process as opposed to the melting process. The heat needed to convert liquid water to steam is called the **heat of vaporization** of water. The heat of vaporization for water is 41 kJ per mole. This value is roughly seven times the heat of fusion and indicates it takes only 1/7 of the energy to melt water compared to vaporizing water. Vaporization takes significantly more energy than melting because to convert liquid water to steam requires completely breaking the hydrogen bonds and separating the water molecules. In the gas phase, the water molecules can

be thought of as independent molecules with minimal intermolecular attraction. The heat of vaporization of a substance is generally several times that of its heat of fusion. This is because the intermolecular forces present in the condensed phases must be overcome before a substance can be converted to gas.

A couple of points should be made about the heating curve of water. The last section of the heating curve represents the situation when all the liquid water has been converted to steam. At this point, as the temperature of the steam begins to rise, superheated steam would be obtained. It should also be realized that phase changes can be considered in terms of a cooling curve. In this case, following the curve from right to left shows that 41 kJ per mole of heat would have to be released to condense steam to liquid water, and 6.0 kJ per mole would have to be released by liquid water in order for it to freeze.

Phase Diagrams

The heating curve for water shown in Figure 6.3 shows that as heat is added to a substance it changes phases from solid to liquid to gas. The heating curve for other substances follows a similar pattern. While phase changes occur when heat is added or subtracted from a substance, pressure can also bring about phase

Figure 6.4
Phase diagram for water

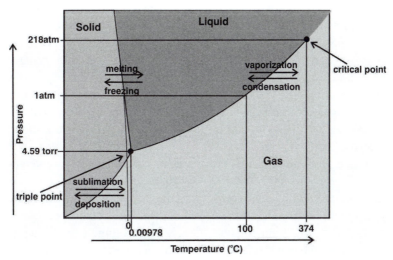

changes. An increase in pressure compresses a substance and favors the solid state, while a decrease in pressure allows a substance to expand and favors the gaseous state. The relationship between temperature, pressure, and phases can be displayed using a phase diagram. In a phase diagram, temperature is plotted on the horizontal axis and pressure along the vertical axis. Regions for solid, liquid, and gas are defined for various combinations of temperature and pressure. Figure 6.4 is the phase diagram for water.

At low temperature and pressure, water exists as a gas and a solid, but not as a liquid. At a temperature slightly above 0°C and a pressure of 4.58 torr, water can exist in all three phases. This point on the phase diagram is called the triple point. As temperature and pressure increase, a point is reached called the critical point. At the critical point and beyond, the substance cannot be liquefied and exists as a dense gaseous fluid called a supercritical fluid. A supercritical fluid has properties of both a liquid and a gas. It can penetrate materials as if it were a gas, but it also acts as a solvent, similar to a liquid.

Supercritical carbon dioxide is used as a substitute for more toxic organic solvents and is used to extract substances in chemical analysis. Supercritical CO_2 is used to extract caffeine to produce decaffeinated coffee.

The boundary between solid and liquid in the phase diagram for water has a slight negative slope. This means that as pressure is applied to ice its melting point decreases and solid water will be converted to liquid. This process takes place when an ice skater glides across the ice. The pressure the skater exerts on the ice is concentrated under the blades of the skates. This pressure is sufficient to momentarily melt the ice as the skater glides across the ice. The water quickly solidifies back to ice when the pressure is reduced after the skater passes.

Thermal Expansion of Substances

In addition to temperature and phase changes, another common phenomenon that occurs when substances are heated is a change in volume. In almost all cases, substances expand when heated. The most notable exception is the expansion of water as it approaches its freezing point. The latter will be examined as a special case after examining the thermal expansion of substances. When a substance absorbs thermal energy, the atoms comprising the substance become more energetic. The energy transferred to a substance produces greater motion in the substance's constituent atoms and molecules, resulting in expansion.

The change in length of an object when it is heated or cooled depends on the object's original length and the change in temperature. If the original length is L and the change in temperature ΔT, then the change in length, ΔL, is given by the equation: $\Delta L = \alpha L \Delta T$. In this equation, α is the coefficient of linear expansion. The unit for the coefficient of linear expansion is per degree Celsius ($1/°C$) and depends on the material. Table 6.3 gives the coefficients of linear expansion for several common substances.

Because materials generally expand or contract in three dimensions when heated or cooled, the change in volume of an object can be related to its original volume and temperature change. The equation governing the change in volume is analogous to that for change in length and is given by $\Delta V = \gamma V \Delta T$, where ΔV is the change in volume, V is the original volume, and ΔT is the temperature change. The symbol γ represents the coefficient of volume expansion. It has the same units as the coefficient of linear expansion, and values of γ are approximately three times that of α (Table 6.3). Values of the coefficients of expansion demonstrate why running hot water over a tight metal cap on a glass jar helps to loosen the cap. The coefficients of common metals are several times that of glass. Additionally, because the thermal conductivities of metals are high (see section on thermal conductivity), they experience a faster temperature change when placed under hot water. Therefore, the greater expansion of the metal cap compared to that of glass helps to loosen a tight cap.

While almost all materials expand when heated, water is one common substance that defies this behavior near its freezing point. The reason for this is due to water's crystalline structure and hydrogen bonding. Since each water molecule has two hydrogen atoms and two lone pairs of electrons, the water molecules can form a three-dimensional network of approximately tetrahedrally bonded atoms. Each oxygen is covalently bonded to two hydrogen atoms and also hydrogen bonded to two oxygen atoms (Figure 6.5). When water exists as ice, the molecules forms a rigid three-dimensional crystal. As the temperature of ice increases to the melting point

Table 6.3
Coefficients of Linear and Volume Expansion

Material	α (x 10^{-6}) per $°$C	γ (x 10^{-6}) per $°$C
Aluminum	24	70
Copper	17	51
Gold	14	42
Concrete	12	36
Glass (normal)	9	26
Glass (Pyrex)	4	11
Rubber	80	240
Mercury	—	180
Water	—	210
Gasoline	—	900
Air	—	3,700

of 0°C, hydrogen bonding provides enough attractive force to maintain the approximate structure of ice. The increase in temperature above water's freezing (melting) point causes thermal expansion, and this would normally lead to an increase in volume and a corresponding decrease in density. Since density is mass divided by volume, a larger volume results in a smaller density. The reason water actually becomes more dense is that at the melting point and up to 4°C, there is enough energy for some of the water molecules to overcome the intermolecular attraction provided by the hydrogen bonding that produces the crystalline structure. These water molecules occupy void space in the existing approximate crystalline structure. Since more molecules of water occupy the same volume, the density increases. This continues until the maximum density is reached at 4°C. At this point, the trapping of additional water molecules in the void space is not great enough to overcome the thermal expansion effect, which lowers the density; therefore, solid ice is less dense than liquid water.

The fact that the maximum density of water is 4°C has significant environmental impacts. Consider the freezing of a freshwater lake in winter. As the temperature decreases, the surface water becomes progressively denser and sinks to the bottom. This process helps carry oxygen from the surface to deeper water. The cycling of a lake's waters also helps to bring nutrients

Figure 6.5
Hydrogen bonding in a water crystal

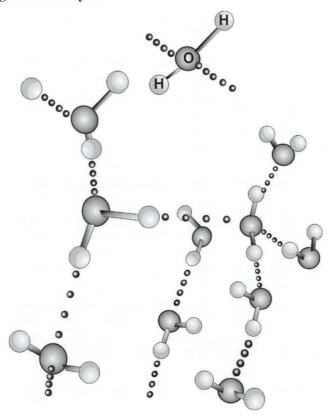

to the surface. This process is sometimes referred to as the fall overturn. Once the lake reaches a temperature of $4°C$, the water is at its maximum density. Further cooling results in less-dense surface water that doesn't sink, eventually forming a layer of ice when temperatures are cold enough. The ice layer effectively insulates the rest of the lake. Since the water was replenished with oxygen during the fall overturn, there is generally not a problem with the lake becoming anoxic. If water behaved like most substances, the entire lake would cool down to its freezing point and then the entire lake would freeze solid.

Heat Transfer

Heat is the movement of thermal energy from a region of higher temperature to a region of lower temperature. The movement of thermal energy takes place by three main processes: conduction, convection, and radiation. Conduction occurs at the atomic level as thermal energy is transferred between adjacent particles making up a substance. For example, when placing a metal spoon in a pot of boiling water, the energetic water molecules transfer their energy to the atoms and electrons in the end of the metal spoon immersed in the water. The atoms and electrons of the spoon become more energetic and in turn transfer their energy up the handle of the spoon, which in turn becomes hotter. Conduction can be thought of as a process in which electrons, atoms, and molecules are jostled and transfer this "bumping" to neighboring particles. Experience tells us that certain substances, such as metals, are particularly effective in conducting thermal energy. The reason for this is related to the atomic structure of metals. Metals are good conductors because their valence electrons hold metal crystals together loosely with metallic

bonds. Rather than valence electrons being held by a single nucleus, they are shared regionally by several nuclei. In this sense, the electrons are referred to as delocalized and are free to move throughout the crystalline structure making up the metal. The relatively unencumbered movement of this pool of delocalized free electrons in metals makes them good conductors of both thermal energy and electricity. Conversely, in other substances the valence electrons are held tightly and associated with a single atom. The lack of mobility makes these substances poor conductors of thermal energy. Poor conductors of thermal energy are known as **insulators.**

The amount of energy conducted through a material depends on the material, the temperature difference, the distance from the heat source, and the cross-sectional area. For example, consider a metal rod that is heated at one end, such as in Figure 6.6. The amount of conductive thermal energy that is transported through a cross section, A, a distance, L, down the length of the rod in a given amount of time, t, is given by the equation.

$$Q = \frac{kA\Delta T}{L} t$$

This equation shows that the energy conducted through the rod is directly proportional to the difference in temperature, cross-sectional area, time, and k. The symbol k represents the coefficient of thermal conductivity and depends on the type of material. Table 6.4 gives k values for a number of common materials. The table demonstrates that the coefficient of thermal conductivity is high for conductors and low for insulators. Table 6.4 shows that the coefficients of thermal conductivity for goose down and air are similar. The insulating properties of many materials are due to their ability to trap air, for example, fiberglass insulation and Styrofoam.

Heat transfer by convection occurs by movement of a substance, which carries energy from one location to another. On a global scale, convection takes place as wind and water currents transport energy across the face of the Earth. Heating systems in buildings use convection by using moving fluids, either air or water, to transport heat from a central energy source such as a furnace or boiler to the rest of the building. Natural convection takes place due to the differential densities of fluids. This takes place between areas of different temperature, pressure, or in the case of the ocean, salinity. The axiom "hot air always rises" is true because when air is heated it expands and its density decreases. As the Sun warms the Earth's surface, the surface will heat the air over it, causing it to rise by natural convection. This natural convection serves to mix the atmosphere near the Earth's surface and is an important aspect of local weather. Convective cells give rise to thunderstorms. In a number of areas, daily thunderstorms occur in the late afternoon as warm humid air rises by natural convection and condenses as precipitation. Another important consequence of natural convection producing a well-mixed atmosphere is distributing pollutants near the Earth's surface. Atmospheric mixing serves to dilute air pollutants. At night and under certain atmospheric conditions, an inversion can occur. During an inversion, cold air rests beneath warm air. Since cold air is denser

Table 6.4
Thermal Conductivity of Various Substances

Material	k (J/m =s=° c)
Aluminum	220
Lead	35
Steel	30
Water	0.60
Glass	0.80
Air	0.024
Goose down	0.025

Figure 6.6
The energy transferred by conduction, Q, in a given time through the cross-sectional area, A, depends on the temperature difference and distance, L, from the energy source

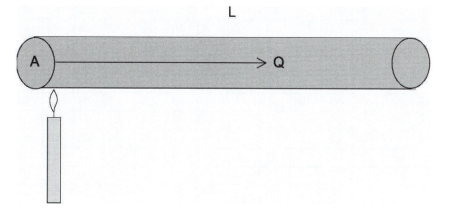

than warm air, atmospheric conditions are stable and natural convection is suppressed. Inversions often lead to elevated levels of air pollutants at ground level.

Forced convection depends on the use of a mechanical device to transport heat from one region to another. Pumps and fans are the most common devices used in forced convection. A car radiator uses the car's water pump to continually transport heat from the engine to the radiator, where a cooling fan and the car's forward motion assist in moving this energy to the atmosphere. Likewise, anytime you use a fan, you are using forced convection to transport heat away from your body. The fact that you feel cooler from convective heat loss when using a fan is analogous to the wind chill index often reported. The wind chill temperature is the calm air temperature that produces the same convective heat loss at that same temperature, accounting for wind. Table 6.5 displays the wind chill index for various temperatures and wind speeds. The wind chill index may also be calculated using the formula

$$\text{wind chill} = 35.74 + 0.6215T - 35.75(V^{0.16}) + 0.4275T(V^{0.16})$$

where T is the air temperature and V is the wind speed in miles per hour.

Thermal radiation is heat transferred by electromagnetic waves (Figure 6.7). Radiation occurs due to the constant motion of charged particles (electrons and protons) in the atoms comprising matter. When a charged particle such as an electron accelerates, it generates electromagnetic radiation. Electromagnetic radiation is emitted in all directions from any substance with a temperature above absolute zero. Radiation emitted from one object is absorbed by other objects, resulting in a change in motion at the atomic level. This transfer of energy is what occurs when heat transfer occurs by radiation. For example, as you stand in front of a heat source such as a wood stove, radiant energy is transferred to your body, causing your skin molecules to move faster and increasing your skin temperature.

While radiation is emitted as electromagnetic waves over a wide range of

Table 6.5
Wind Chill Chart

		Temperature (° F)													
		40	35	30	25	20	15	10	5	0	−5	−10	−15	−20	−25	−30
Wind Speed (mph)	5	36	31	25	19	13	7	1	−5	−11	−16	−22	−28	−34	−40	−46
	10	34	27	21	15	9	3	−4	−10	−16	−22	−28	−35	−41	−47	−53
	15	32	25	19	13	6	0	−7	−13	−19	−26	−32	−39	−45	−51	−58
	20	30	24	17	11	4	−2	−9	−15	−22	−29	−35	−42	−48	−55	−61
	25	29	23	16	9	3	−4	−11	−17	−24	−31	−37	−44	−51	−58	−64
	30	28	22	15	8	1	−5	−12	−19	−26	−33	−39	−46	−53	−60	−67
	35	28	21	14	7	0	−7	−14	−21	−27	−34	−41	−48	−55	−62	−69
	40	27	20	13	6	−1	−8	−15	−22	−29	−36	−43	−50	−57	−64	−71
	45	26	29	12	5	−2	−9	−16	−23	−30	−37	−44	−51	−58	−65	−72
	50	26	19	12	4	−3	−10	−17	−24	−31	−38	−45	−52	−60	−67	−74
	55	25	18	11	4	−3	−11	−18	−25	−32	−39	−46	−54	−61	−68	−75

wavelengths, most of the radiation is emitted over a limited range. The radiation emitted at any particular wavelength depends on the surface temperature of the substance. The peak wavelength emitted moves to shorter wavelengths and increases in intensity as the temperature increases (Figure 6.8). Figure 6.8 shows that most of the radiation coming from the Sun is in the visible range. Conversely, for objects at normal temperatures experienced on the Earth, most of the thermal radiation falls into the infrared range. This explains why objects can't be seen in a completely dark room. Infrared sensors or cameras can be used to "see" the object. As the temperature of an object at room temperature increases, more of the radiation it emits falls into the visible range. Once an object has reached a temperature of approximately 500 K, enough visible light is emitted from the object to detect a faint red glow. As the temperature continues to

Figure 6.7
Electromagnetic spectrum

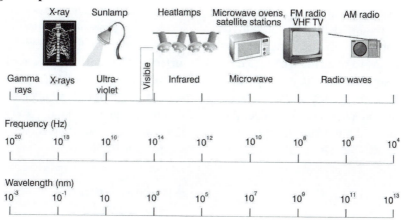

Figure 6.8
Peak wavelengths

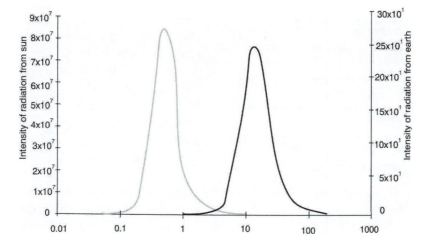

increase, the object's glow will change to the yellow and blue hues characterizing the shorter visible wavelengths associated with higher temperatures.

As an object continually emits radiation to its surroundings, it also continually absorbs radiation from its surroundings. When an object is at the same temperature as its surroundings, it is in thermal equilibrium, and the amount of radiation absorbed equals the amount of radiation emitted. Since our body temperature, approximately 37°C, is greater than our surroundings (the exception being if we are outside on an extremely hot day or in a hot tub, sauna, etc.), the absorption of radiation is important in reducing our metabolic requirements. The amount of radiation absorbed by an object depends on several factors. Most of us realize that a dark object will absorb more radiant energy than a light object. This is why we wear light-colored clothes on a hot day. An object that absorbs all the radiation falling upon it is termed a blackbody. Good absorbers of radiation are also good emitters of radiation. Therefore, a blackbody is not only a perfect absorber of radiation, but also a perfect emitter. The amount of radiation energy, Q, emitted by an object that is assumed to be a blackbody is given by the Stefan-Boltzmann law: $Q = \sigma T^4 A t$. In this equation, the radiation emitted is expressed in joules. The symbol σ is the Stefan-Boltzmann constant and has a value of 5.67×10^{-8} J/s-m²-K⁴. The symbols T, A, and t represent the absolute surface temperature of the object, the surface area of the object, and time, respectively.

Most objects are not perfect emitters, and the radiation from an object also depends on characteristics of the object's surface. For example, rough surfaces absorb and emit more radiant energy than smooth surfaces. The emissivity is a dimensionless number between 0 and 1 that equals the ratio of the amount of radiation an object radiates to how much it would radiate as a blackbody. The emissivities of several common materials are given in Table 6.6.

Applications of Heat Transfer

The use of insulation in buildings, clothing, and other situations reduces heat transfer by conduction, convection, and radiation. In buildings, materials such as fiberglass, cellulose, and urethane foam placed in ceilings, floors, and walls reduce heat transfer between the interior and exterior of the building. The materials retard the convective movement of air in the exterior shell of the structure. Bats of insulation often have a shiny aluminum backing that reflects radiant energy from the interior of an unfinished interior such as a garage. The material and the "dead" air it contains produce a material with a low thermal conductivity. Insulating material is rated using an R value, where R stands for the resistance

Table 6.6
Emissivities of Various Materials

Material	Emissivity
Polished Steel	0.07
Polished aluminum	0.05
Brick	0.45
Graphite	0.75
Wood	0.90
Paper	0.95
Water	0.96
Human Skin	0.98

to thermal energy transfer. *R* is the reciprocal of the coefficient of thermal conductivity when the latter is expressed in the units $\frac{BTU}{ft^2 - h°F}$. Since the coefficient of thermal conductivity and *R* are inversely related, a low *k* value gives a high *R*. Building codes across the country require a specific *R* rating for insulation, depending on where the insulation is placed. For example, in the central United States, R-38 may be required for ceilings and R-22 for walls. Fiberglass insulation has an *R* rating of approximately 2.2 per inch and cellulose 5.0 per inch. A thermos bottle uses principles similar to those of insulating a house. A thermos has a double wall with trapped air to reduce convection and produce a low conductive air space to retard the flow of heat. The silver interior of a thermos reflects heat back into the interior of the bottle. The shiny material surface used in insulation and thermos bottles is also used in emergency blankets. These blankets are made of a shiny Mylar material that reflects heat radiated by the human body back toward it.

When building on permafrost, it is essential to keep the ground frozen. If a structure or road is place over permafrost and the ground beneath thaws, the road may fail. An area where this is important is the trans-Alaskan pipeline. Much of the pipeline is built over permafrost, and the supports for the pipeline are anchored in permafrost. In order to keep the ground frozen, any excess heat must be transported away from the ground.

To accomplish this, the supports use a venting system with a Freon-type substance that has a low boiling point. The Freon absorbs heat from the surrounding ground, and this causes it to vaporize. The vapor moves up through the support and loses its heat to the surrounding air through fan-shaped appendages. As it loses heat, it condenses back into a liquid and falls back to the ground inside the support, where the cycle is repeated. The system, called a thermosiphon, acts as a convection pump to transfer heat from the permafrost to the atmosphere.

Summary

One of the most prevalent physical processes that occur in our life is the transfer of thermal energy or heat. Heat processes control the weather, our metabolism, transportation, and numerous other daily activities. The radiant energy received from the Sun is responsible for our very existence. The interaction of heat and matter is an important design consideration. Cooking utensils are made of glass and metals, winter clothing consists of insulating fabrics, and summer clothing is light colored and breathable. Our ability to understand the interaction of matter and thermal energy has contributed to an overall increase in the quality of life. As this understanding increases, greater advantages will be gained in wide-ranging areas. A few of these include energy conservation, new products, medical advances (thermography), and remote sensing (infrared imaging).

7

Matter

Introduction

Physics involves the study of energy and matter. Matter is anything that occupies space and possesses mass. In the last chapter, the interaction of heat and matter were examined. It was seen that matter normally exists as one of three common phases: solid, liquid, gas. In this chapter these three phases will be examined in greater detail. The first part of the chapter will explore solids and how solids behave when subjected to forces. This will be followed by an examination of fluids and the basic principles of both the static and dynamic state.

Solids

Solids are the most obvious state of matter for us; they are ubiquitous and come in all shapes and sizes. Solids are classified as either **crystalline** solids or **amorphous** solids. A crystalline solid displays a regular, repeating pattern of its constituent particles throughout the solid. At the microscopic level, crystalline solids appear as crystals. Amorphous solids do not display a regular repeating pattern of constituent particles. Crystalline solids are composed of atoms,

molecules, or ions that occupy specific positions in a repeating pattern. The positions occupied by the particles are referred to as **lattice points.** The most basic repeating unit making up the crystalline solid is known as the **unit cell.** The unit cell repeats itself throughout the crystalline solid (Figure 7.1).

Crystalline solids can be classified according to the type of particles occupying the lattice points and the type of intermolecular forces present in the solid. Ions occupy the lattice points in an **ionic solid,** and the crystal is held together by the electrostatic attraction between cations and anions. Common table salt is an example of a crystalline solid. In table salt, positive sodium ions and negative chloride ions occupy lattice points in a repeating crystalline pattern. In **molecular solids,** molecules occupy the lattice points, and van der Waals forces and/or hydrogen bonding predominates in this type of crystalline solid. Common table sugar, sucrose, is an example of a common molecular solid. **Covalent crystals** are generally composed of atoms linked together by covalent bonds in a three-dimensional array. Carbon in the form of graphite or diamond is an example of a covalent solid. The final type of solid is

Figure 7.1
A crystalline solid is composed of a repeating unit known as the unit cell. Lattice points are occupied by atoms, molecules, or ions.

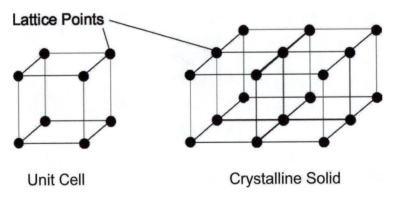

Lattice Points

Unit Cell   Crystalline Solid

a **metallic solid,** in which metallic bonding characterized by a delocalized sea of electrons holds together metal cations.

X-ray Diffraction

Much of our knowledge on the structure of crystalline solids comes from **x-ray crystallography** or x-ray diffraction. **Diffraction** is the scattering or bending of a wave as it passes an obstacle. X-rays are a form of electromagnetic radiation that travel as electromagnetic waves. They are particularly useful for probing crystalline solids because their wavelengths are roughly on the same scale as that of atoms, 10^{-10} m. Visible light can't be used for probing crystalline solids because the wavelengths of visible light are far too large to produce an image. A simple analogy may help to put this concept in perspective. Say you wanted to pick up a very tiny grain of rice with your fingers. It would be impossible to use your thumb and index finger in the usual manner to do this, because your fingers are just too large. However, you could easily pick up the grain with a pair of fine tweezers. Just as your fingers are too large to perform certain tasks, visible

light is too large to probe at the atomic level. X-rays are the right size to "observe" crystalline solids.

Max Theodor Felix von Laue (1879–1960) first predicted X-rays could be diffracted by a crystal in 1912. He received the Nobel Prize in 1914 for his discovery of x-ray crystallography. One year later, William Henry Bragg (1862–1942) and his son William Lawrence Bragg (1890–1951) shared the Nobel Prize for expanding on von Laue's work. In x-ray crystallography, an x-ray beam is focused through a crystal (Figure 7.2). The particles making up the crystal scatter the x-rays, resulting in a pattern of **constructive interference** and **destructive interference.** Constructive interference results when waves are in phase and reinforce each other, while in destructive interference the waves are out of phase and cancel each other (Figure 7.3). Because the particles (atoms, molecules, or ions) making the crystal have an ordered arrangement, x-rays scattered by the particles produce an ordered pattern. The pattern can be captured on a photographic plate or film, with dark areas representing constructive interference and light areas destructive interference. The diffraction pattern can then

Figure 7.2
Picture of x-ray diffraction pattern

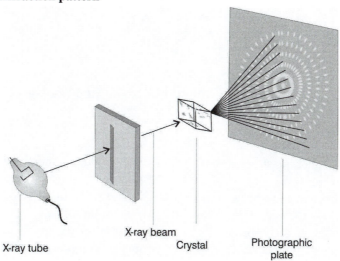

X-ray beam

X-ray tube

Crystal

Photographic
plate

Figure 7.3
**In constructive interference, waves reinforce each other. In destructive interference, waves
cancel each other, resulting in no wave pattern, as indicated by the horizontal line.**

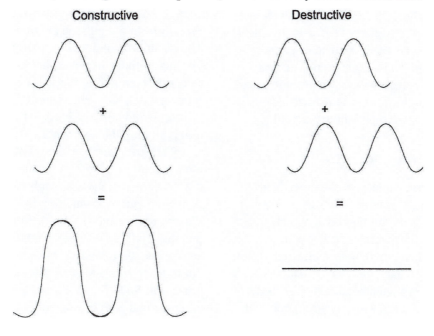

Constructive

Destructive

Figure 7.4
Three simple unit cells for crystalline solids

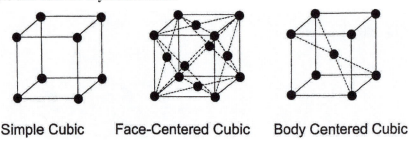

Simple Cubic   Face-Centered Cubic   Body Centered Cubic

be used to help decipher the structure of the crystal.

The x-ray diffraction pattern reveals how the particles are arranged in a crystalline structure. A number of common arrangements exist among crystalline solids. These can be classified according to the arrangement of particles making up the unit cell. Three common arrangements are simple cubic, face-centered cubic, and body-centered cubic (Figure 7.4). In sodium chloride, the Na^+ and Cl^- ions are positioned in a face-centered cubic arrangement. The alkali metals follow a body-centered cubic arrangement. In addition to the arrangement of particles in the crystalline solid, interpretation of the x-ray diffraction pattern allows scientists to determine the distance between particles and how much void space is present in the solid.

Amorphous Solids

The term "amorphous" comes from the Greek word meaning "disordered." Amorphous solids do not display a regular three-dimensional arrangement of particles. The most common amorphous solid is glass. We normally use the term "glass" in association with silica-based materials, although the term is sometimes used with any amorphous solid including plastics and metals. The technical definition for glass is an optically transparent fusion product of inorganic material that has cooled to a rigid state without crystallizing. Our discussion in this section will primarily focus on silica glass.

Both crystalline quartz and silica glass are primarily composed of silica, SiO_2. The difference between quartz and glass is that quartz displays **long-range order** (Figure 7.5). In quartz, the silicon is covalently bonded to the oxygen in a tetrahedral arrangement. When glass is produced, SiO_2 is heated to an elevated temperature and then rapidly cooled. The rapid cooling doesn't allow the SiO_2 to form a regular crystalline structure. The result is a solid that behaves like a very viscous liquid when it is heated. Glass is sometimes referred to as a solid solution and flows at a very slow rate. This can sometimes be seen in old window glass, where the bottom is slightly thicker than the top.

The first silica glass was produced around 3500 B.C.E. in Mesopotamia (present-day Syria and Iraq), although there is evidence of early production in Egypt and Phoenicia (Lebanon). The melting point of pure SiO_2 is 1713°C, but the mixing of other substances with the SiO_2 lowers its melting point. Egyptians added natron (sodium carbonate) to SiO_2. Some glasses are produced at temperatures as low as 600°C. As the art of glass making developed, individuals discovered how to produce different glasses

Figure 7.5

Both quartz and silica glass are primarily composed of silica, SiO_2. Quartz is a crystalline solid and has an ordered structure, but glass is an amorphous solid that is disordered. The structures are actually arranged in a three-dimensional tetrahedral pattern. They are shown here in a two-dimensional representation.

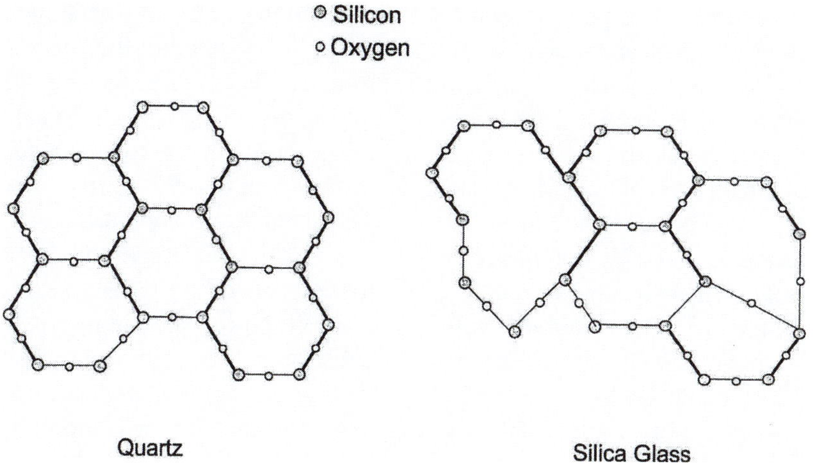

◎ Silicon
○ Oxygen

Quartz   Silica Glass

by adding various substances to the silica melt. The addition of calcium strengthened the glass. Other substances imparted color to the glass. Iron and sulfur gave glass a brown color, copper produced a light blue, and cobalt a dark blue. Manganese was added to produce a transparent glass, and antimony to clear the glass of bubbles. Most modern glass produced is soda-lime glass and consists of approximately 70% SiO_2, 15% Na_2O (soda), and 5% CaO (lime). Borosilicate glass is produced by adding about 13% B_2O_3. Borosilicate glass has a low coefficient of thermal expansion and, therefore, is very heat resistant. It is used extensively in laboratory glassware and in cooking, where it is sold under the brand name Pyrex.

Elasticity of Solids

As oppose to fluids, discussed in the next section, solids tend to maintain their shape when forces are applied to them. Still, solids may stretch, twist, or be compressed by forces. When these forces are removed and the solid returns to its original size and shape, it is called elastic. A prime example of this occurs when a spring is stretch or compressed. The relationship between the force applied to a spring and its displacement from equilibrium is called **Hooke's law** in honor of Robert Hooke. Hooke's law can be written as $F = kx$, where F is the applied force, k is the spring constant, and x is the displacement. Hooke's law applies for small displacements. Once a spring has been stretched beyond a particular point, it won't return to its original length. This point is known as the **elastic limit.**

Stretching a material puts the material under tension, and compressing the material puts it under compression. Tension and compression are caused by applying **stress** to a solid. Stress is defined as the force acting on the solid divided by the area over which it acts. For example, hanging weights from a spring stretches the spring due to the stress applied to the spring. The spring is deformed by the hanging weight. The deformation caused by the stress is called the stain and is defined as the change in dimension divided by the original dimension. For elastic materials, the strain is directly proportional to the stress. The proportionality constant between stress and strain is the **modulus of elasticity:** stress = (modulus of elasticity)(strain). Placing a solid under tension or compression will cause its length to change according to this equation. The change in length can be calculated by knowing the stress and the modulus of elasticity. The latter is known as Young's modulus, Y, and varies for different materials. Robert Young (1773–1829) was an English physicist. Table 7.1 gives Young's modulus for several common materials. Stress equals Young's modulus times strain. Using the definition of strain and stress give the equation: $F = Y(\Delta L / L_0)A$. This equation defines the elastic property of a solid when it is stretched or compressed.

In addition to compression and tension, a solid may be deformed with a shearing stress. A shearing stress occurs when pushing on the top of a deck of cards laying on a rough surface. When pushing on the top of the deck, the cards fan out because the bottom card sticks to the surface due to friction (Figure 7.6). When a shearing stress is applied to a solid, the shear stress is the force applied parallel to the surface divided by the area of the surface. In the card example, it would be the force applied to the top of the deck divided by the area of a card. The shear strain is defined as the displacement of the surface in the direction of the force divided by the thickness of the solid undergoing the shearing stress, $\Delta x/L_0$. The equation governing the relationship between the shearing stress and shearing strain is analogous to that for compression and tension: $F = S(\Delta X / L_0)A$. In this equation, S is the shear modulus (Table 7.2).

When a balloon is inflated, it expands it all directions. If the inflated balloon were subjected to increased pressure, it would be compressed. Similar to a balloon, solids can change volume as they expand and compress. This occurs when an object is subjected to changes in pressure. The relationship between the pressure and volume is $\Delta P = -B(\Delta V / V_0,)$ where ΔP is pressure change, ΔV and V_0 are the change in volume and the original volume, respectively, and B is the bulk modulus. The negative sign indicates that as pressure

Table 7.1
Young's Modulus for Various Materials

Material	Y (N/m²)
Steel	20×10^{10}
Bone (compression)	1.6×10^{10}
Bone (tension)	0.94×10^{10}
Aluminum	6.9×10^{10}
Wood	1.3×10^{10}
Glass	6.5×10^{10}
Nylon	0.39×10^{10}
Cotton	0.80×10^{10}

Table 7.2
Shear Modulus of Various Solid Materials

Material	S (N/m²)
Steel	8.1×10^{10}
Bone	8.0×10^{10}
Aluminum	2.4×10^{10}
Wood	0.9×10^{10}

increases volume will decrease. Values of the bulk modulus are given in Table 7.3.

Fluids

Fluids have the ability to flow and include both liquids and gases. An important characteristic used to describe fluids is density. The density of liquids is generally less than solids, and that of gases is less than liquids. Table 7.4 gives the densities of several common liquids and gases and includes a couple of representative solids for comparison. Density is often characterized by reporting a substance's specific gravity. The specific gravity of a substance is the ratio of its density to the density of water. Specific gravity is expressed in decimal form, for example, the specific gravity of iron is 7.9. Table 7.4 demonstrates that not all solids are denser than liquids. Solids such as ice and wood are common solids that are denser than water. Liquid mercury is much denser than many solids.

Viscosity is another important property of fluids. Although the ability to flow distinguishes fluids from solids, this ability varies greatly depending on the fluid. Viscosity is a measure of a fluid's resistance to flow. Viscosity can be thought of as the internal friction characteristic of the fluid. Anyone familiar with the saying "slow as molasses" can appreciate the difference in the ability of molasses to flow compared to other liquids such as water. The viscosity of fluids is measured in newtons-seconds per meter^2. A common unit for viscosity is the poise (P). One poise is equivalent to 1 g/cm-s, so 10 P equals 1 N-s/m^2. The poise is named in honor of Jean Louis Marie Poiseuille (1799–1869). Poiseuille was a French physician who made quantitative measurements of viscosity in his studies of the movement of blood through vessels. Table 7.5 gives the viscosities of several common fluids and shows that the viscosity of liquids is several orders of magnitude greater than that of gases.

Table 7.3
Bulk Modus of Various Solids

Material	B (N/m^2)
Steel	16×10^{10}
Water	0.22×10^{10}
Aluminum	7.0×10^{10}
Glass	2.0×10^{10}

Figure 7.6
A force, F, applied to a solid causes a shearing stress resulting in a shearing stain

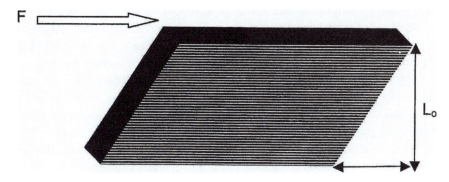

Table 7.4
Density of Common Substances

Substance	Density (kg/m³)
Wood (pine)	500
Ice	920
Iron	7,900
Gasoline	675
Ethyl alcohol	800
Water	1,000
Blood	1,060
Mercury	13,600
Helium	0.18
Air at 20°C	1.21
Carbon dioxide	1.98
Universe	10^{-21}

Table 7.5
Viscosity of Common Substances

Substance	Viscosity (N−s/m²)
Air	1.8×10^{-5}
Water	1.0×10^{-3}
Oil	2.0×10^{-1}
Glycerin	1.0
Honey	5.0

The ability of a liquid to flow is related to several factors including intermolecular forces, size of particles, temperature, and structure. The stronger the intermolecular forces in liquids, the higher the viscosity. Small particles can slide pass neighboring particles easier, and for this reason substances composed of smaller particles have lower viscosities. Another factor affecting viscosity is the size and shape of molecules. Structures composed of long chains have higher molecular masses and a greater probability of molecules entangling. The dependence of size and shape on viscosity can be pictured by contrasting the difference in dumping out a box of marbles versus dumping out a box of rubber bands. The marbles flow readily out of the box, while rubber bands become entangled and don't flow.

Anyone who has heated honey or syrup is aware that viscosity decreases with increasing temperature. Viscosity is a primary concern in lubricants, and a practical example of viscosity deals with the labeling of motor oils. Motor oils have an arbitrary viscosity rating given by their SAE (Society of Automotive Engineers) numbers. A higher number indicates a higher viscosity. This means that oils with higher SAE numbers are thicker. Most common motor oils are produced to have multiviscous characteristics in order to meet the needs of both cold start-up conditions and the elevated temperatures during operation. A typical SAE number would be 5W-30 or 10W-40. The first number can be considered the base number appropriate for cold conditions (W stands for winter). Cold conditions are considered to be 0° F. The second number is the equivalent viscosity at an elevated temperature, say 250° F. Multiviscous oils are used to meet the typical driving cycle. When an engine is cold, it needs a lower-viscosity (thinner) oil to lubricate the engine effectively. As the engine warms up, the oil inside also warms and becomes even less viscous. In order to compensate for the lowering of the viscosity with increased temperature, multiviscous oils contain carbon polymer additives. These polymers have a coiled shape at low temperatures and uncoil as the engine temperature increases. The uncoiled polymers act to increase the viscosity, thereby compensating for the heating effect. The second SAE number is an equivalent viscosity at the higher temperature.

Another characteristic dependent on the intermolecular forces is the surface tension of the liquid. Surface tension results from the unbalanced forces on molecules at the surface of a liquid. Figure 7.7 shows how surface tension results from these unbalanced forces. Consider water as the liquid in Figure 7.7. A water molecule in the interior of the liquid is surrounded on all sides by other water molecules. Attractive intermolecular forces pull the molecule equally in all directions, and these forces balance out. A water molecule on the surface experiences an unbalanced force toward the interior of the fluid. This unbalanced force pulls on the surface of the water, putting it under tension. This is similar to the tightening of the head of a drum. The tension means the surface of the water acts like a thin film. If you carefully use tweezers to place a clean needle on the surface of water, surface tension will allow the needle to float even though the needle is denser than water.

Surface tension is directly related to the magnitude of intermolecular forces in a liquid. The greater the intermolecular force, the greater the surface tension. For this reason, water has a high surface tension. Because of the surface tension, water drops will assume a spherical shape. This shape minimizes the surface area of the drop. Related to surface tension are the properties of **cohesion** and **adhesion.** Cohesion refers to the attraction between particles of the same substance. That is, it is just the intermolecular force between molecules of the liquid. Adhesion

Figure 7.7
Surface tension results from the fact that surface molecules are pulled toward the interior of the liquid, as compared to interior molecules, where the forces are balanced.

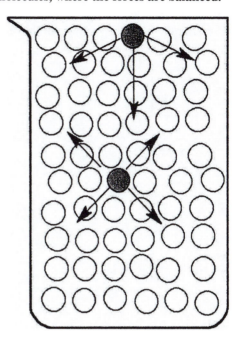

is the force between particles of different substances. Adhesion results in the ability of water to rise in thin tubes by capillary action. The adhesion between water and glass is greater than the cohesion between water molecules. The glass pulls water up the interior of the tube, and the cohesive force between water molecules creates a concave **meniscus** (Figure 7.8). The water will rise until the cohesive force balances the weight of the water pulled up on the side of the tube. Since smaller-diameter tubes contain less volume of water, water rises higher as the tube diameter decreases.

The interaction of cohesive and adhesive forces determines whether a liquid will spread out over a surface or bead up. Water will tend to spread out over a glass surface because the adhesion between glass and water is greater than the cohesion between

Figure 7.8
In capillary action, the adhesion between the walls of the tube and the liquid is greater than the cohesion between liquid molecules. The liquid rises up the tube and forms a meniscus that is concave upward. The smaller the tube, the greater will be the height the liquid rises along the sides.

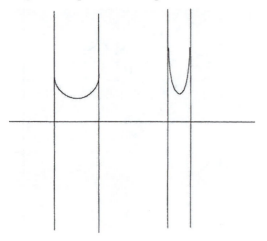

water molecules. When a waxy substance is applied to the glass, because the adhesion between the wax and water is less than the cohesion of water molecules, the water will bead up. This is the basis of certain windshield treatments such as Rain-X. When these treatments are applied to a windshield, rain drops bead up and flow readily off the windshield. Conversely, in other cases substances may be used such that the adhesive forces are greater than cohesive forces, causing the water to spread. Wetting agents used in dishwashers are a prime example. Wetting agents decrease the cohesive forces between water molecules, and this helps water to spread over the surface of an object due to adhesive force.

Atmospheric and Hydrostatic Pressure

One of the most important properties characterizing a fluid is its pressure. Pressure is defined as the force exerted per unit area. Atmospheric pressure is the force exerted by the weight of the atmosphere on the Earth's surface. A common device used to measure atmospheric pressure is the barometer. The barometer was invented in 1643 by Evangelista Torricelli (1608–1647). Torricelli, who was a student of Galileo, was presented with the problem of why water could not be raised more than 32 feet with a suction pump. Galileo explained the rise of water up a pipe by invoking the statement "nature abhors a vacuum." When a pump created a vacuum, Galileo believed water rose to fill the void created. It was noted, though, that the limit to the maximum height to which water could be raised was approximately 32 feet. Torricelli reasoned correctly that the maximum height to which water could be pumped was due to atmospheric pressure. In his study of the pump problem, Torricelli took a sealed tube approximately 4 feet in length and filled it with mercury. He inverted the tube in a pool of mercury. Torricelli noticed that the mercury in the tube fell, creating a vacuum at the top of the tube. The final height of the mercury in the tube was roughly 28 inches. Since mercury has 13.5 times the density of water, Torricelli proposed that water should rise about 13.5 × 28 inches, or 32 feet. Torricelli determined that atmospheric pressure placed a limit on how high a liquid could rise in the tube. His inverted mercury-filled tube was the first barometer (Figure 7.9).

Many units are used to express pressure. Since pressure is defined as force per unit area, a common unit used in the United States is pounds per square inch (psi). This unit is commonly used for tire inflation pressure. The atmospheric pressure at sea level is about 14.7 psi. In the metric system the basic unit for force is the newton, abbreviated N, and area is measured in squared meters, m^2. Therefore, the metric unit for pressure is

Figure 7.9
In a mercury thermometer,
atmospheric pressure, *P*, forces mercury to
rise in the tube to a height of approximately
30 inches

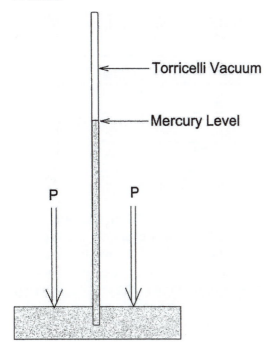

Torricelli Vacuum

Mercury Level

P   P

At the top of Mt. Everest the atmospheric pressure is only one-fourth of that at sea level. The pressure in Denver is about three-fourths of that at sea level. While changes in elevation have pronounced effects on atmospheric pressure, temperature and moisture also affect atmospheric pressure. Cold and dry air is denser than warm and wet air. Changes in the moisture and temperature of air masses produce high and low pressure systems that move across the continent, dictating our weather.

We can understand how atmospheric pressure is just the weight of the atmosphere pushing down on the Earth's surface, but how does this apply to the basic definition of pressure on a confined gas? In order to expand the concept of pressure and provide a basic framework for understanding the behavior of gases, a simple model for a confined gas is employed. This model is known as the **kinetic molecular theory**:

1. A gas consists of small particles, either individual atoms or molecules, moving around randomly.

2. The total volume of the gas particles is so small compared to the total volume the gas occupies that we can consider the total particle volume to be zero. This means that a gas consists almost entirely of empty space.

3. The gas particles act independently of one another. A particle is not attracted to nor repelled by any other particle.

4. Collision between gas particles and between gas particles and the walls of the container are elastic. This means that the total **kinetic energy** of the gas particles is constant as long as the temperature is constant.

5. The average kinetic energy of the gas particles is directly proportional to the absolute temperature of the gas.

newtons per squared meter, N/m^2, which is also known as a **pascal**, abbreviated Pa. As noted, atmospheric pressure can also be expressed by measuring the height of mercury in a barometer. This is often reported in inches in the United States and is how atmospheric pressure is recorded in daily weather reports. Standard atmospheric pressure is 29.92 inches of mercury. In the metric system, the height of mercury is given in millimeters of mercury. Another common unit used to express pressure is atmospheres (atm). One atmosphere is equal to normal atmospheric pressure at sea level. The various units for pressure and the value of standard atmospheric pressure are summarized in Table 7.6.

The values in column three of Table 7.6 give the atmospheric pressure at sea level.

The kinetic molecular theory can be used to explain pressure. Consider a sealed

Table 7.6
Pressure Units and Values for Standard Atmospheric Pressure

Pressure Units	Abbreviation	Standard
Atmosphere	atm	1.00
Millimeters of mercury	mm Hg	760
Inches of mercury	in Hg	29.92
Pounds per square inch	psi	14.70
Newtons per square meter = pascal	N/m² = Pa	101,300
Millibar	mb	1,012.3

syringe as a container and assume it contains a volume of air. To simplify the discussion it can also be assumed that air consists of 80% nitrogen and 20% oxygen (Figure 7.10). The nitrogen and oxygen molecules move randomly in the barrel of the syringe. The molecules collide with the walls of the syringe barrel and the face of the syringe's plunger (as well as with each other). The collisions exert a constant force on the inside surface area of the walls of the syringe barrel. The force exerted per unit area of the syringe's internal surface is the pressure of the gas. Because the molecules move randomly throughout the syringe barrel, the pressure is the same throughout the syringe. If we assume the pressure outside the syringe is 1 atmosphere and that the plunger is not moving, then the pressure inside the syringe must also be 1 atmosphere. The pressure of a confined gas is nothing more than the force caused by the constant bombardment of the gas particles on the sides of the container.

Just as the accumulated weight of the atmosphere exerts pressure over every point on the Earth's surface, water, or any other liquid, exerts pressure due to its weight. We experience this pressure when we dive under water just a few feet. The pressure exerted by the weight of water over a surface is called hydrostatic pressure. Hydrostatic pressure, P, can be calculated using the equation $P = \rho g h$, where ρ is the density of water, g is the acceleration due to gravity, and h is the water depth. For example, at a depth of 10 m the hydrostatic pressure is equal

to $1{,}000 \dfrac{\text{kg}}{\text{m}^3} \times 9.8 \dfrac{\text{m}}{\text{s}^2} \times 10\ \text{m} = 980{,}000 \dfrac{\text{N}}{\text{m}^2}$,

which is approximately 1 atm. This hydrostatic pressure is the pressure due to the water. The actual pressure at 10 m depth in a water body such as a lake would actually be close to 2 atm because the pressure at the surface is 1 atm due to atmospheric pressure.

While the equation $P = \rho g h$ was applied to water, it can be used for any fluid. It should be noted that when using this equation, the density of the fluid is assumed to be constant over the entire depth of the fluid. In most cases, the density of the fluid will vary vertically. In the case of water, the variation is small, and only a small amount of error will be introduced by assuming a constant density. Conversely, the density of air decreases drastically from the Earth's surface with alti-

Figure 7.10
A syringe filled with air, which is assumed to be 80% nitrogen and 20% oxygen

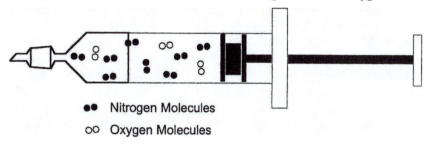

●● Nitrogen Molecules

○○ Oxygen Molecules

tude. In this case the equation should only be used for small vertical segments, or an integral method should be used to determine the pressure.

The Gas Laws

The behavior of a gas under different conditions can be predicted with several basic gas laws. Using a syringe as a container, the gas laws can be explained. These laws apply to what is referred to as an **ideal** or **perfect** gas. An ideal or perfect gas can be thought of as a gas that conforms to the kinetic molecular theory. In reality, gas molecules do have volume and exert forces on each other. Under normal conditions of temperature and pressure, though, the kinetic molecular theory explains the behavior of gases quite well. When a gas is at very low temperatures and/or under extremely high pressure, it no longer behaves ideally. Under these conditions, the gas laws are not valid.

The definition of pressure assumed that the volume of air contained in the syringe was constant. What would happen when the plunger is pushed, while making sure that the opening remained sealed? Pushing in on the plunger decreases the volume of the syringe's barrel containing the gas. Since the volume has decreased, the inside surface area has also decreased. This means that the frequency of collisions of gas molecules per unit area of the syringe has increased. The

increased number of collisions per unit area translates into an increase in pressure on the barrel of the syringe. The increase in the number of collisions can be compared to a hard rain falling on the sidewalk compared to a light drizzle. The hard rain will exert more pressure per unit area than does the light drizzle. When the plunger is pushed in, the pressure increases to some value above 1 atmosphere. As the syringe's barrel volume decreases when pushing the plunger, the pressure of the confined gas increases. The relationship between volume and pressure is known as **Boyle's law.** Boyle's law is named after Robert Boyle (1627–1691). Boyle's law says that the pressure and volume of an ideal gas are inversely related; as one goes up the other goes down. Boyle's law can be stated mathematically as

$$\text{pressure} \times \text{volume} = \text{constant or } PV = k$$

If the gas in the syringe was originally at 1 atm and the volume was 50 mL, then PV would equal 50 atm-mL. If the plunger is depressed to decrease the volume to 25 mL, then the pressure would have to increase to 2 atm in order for PV to remain constant. It should be emphasized that Boyle's law only involves two variables that change: pressure and volume. It is assumed that the temperature of the gas and the number of molecules of gas in the syringe remain constant.

Anyone who has watched a hot air balloon being filled realizes that as air is heated it expands. The direct relationship between temperature and volume in an ideal gas is known as **Charles' law.** Jacques Alexandre Charles (1746–1823) was an avid balloonist. Charles made the first hydrogen-filled balloon flight in 1783 and formulated the law that bears his name in conjunction with his ballooning research. Charles' law can be stated mathematically as

$$\frac{\text{volume}}{\text{temperature}} = \text{constant} \quad \text{or} \quad \frac{V}{T} = k$$

When applying Charles' law, absolute temperature must be used. An example using a syringe can be used to illustrate how much the volume of a gas would increase if its temperature was raised from 20°C to 100°C. Assume the original gas volume at 20°C is 25 mL. Converting temperatures to kelvins (K) and applying Charles' law, the calculations are

$$20\,°C + 273 = 293 \text{ K}$$
$$100\,°C + 273 = 373 \text{ K}$$

$$\frac{25\text{mL}}{293\text{K}} = \frac{V}{373\text{K}}$$

Solving for V, the volume at 100 °C is approximately 32 mL.

Increasing the temperature of a confined gas increases the kinetic energy of the gas molecules and results in more collisions of the molecules with the walls of its container. The increase in frequency of the collisions results in an increase in pressure. Relief valves on items such as pressure cookers and hot water heaters are safety devices that decrease the pressure before it reaches a dangerous level. The relationship between temperature and pressure is given by Gay-Lussac's law, which states that pressure is directly proportional to temperature. Joseph Louis Gay-Lussac (1778–1850) published work on gases at the beginning of the nineteenth century. He was actually the first to publish Charles law who had first put forth the relationship between temperature and volume 15 years before Gay-Lussac's publication. Like Charles, Gay-Lussac was a balloonist who made ascents in hydrogen filled balloons to conduct scientific investigations. He set an altitude record of 23,000 feet in 1804 that stood for half a century. Gay-Lussac's law expressed mathematically is

$$\frac{\text{pressure}}{\text{temperature}} = \text{constant} \quad \text{or} \quad \frac{P}{T} = k$$

Again, temperature must be expressed in kelvins. If the pressure and temperature of the air in a syringe are originally 1.0 atm and 293 K and the syringe is placed in boiling water, the pressure will increase to approximately 1.3 atm. The calculations are

$$\frac{1.0 \text{ atm}}{293 \text{ K}} = \frac{\text{pressure final}}{373 \text{ K}}$$

$$\text{pressure final} = 1.0 \text{ atm} \times \frac{373 \text{ K}}{293 \text{ K}} = 1.3 \text{ atm}$$

The relationship between temperature and pressure provides a method for determining the value for absolute zero. By measuring the pressure of a gas sealed in a constant-volume container at different temperatures and extrapolating to a pressure of 0 atm gives the value for absolute zero (Figure 7.11).

Boyle's, Charles', and Gay-Lussac's laws give the relationships between pressure, volume, and temperature for an ideal gas. In the examples used, the amount of gas in the syringe was considered constant. The amount of gas is measured in moles

Figure 7.11
The value of absolute zero may be determined by extrapolating a plot of pressure versus temperature to a value of 0 atm. This gives a value for absolute zero of –273.16 °C.

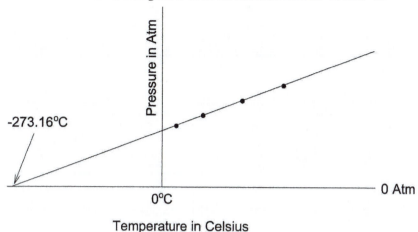

(the standard symbol for moles is n). The volume of a gas is directly proportional to the number of molecules present and, therefore, the number of moles of gas. This relationship is known as **Avogadro's law**. If temperature and pressure are held constant, and somehow more molecules are added to the syringe barrel, the volume would increase. A better example to illustrate Avogadro's law is to think about blowing up a balloon. As the balloon is inflated, more gas molecules (principally carbon dioxide) fill the balloon, and its volume increases. It is assumed the pressure and temperature stay constant as the balloon inflates, and that the increase in volume is due to the increase in the number of moles of gas entering the balloon. Because the relationship between pressure and moles is direct, the relationship is expressed mathematically as

$$\frac{\text{volume}}{\text{moles}} = \text{constant} \quad \text{or} \quad \frac{V}{n} = k$$

The four fundamental gas laws are summarized in Table 7.7.

The Ideal Gas Law

The gas laws discussed in the previous section are limited because they only allow the examination of two variables at a time. Fortunately, all four laws can be combined into one general law called the **ideal gas law.** The ideal gas law relates the four quantities pressure, volume, moles, and temperature. The ideal gas law is given by the equation

$$\text{pressure} \times \text{volume} = \text{moles} \times R \times \text{temperature}$$

$$PV = nRT$$

In this equation, R is called the ideal gas law constant. Its value depends on the units used, but assuming pressure is measured in atmospheres, volume in liters, and temperature in kelvins, its value is 0.082 atm-L/mol-K. When using metric units, the value of R is 8.314 J/mol-K Other forms of the ideal gas law are

$$\frac{PV}{nT} = R \quad \text{or} \quad \frac{P_1 V_1}{n_1 T_1} = \frac{P_2 V_2}{n_2 T_2}$$

Table 7.7
Summary of Four Laws for an Ideal Gas

Gas Law	Variables	Relationship	Equation
Boyle	Pressure, volume	Inverse	$P_1V_1 = P_2V_2$
Charles	Volume, temperature	Direct	$\dfrac{V_1}{T_1} = \dfrac{V_2}{T_2}$
Gay-Lussac	Temperature, pressure	Direct	$\dfrac{P_1}{T_1} = \dfrac{P_2}{T_2}$
Avogadro	Volume, moles	Direct	$\dfrac{V_1}{n_1} = \dfrac{V_2}{n_2}$

The four gas laws in the previous section are all special cases of the ideal gas law.

Pascal's Principle

You know that when you squeeze one part of a balloon the rest of the balloon will expand. This is a demonstration of **Pascal's principle,** named in honor of Blaise Pascal (1623–1662), which states that when an external pressure is applied to a confined fluid, this pressure is transmitted to every point within the fluid. For example, if 10 lb/in^2 are added to a tire, then the pressure at every point within the tire increases by 10 lb/in^2. Pascal's principle is the basis of hydraulics, where forces are transmitted through fluids. Small hydraulic jacks are often used to change tires, and large hydraulic lifts are employed in service stations to lift vehicles several feet. The use of a hydraulic lift is illustrated in Figure 7.12. Referring to Figure 7.12, a relatively small force, Force 1, is applied to an area, A1. The pressure transmitted to the enclosed fluid

within the lift would equal Force 1/A1. This pressure is transmitted undiminished to the opposite side of the lift. The pressure exerted on area A2 would equal the pressure exerted on area A1, therefore,

$$\text{pressure} = \frac{\text{Force 1}}{\text{A1}} = \frac{\text{Force 2}}{\text{A2}}$$

By rearranging this equation, it is seen that Force 1 is multiplied by a factor of A2/A1. While a greater force is obtained with a hydraulic lift, it should be remembered that work equals force times distance. Neglecting friction and the weight of the hydraulic fluid, the input force must be exerted over a proportionately greater distance. For example, in order to lift a 3,000 lb vehicle 1 m with a force of 600 lb would require the 600 lb force to act over 5 m. Pascal's principle applied to hydraulics can be observed in numerous applications. Landing gears, heavy equipment (backhoes, tractors), and industrial machinery are a few common examples where this can be seen.

Figure 7.12
In a hydraulic lift, a small force, Force 1, is multiplied and used to lift a heavy object. While the force is multiplied, it must be exerted over a proportionately greater distance than the object is lifted.

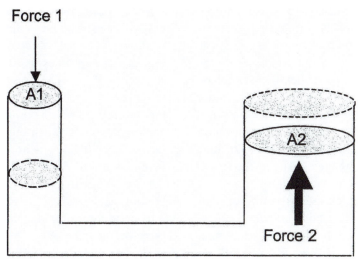

One of the most common applications of Pascal's principle is for braking systems. When the brake pedal on a car is pushed down, brake fluid is delivered to the wheels through a series of hoses and tubes collectively called brake lines. Pascal principle's results in a great mechanical advantage, so very little foot pressure results in a relatively large force applied at the wheels. The brake pedal itself is linked to a piston (or pistons) in the brake's master cylinder. The master cylinder serves as a reservoir for the brake fluid. Pushing the brake pedal forces the piston to exert pressure on the fluid in the brake lines, and this in turn forces the calipers or brake drums on the wheels to close, providing the braking action.

Buoyancy and the Archimedes Principle

When an object, such as a rock, is dropped in a container of water, the water level increases. The rock displaces a volume of water equal to its volume. The rock makes room for itself in the container by forcing water out of its way and causing the volume of water to rise in the container. According to Newton's third law, if the rock exerts a force on the water, the water will exert an equal and opposite force on the rock. This force is the buoyant force and can be found by the Archimedes principle. Archimedes' principle is named in honor of Archimedes (287–212 B.C.E.), who supposedly discovered it while attempting to determine whether a king's gold crown was counterfeit. The story of **Archimedes** begins when King Hiero II ordered a crown that was suppose to be pure gold. The King, suspecting that the crown was not pure, asked Archimedes to determine if the crown was indeed pure or if it contained a mixture of gold and silver. While at the local baths, upon immersing himself in the tub, Archimedes noticed how the water overflowed. Archimedes was struck with a solution to the King's gold problem, exclaiming *Eureka!* ("I have found it"). Archimedes could put a mass of gold equal to the mass of the crown into a container filled to the brim with water. The amount of overflow from this test could then be compared to the amount of overflow when the crown was placed in the container. If the two were equal, Archimedes could conclude that the crown was pure gold. If less water overflowed when the crown was placed in the container, then the crown was not pure gold. Archimedes' method involved indirectly determining the density of the crown and comparing it to that of gold.

While the Archimedes story survives to this day, the method described above was far beyond Archimedes' measurement capabilities. The difference in overflows would be impossible to measure. Archimedes actually used a special balance to solve the King's problem. The two weighing pans on Archimedes' balance were suspended in water. By suspending the pans in water and using a mass of gold equal to that of the crown, Archimedes accounted for the volumes of the metal used to make the crown and that of pure gold. That is, since the masses were equal and the density of the crown should equal that of pure gold, then the volumes of the gold and crown should also be equal. Archimedes could determine if a metal was pure or alloyed because the balance would tip toward the weight of pure gold if the crown was counterfeit and contained silver.

Archimedes' principle states that an object partially or completely immersed in a fluid will be buoyed up by a force equal to the weight of the volume of the fluid displaced. Archimedes' principle can be derived by considering a cube immersed in a fluid, as depicted in Figure 7.13. The fluid exerts pressure on all faces of the cube. The forces exerted on each of the vertical faces are equal. Since the forces on the left and right faces, as well as the front and back faces, are equal and opposite, they cancel. The net force on the cube is due to the difference in pressures exerted on the top and bottom faces. The pressure on the top face is shown as P_{top}. The force exerted on the top face is P_{top} times its area, which is x^2: force$_{top}$ = $P_{top}x^2$. The pressure on the bottom face, P_{bottom}, is equal to the pressure on the top face plus the pressure due to difference in depths between the top and bottom faces. Therefore, the pressure on the bottom face is equal to $P_{top} + \rho gx$, where ρ is the density of the fluid. The force exerted on the bottom face equals $(P_{top} + \rho gx)x^2$. The difference in forces on the two faces is the net force on the cube: $P_{top}x^2 + \rho gx^3 - P_{top}x^2 = \rho gx^3$. The quantity x^3 is the volume of the cube and equals the volume of fluid displaced. The product ρgx^3 is the weight of the fluid displaced, demonstrating that the net upward force exerted on a submerged object is equal to the weight of fluid displaced (Archimedes' principle). Although Archimedes' principle was derived from a regular cube, the relationship can be shown to hold for any submerged object and for objects only partially submerged.

Archimedes' principle explains why objects more dense than water, such as iron ships, float. The hulls of ships are shaped to displace an ample amount of water to produce enough buoyant force, allowing the vessel to float. Similarly, when a hot air balloon is inflated, it displaces enough air in the atmosphere to allow the balloon to float. Buoyant forces are often neglected in familiar situations. For example, when an object is weighed on Earth, rarely is the buoyant force of the atmosphere on the object considered. When precise weighing is required, the buoyant force of the atmosphere needs to be taken into account and corrections made.

Figure 7.13

A cube immersed in fluid experiences a net upward force equal to the weight of the fluid it displaces

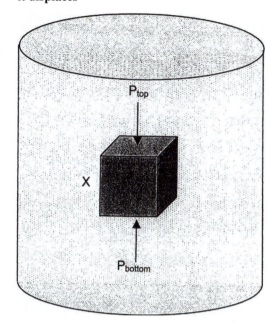

Fluid Dynamics

The material presented thus far in this chapter has focused on fluids at rest. A defining characteristic of a fluid is its ability to flow, and this ability extends the study of fluids to the motion of fluids. The general study of the motion of fluids is called fluid dynamics. When the fluid is water, the term hydrodynamics applies.

Experience has taught us that fluid flow varies. We hardly notice a light breeze but become alert when the weather forecast is for high winds, tornadoes, or hurricanes. Similarly, a meandering stream can transform itself into a torrential flow during heavy rains. Two main types of flow are **streamline (laminar)** flow and **turbulent** flow (Figure 7.14). Streamline flow is characterized by each particle within the fluid following in steady succession of those particles that preceded it along a streamline. Turbulent flow, as the name indicates, consists of disorderly motion made up of eddies of various sizes moving in various directions. Streamline flow would be analogous to a steady stream of traffic moving down a highway, where each car represents a particle. The cars, for the most part, move steadily along in organized lanes that are similar to streamlines. Each car follows the one ahead of it down the road. Turbulent flow can be compared to the movement of cars in a large parking lot, such as at a mall. Cars are moving in all directions at any one time. While there is some organization to the traffic flow, the pattern is highly random, and the movement of any one car cannot be predicted. The material of the rest of this chapter will focus mainly on streamline flow, in which frictional forces are negligible.

Consider a liquid such as water moving through a pipe that narrows. The density of water can be considered constant as it enters the constriction because water can be assumed to be incompressible (water's compressibility is so small under normal conditions it can be assumed to be zero). If the flow in the pipe is to remain steady, the volume of water that passes any section along the pipe must be constant. This means that the water's velocity must increase when the pipe narrows. As water enters the narrow part of the pipe, its velocity must increase. This is similar to using a squeeze nozzle on a garden hose. Squeezing the nozzle forces the water through a constriction narrower than the hose, and its velocity increases. The conservation of volume of a liquid through a vessel can be expressed in terms of the continuity equation $v_1 A_1 = v_2 A_2$, where v_1 and v_2 are the respective velocities through cross-sectional areas A_1 and A_2, respectively. The continuity equation is not valid for gases because gases are compressible.

Figure 7.14
Laminar and turbulent flow

LAMINAR FLOW

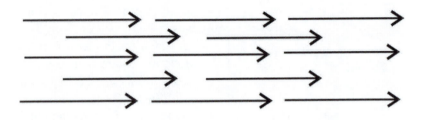

TURBULENT FLOW

An important relationship in fluid dynamics is the conservation of energy principle (discussed in chapter 5) applied to a fluid. Daniel Bernoulli (1700–1782), a Swiss mathematician, first developed ideas on fluid flow that preceded energy and thermodynamic principles developed in the latter half of the eighteenth century. The **Bernoulli principle** states that as the velocity of a fluid increases, the pressure decreases. This is a consequence of the conservation of energy for a fluid, as mathematically expressed with the Bernoulli equation. The Bernoulli equation says that the sum of kinetic energy, potential energy, and work is constant for an incompressible, steady flow situation in the absence of friction:

$$\text{kinetic energy} + \text{potential energy} + \text{work} = \text{constant}$$

$$\frac{1}{2}mv^2 + mgh + pV = \text{constant}$$

In this equation, m is mass, v is velocity, g is the acceleration due to gravity, h is elevation with respect to a reference level, p is pressure, and V is volume of the fluid. If ρV (density $\times$ volume) is substituted for m and the equation divided through by V, then the Bernoulli equation can be expressed as

$$\frac{1}{2}\rho v^2 + \rho gh + p = \text{constant}$$

In fluid flow, when the elevation is constant, there is no change in potential energy. In this situation, Bernoulli's equation shows that when velocity increases, the pressure of the fluid decreases.

Applications of the Bernoulli principle can be observed in a number of common objects. In a hand-pumped spray bottle, a burst of air is force through a small opening. The increase in velocity of the air across the orifice in the spray mechanism lowers the pressure at the orifice and draws liquid up a tube through it creating a spray. In a similar manner, spray painters use compressed air forced through an orifice to spray paint. The Bernoulli principle is also demonstrated when turning water on in a shower stall and observing that a shower curtain moves inward. In this case, the velocity of the water within the shower lowers the pressure inside

Figure 7.15
As an airplane's wing moves through the air, flow over the wing produces lift

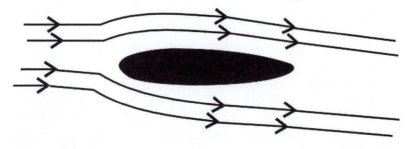

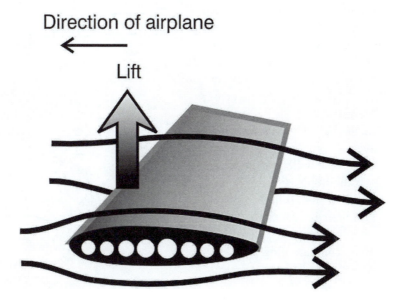

the curtain slightly, while the pressure outside the curtain remains at atmospheric pressure.

Blowing winds lowers the pressure over structures. Strong winds, especially those associated with tornadoes and hurricanes, have the ability to blow out windows and blow off roofs due to the decrease in pressure outside buildings. Leaving windows open in a storm helps to equalize the pressure between the inside and outside of buildings and reduces the potential of a disaster.

The Bernoulli principle provides the lift that enables winged aircraft to fly. Wings on aircraft are shaped in the form of an airfoil, as depicted in Figure 7.15. As the wing cuts through the air, the air is force over and under the wing. Since the air travels a greater distance moving over the top of the wing as compared to below the wing, the velocity above the wing is greater. According to the Bernoulli principle, the pressure below the wing in the region of lower air velocity will be higher. The higher pressure on the underside of the wing helps create lift. In addition to the shape of the wing, the angle of attack also serves to provide lift.

The Bernoulli principle is also responsible for the curve of a baseball and the hook or slice of golf balls. The curve is created by the movement of the ball through the air coupled with the spin of the baseball. The seams on a baseball or dimples of a golf ball create drag, and the layer of air immediately surrounding the ball moves with the spin of the ball. On one side of a baseball, the layer of spinning air moves in the same direction as the ball's translational motion, while on the opposite side the air layer opposes the translational motion of the ball. The result is that the relative velocity of air is lower on one side of the ball than the other, thus creating a difference in pressure and a net force that causes the ball to curve (Figure 7.16).

Material Science

The material in this chapter has focused on the solid, liquid, and gas states. Up until recent times there has been a clear distinction between these states. Recent advances in modern science have led to materials that blur the distinction between states of matter. Material science is a relatively new interdisciplinary area of study that applies knowledge from physics, chemistry, and engineering. Material scientists study the properties, structure, and behavior of materials. Material science can further be divided into branches that deal with areas such as ceramics, alloys, polymers, superconductors, semiconductors, corrosion, and surface films. Material scientists continue to modify our traditional view of the

Figure 7.16
Curve ball. The spin of a baseball causes a change in pressure on opposite sides of the ball, resulting in a curveball.

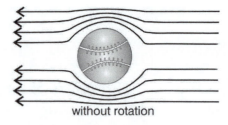

without rotation

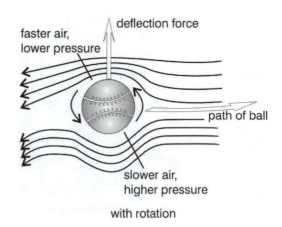

faster air, lower pressure

deflection force

path of ball

slower air, higher pressure

with rotation

different states of matter. A good example of a material that straddles two states of matter is **liquid crystals.** The term liquid crystals sounds contradictory but is an appropriate name for certain organic molecules. There are close to 100,000 different organic compounds identified as liquid crystals. They are slender, rodlike molecules that have mobility like a liquid, but display order by having their axes line up parallel to each other. An analogy might be that a liquid is like the people walking in one direction down a busy sidewalk, while a liquid crystal is like a marching band. On the sidewalk, people flow in a definite direction, but they are not aligned. People in the marching band are aligned in columns. There is a definite structure to the marching band, like a solid, but members of the band still flow.

Liquid crystals are everywhere: watches, calculators, digital clocks, laptop computers, and so on. Devices with liquid crystals are so commonplace that it is hard to believe that the first digital watch wasn't marketed until 1973. If a liquid crystal display is examined, a figure eight arrangement consisting of seven bars is seen. The liquid crystals in these displays are controlled by small electric fields; the crystals are very sensitive to these electric fields and orient themselves according to the electric field. A liquid crystal display works by rotating **polarized light** so that it either passes through or is blocked by a second polarizer. One can think of unpolarized light as light waves that vibrate in all directions. In this condition light is said to be incoherent, or in simple terms, it is all scrabbled up. A polarizer lines up the light waves in one direction, making the waves coherent. In short, it unscrambles the light waves. A polarizer is like an optical grating because it allows light that has the right orientation to pass through it. This is like a narrow passage only allowing those people who are turned sideways to squeeze through. When liquid crystals cause the light to be blocked by the second polarizer, the display registers a dark image (Figure 7.17). These dark images are displayed as numbers or letters in the liquid crystal display. By mixing other chemicals with liquid crystal molecules, color displays can be made.

Liquid crystals are just one of the many advances made in materials science during the last 30 years. There are many other discoveries of new substances with interesting properties. One that captured widespread attention starting in 1985 was **buckminsterfullerenes,** or "bucky balls." This naturally occurring form of carbon was named after the American architect F. Buckminster Fuller (1895–1983), who designed the geodesic dome. Buckminsterfullerene is said to be the most spherical molecule known. The first buckminsterfullerene identified was C-60, by Richard Smalley (1943–), Robert Curl (1933–), and Harold Kroto (1939–). These three shared the 1996 Nobel Prize in chemistry for their pioneering discovery. C-60 has the shape of a soccer ball, with the 60 carbons making up 12 pentagonal and 20 hexagonal faces (Figure 7.18). Since the original discovery of C-60, numerous other fullerenes of various shapes have been produced, ranging from C-28 to C-240. Research into their potential use is currently underway, but their expense, which is several times that of gold, currently limits widespread use.

Other advances in material science have helped humans mimic nature in the production of certain materials. For example, in the last half of the twentieth century we have learned how to produce synthetic diamonds. General Electric first produced diamonds commercially in the 1950s by subjecting graphite to temperatures of $2500°C$ and pressures approaching 100,000 atm. Currently, there are well over a hundred companies that produce synthetic diamonds.

Figure 7.17
In a liquid crystal display, unpolarized light passes through a polarizer. The polarized light is either unaffected (1) or rotated (2) by the liquid crystals, depending on the voltage applied to the liquid crystals. Light either passes through the second polarizer and is reflected by the mirror producing a light image or is blocked by the second polarizer, in which case a dark image appears.

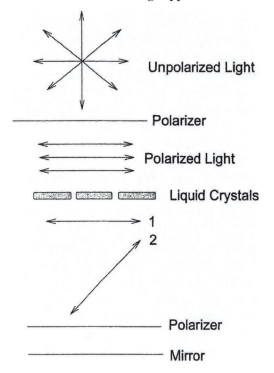

Figure 7.18
Buckminsterfullerene

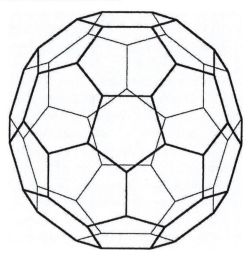

Other advances in material science are being made in a number of areas. Materials collectively known as "smart" or "intelligent" materials are materials that have the ability to adapt to their surroundings. A common example is lenses made from material that darken in bright light, used in eyeglasses. Memory metals have the ability to return to a preformed shape at a certain temperature. These metals are used in orthodontic devices and guide wires. A popular memory metal is the nickel-titanium alloy called Nitinol. Memory metals assume a "parent" shape above a specific transition temperature. Below the transition temperature the metal can be formed into various shapes. When the memory metal is heated above the transition temperature, the metal goes back to its parent shape as though it has remembered it. Shape memory polymers have recently been introduced that work in a way similar to metals. Advances in memory materials may make it possible to repair bent objects by simply heating the object. Another group of materials is known as self-healing. To understand how self-healing materials are designed, think of how a cut heals. When you cut yourself, a number of signals muster a chemical response to the wound. Over several days the wound heals. Self-healing plastics are currently being studied. Materials such as plastics develop numerous microcracks through everyday use. As time goes on these microcracks propagate and grow, eventually leading to failure of the product. One concept is to strategically position liquid plastic–filled capsules in self-healing plastics. When the plastic object is dropped or handled in a certain way, the capsules break open and plastic flows into the microcracks, where it hardens. This self-healing process would serve to lengthen the normal life of materials. Stealth weaponry is yet another example of how materials can be modified to behave in a totally new manner. Stealth airplanes are built with materials designed to absorb radar waves, rather than reflect a signal back to the receiver.

Research in material science continues to modify the substances that make up our world. As the twenty-first century unfolds, new materials will improve our lives. Just consider biomaterials used for bone replacement, drug delivery microchips, and skin grafts; composites used in cars, planes, and other transportation vehicles; building materials that are lighter, provide insulation, and have superior strength. All we have to do is take a good look around us to realize that many of the common materials present today were not available even 50 years ago, for example, Teflon, fiberglass, and semi-conductors. Material scientists continue to improve old materials and create new ones. In this manner they are the modern version of the ancient alchemists. Like alchemists, material scientists strive to perfect the basic elements into something more perfect, although without the mystical connection. As these new alchemists continue to perfect materials, chemistry will play a central role in their quest.

Summary

Matter generally exists in one of three main forms. While solids, liquids, and gases share several common properties, all these three states of matter have unique properties that distinguish them. Knowledge of these properties has contributed immensely to modern technology. Skyscrapers currently approaching 500 meters depend on knowing how stress and strain in solids contribute to the strength of materials. Hydraulics are used to move and reshape the Earth into a web of roadways that span the country, while fluid mechanics has taken flight from Kitty Hawk to outer space. Knowledge of matter has been used to develop different materials that blend the traditional qualities of solid, liquid, and gas into a wide range of composites that have further improved life. This trend should continue and lead to a greater variety of materials and expand our view of matter.

8

Vibrations, Waves, and Sound

Introduction

A vibration is an oscillatory repetitive motion that occurs at a specific location over time. Common examples of vibrations are a pendulum swinging back and forth, a weight bobbing on a spring, or a plucked string on a musical instrument. Vibrations lead to the propagation of waves. A vibrating string creates a sound wave, electromagnetic waves are produced by the vibrations of electric charge, and seismic waves come from vibrations from within the Earth. All waves can be described by common characteristics such as **wavelength, frequency,** and **period.** This chapter begins the study of vibrations by examining simple harmonic motion and follows this with an examination of the general principles that apply to all waves. Different types of waves will be examined, concluding with a more comprehensive examination of sound.

Simple Harmonic Motion

A simple model used to examine vibrations is that of a simple harmonic oscillator.

A simple harmonic oscillator is a system consisting of a "to and fro" motion around an equilibrium position. A condition for simple harmonic motion is that the restoring force is proportional to the displacement from the equilibrium position. For example, consider a weight attached to a vertical spring (Figure 8.1). If the spring is stretched by pulling on the weight and then released, the weight will bob up and down around the initial equilibrium position. As the weight moves down past the equilibrium position, stretching the spring, it will decelerate in proportion to a restoring force acting upward that attempts to return the weight to its equilibrium position. Once the weight "bottoms out," it will then move upward through the equilibrium position, compressing the spring. In the absence of friction, this up and down motion would continue forever, but in actuality, it eventually damps out due to friction.

Several parameters are used to describe simple harmonic motion (as well as wave motion). The maximum distance from the equilibrium position is the **amplitude.** The number of vibrations in a given interval of

Figure 8.1
Spring simple harmonic motion. A weight on a spring moving up and down produces simple harmonic motion.

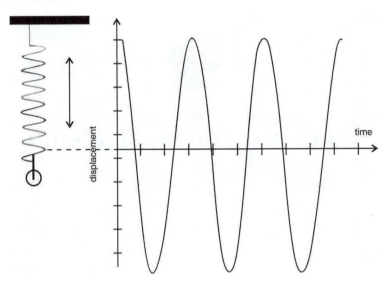

time is the frequency. Frequency is measured in the units of cycles per second. One cycle per second is equal to 1 hertz (abbreviated Hz). The time it takes for one cycle, or vibration, is the period. Frequency and period are inversely related by the equation: period = 1/frequency. Therefore, if the frequency is 10 Hz, then the period equals 1/10 s.

A common example of simple harmonic motion is a pendulum swinging through a small angle of several degrees or less. Galileo is credited with discovering that a simple pendulum's period is independent of the arc length but depends on the length of the pendulum. Galileo realized that pendulums could be used as timepieces and spent his last years trying to perfect an accurate clock based on the pendulum. The invention of the pendulum clock occurred a decade after Galileo's death and is credited to the Dutch physicist Christiaan Huygens.

For a simple pendulum, it can be shown that the frequency depends only on its length and the acceleration due to gravity (at small angles), according to the equation

$$\text{frequency} = \frac{1}{2\pi}\sqrt{\frac{g}{L}}$$

In this equation, g is the acceleration due to gravity and L is the length of the pendulum. Since the frequency of a pendulum depends on g, pendulums have historically been employed to determine the density of the Earth and to identify geologic formations. Locations on the Earth's surface where the acceleration of gravity is greater produce higher frequencies (shorter periods) in a free-swinging pendulum. Using very accurately calibrated pendulums, explorers navigating the globe during the 1800s took readings, and this led to the discovery that the Earth was not a perfect sphere but was an oblate spheroid. An oblate spheroid shape means the Earth is flatter at the poles, in other words, the radii to the North and South poles were smaller than the equatorial radius. Data showed that pendulums experienced more vibrations in a given period toward the pole and fewer toward the equator. This was interpreted as a flattening of the Earth toward the pole. This meant the Earth's radius was greater at the equator where the value of g was smaller.

Vibrations can be classified as free (natural) or forced. Free vibrations take place when an object is displaced and released, for example, when a weight hanging on a string is displaced slightly to the side and then released, the weight will swing with a natural period dependent on the string's length. Many natural objects, from atoms to oceans, have a natural period of vibration. Forced vibrations are driven by an outside agent, and the vibration matches that of the agent applying the force. For example, when one person pushes another person on a swing, the frequency depends on the person doing the pushing. When the frequency of a forced vibration matches the natural vibration of an object, **resonance** occurs. Resonance leads to an increase in frequency. A simple example of resonance occurs when swinging. The swing has a natural period that depends on its length. In order to increase the amplitude, which in the case of a swing means a greater height, the person swinging should pump his legs at a frequency that matches the natural period of the swing. Resonance also explains the great tidal heights in places such as the Bay of Fundy. Natural water bodies have a natural period of vibration that depends on the shape of the basin. When the dimensions of the basin produce a natural period that matches the tidal forces of the Moon and Sun (roughly 12 hours and 24 hours, respectively), then the amplitude of the tide is amplified to a much greater height. The natural period of a water body is also evident when trying to carry a shallow pan of water. The water in the pan tends to resonate at a frequency that matches the frequency of walking, making it difficult to carry the water without it sloshing over the sides of the pan.

Waves

While vibrations consist of oscillations taking place over time at a specific location, waves can be thought of as a vibration that propagates through space. As the vibration travels through space, it creates a disturbance that carries energy from one location to another. There is no net displacement of matter as a wave passes. This is observed when watching an object floating in water as waves pass. The object bobs up and down and remains in a relatively fixed location as the wave passes. Wind and other forces may cause the object to move laterally, but the wave itself causes a circular or elliptical motion with no net lateral displacement.

The creation of waves from vibrations is observed when the end of a long rope is vibrated in an up and down motion causing a wave to travel down the length of the rope. If a particular point on the rope is examined as the wave travels down the rope, it is found that this point vibrates up and down with simple harmonic motion. The up and down disturbance of the point is perpendicular to the direction of the wave, and this type of wave is referred to as a **transverse wave.** Water waves are probably the most common example of transverse waves. There are several other common categories of waves. Moving a slinky back and forth creates a **longitudinal wave** in the slinky. As the slinky is moved up and down, different regions of the slinky are stretched and compressed in areas of **rarefaction** and **compression,** respectively. Longitudinal waves vibrate in a direction parallel to the wave motion. Water waves continually washing up on a beach are **periodic waves** that result from a sustained force or repeated motion. **Impulsive waves** are the result of an isolated single disturbance. The ripples from a rock thrown into a pond and the sound from the crash of cymbals are examples of impulsive waves.

Similar to simple harmonic motion, waves can be characterized by an amplitude, period, and frequency. Additionally, the wave has a wavelength, which is the distance between any two consecutive points in phase, for example, two consecutive crests or two consecutive troughs (Figure 8.2). Since speed is defined as distance divided by time, the speed of a wave is determined by dividing the wavelength by the period. Because the reciprocal of the period is the frequency, wave speed can also be determined by multiplying the wavelength by the frequency:

$$\text{wave speed} = \frac{\text{wavelength}}{\text{period}}$$
$$= \text{wavelength} \times \text{frequency}$$

More than one wave can share the same space. When this occurs, the waves interact with each other, and wave interference occurs. The result of wave interference can be determined by the **superposition principle.** The superposition principle states that when waves travel through the same space, that the individual disturbances from the separate waves add to produce a net disturbance. The superposition principle means that when two crests meet, the result is a crest with an amplitude equal to the sum of the two wave crest amplitudes. This is an example of **constructive interference.** Constructive interference takes place when waves meet and the resultant wave's amplitude is greater than the amplitude of the waves that meet. Conversely, **destructive interference** takes place when the amplitude is smaller than the component waves, for example, when the crest of one wave meets the trough of another (Figure 8.3).

One consequence of superposition can be the production of a standing wave. If one end of a rope is attached to a wall and the free end moved up and down to create a series of waves, these waves will travel down the length of the rope and be reflected off the wall. The reflected waves will interact

Figure 8.2
Wave features. A wave can be characterized by its wavelength, amplitude, and period (not shown).

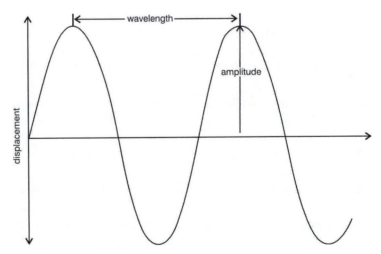

with the incident waves through interference. At particular frequencies of vibration, the interference between the incident and reflected wave will produce a wave that doesn't travel down the rope, but results in a wave pattern that oscillates in place. That is, it oscillates at a fixed or standing position (Figure 8.4). Positions along the rope where there is no vertical displacement are referred to as **nodes.** Positions of maximum displacements are called **antinodes.**

One of the most common examples of standing waves is associated with musical instruments. When a string is plucked on a musical instrument, such as a banjo, the sting vibrates according to a number of natural frequencies. The most fundamental frequency consists of the entire length of the string moving up and down in a standing wave pattern, with the nodes at each end of the string and an antinode in the center. The wavelength of this standing wave is twice the length of the string. This fundamental frequency is called the **fundamental** or **first harmonic.** The string also vibrates in a manner such that a standing wave produces a node at each end and at the middle of the string. This is referred to as the second harmonic or the first overtone. Higher overtones exist with increasingly more nodes (Figure 8.5). Standing waves and harmonics are an important aspect of music, and more will be said about this subject later in the chapter.

Another wave phenomenon that has important application in physics is the **Doppler effect,** named in honor of Christian

Figure 8.3
Superposition principle. Wave 1 and wave 2 add together to produce a resultant wave

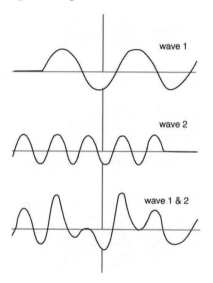

Figure 8.4
Standing wave

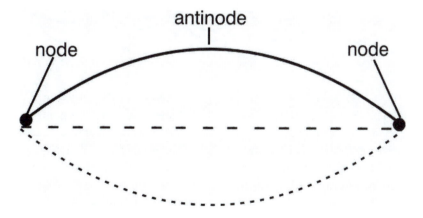

Figure 8.5
Harmonics. A vibrating string has various modes of vibration producing harmonics or overtones.

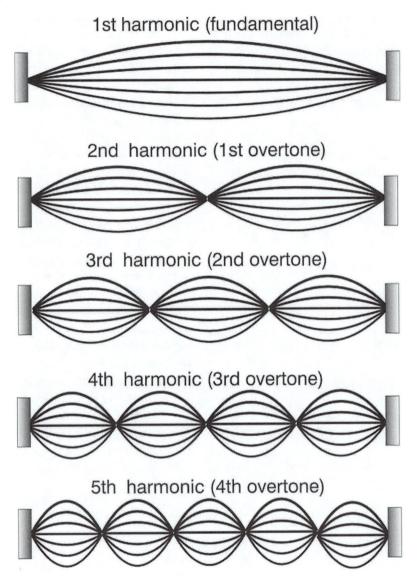

1st harmonic (fundamental)

2nd harmonic (1st overtone)

3rd harmonic (2nd overtone)

4th harmonic (3rd overtone)

5th harmonic (4th overtone)

Doppler (1803–1853), who described the effect in 1842. The Doppler effect describes the apparent change in frequency of waves due to the relative motion between the source of the waves and the observer. The Doppler effect can be compared to traveling down a highway and noting the number of vehicles that pass in each direction. The number of vehicles that pass you going in the opposite direction is much greater than the number of vehicles you pass going in the same direction. Likewise, the number of waves an observer detects when the wave source and observer are approaching each other is greater than

Figure 8.6
Doppler effect. As a sound source approaches an observer, the sound's frequency increases while it decreases as it moves away from an observer.

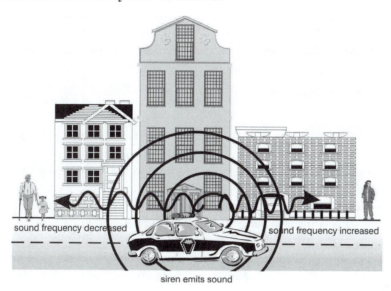

sound frequency decreased

sound frequency increased

siren emits sound

when the observer and wave source are moving away from each other. The Doppler effect results in greater wave frequency when the relative motion of the source and observer is toward each other, and lower frequency when the relative motion is away from each other (Figure 8.6).

The Doppler effect results in the change in pitch of a sound from higher to lower as a vehicle travels past a person. It is also used to estimate the speed and distance to distant galaxies. If the universe is expanding, all objects in the universe tend to be moving away from each other. The light emitted from these galaxies would, therefore, be shifted toward a longer wavelength or toward the red end of the visible light spectrum. The amount of red shift is proportional to how fast the galaxies are receding from observers on Earth. Astronomical measurements have demonstrated that the most distant galaxies show the greatest red shift and, hence, are moving away from Earth at the greatest rate.

Another area where the Doppler effect has practical applications is in radar. The word "radar" is an acronym for **ra**dio **d**etection **a**nd **r**anging. Police use radar and the Doppler effect to measure the speed of vehicles. Coaches monitor pitches with radar guns to determine if pitchers are losing velocity and becoming fatigued. Another widespread use of radar is for weather forecasting. Doppler radar images are standard on the nightly weather forecast. Doppler radar works by emitting waves with at a specific frequency from a rotating emitter. As the waves encounter precipitation in the atmosphere (in the form of rain or snow) carried by winds, the signal is reflected back to a receiver that produces color-enhanced radar images. The standard color pattern displays movement toward the radar as shades of green and movement away from the radar as shades of red. The intensity of the signal represents the strength of the weather pattern. Thunderstorms and severe weather such as tornadoes can be identified by trained weather personnel by examining the color patterns and the strength of the signal. Currently, approximately 150 Doppler radar sites are located throughout the United States. Images from these sites can be accessed through the National Weather Service's Web site (http://weather.noaa.gov/radar/national.html).

Sound

From our time of birth, we are exposed to sounds. In fact, we are exposed to sounds 24 hours a day our entire life, and sound is one of the most ubiquitous phenomena in our world. It is truly a rare occasion when we cannot perceive some type of sound. Sound is a longitudinal wave that travels through a medium. Typically this medium is air, but sound also travels through liquids and solids. Sound and music were one of the first areas to which the ancient Greeks devoted appreciable scientific study. As mentioned in chapter 1, Pythagoras founded a school of philosophy heavily based on the mathematical relationships derived from vibrating strings. The Pythagoreans applied their ideas, derived from the harmonic sounds of music, to the universe as a whole. They believed the universe was arranged according to the same fundamental mathematical relationships that explained music. Pythagoras' disciples continued the study of sound and music. One of them, Archytas of Tarentum (428–350 B.C.E.), produced a work on the science of sound. Aristotle correctly inferred that the sound of a vibrating string was due to the disturbance of the string being transferred to the surrounding air. That sound was a wave phenomenon grew from a comparison to water waves. The analogy to water waves was emphasized by the Greek philosopher Chrysippus (280–206 B.C.E.). This idea was more fully developed by the Roman philosopher Boethius (480–524 C.E.), who compared the propagation of sound to the ripples of waves when a rock is thrown into a pool of water. Isaac Newton proposed a mathematical theory of sound in his *Principia*, in which he considered sound as pressure waves transferred to surrounding particles.

As a longitudinal wave or pressure wave, sound contains alternating regions of compression and rarefaction. The areas of compression and rarefaction associated with a sound wave correspond to areas of high and low air pressure, respectively, as compared to the surrounding air. The nature of sound can be described using standard wave characteristics. The loudness of a sound is directly related to the amplitude of the wave. While people relate to the loudness of a sound, the correct physical term to describe loudness is **sound intensity.** The intensity of a sound wave is related to the amount of wave energy that crosses an area perpendicular to the wave direction in a specific period of time. Using standard metric units, energy is expressed in joules, area in meters^2, and time in seconds. Therefore, the intensity of a sound wave is given in the units of J/m^2-s. A joule divided by a second is equivalent to a watt, which is a unit of power. Therefore, sound intensity is the amount of power transported through an area and is given in units of watts per meter^2 (W/m^2). As a wave is generated from its source, it spreads out in three dimensions. As it spreads out, its intensity decreases according to the **inverse square rule.** This rule relates the intensity of a sound to the distance from its source. According to the inverse square rule, the sound intensity is inversely related to the square of the distance from the source. For example, if a firecracker explodes, the sound intensity will be 1/4 the value at a distance of 10 feet compared to 5 feet, that is, doubling the distance reduces the intensity by 1/4. At a distance of 15 feet, the sound intensity (loudness) will be only 1/9 of that at 5 feet.

The threshold of human hearing is approximately 1×10^{-12} W/m². This corresponds to a change in air pressure of less than one billionth of an atmosphere. Pain occurs for most humans when the intensity approaches 10 W/m². Rather than use intensity to describe loudness, intensity level is employed. Intensity level is defined by the equation:

$$\text{intensity level} = 10 \log \frac{\text{intensity}}{\text{intensity}_0}$$

In this equation, intensity_0 is the reference intensity equal to 1×10^{-12} W/m². According to this equation, a television with a sound intensity of 0.5×10^{-4} W/m² would have an intensity level of 77. Even though the equation for intensity level is dimensionless, it is assigned the unit of bel in honor of Alexander Graham Bell (1847–1922). The factor 10 in the equation makes the unit for sound intensity 1/10 bel, or a decibel, abbreviated db ("deci-" being the prefix for 1/10). The threshold value of human hearing has a sound intensity of 0 db, and Table 8.1 lists the sound intensities of various activities. Sound intensity is a logarithmic scale, and it should be noted that a change in 10 db equates to a 10-fold increase in sound intensity.

Pitch is another characteristic of sound subjectively perceived by humans. Pitch is physically related to the frequency of the

Table 8.1
Sound Intensities of Common Activities

Activity	Intensity (W/m²)	Intensity Level (db)
Ruptured eardrum	10^4	160
Jumbo jet takeoff on runway	10^3	150
Rock concert within several feet of speakers	1	120
Rock concert front row	10^{-1}	110
Walkman turned up	10^{-2}	100
Power saw	10^{-4}	80
Heavy street traffic	10^{-5}	70
Conversation	10^{-6}	60
Quiet library	10^{-8}	40
Whisper	10^{-10}	20
Cat purr	10^{-11}	10
Human threshold of hearing	10^{-12}	0

sound wave. High-pitched sounds are produced by more rapidly vibrating strings of musical instruments. Humans can detect sounds ranging from 20 Hz to 20,000 Hz, although with age the ability to detect higher frequencies decreases. A 50-year-old person's hearing may range only to 15,000 Hz. Sounds above the normal hearing range of humans are referred to as ultrasonic, and those below the normal hearing range are called infrasonic. Many organisms have the ability to detect sounds outside the normal hearing range of humans. Dogs and cats can hear sounds above 20,000 Hz. Animals such as bats and dolphins have the ability to emit high-frequency sounds and use their echoes to orient themselves in their environment. This ability is referred to as **echolocation.** A human application of echolocation takes place in the form of an ultrasound examination. This technique produces an image reflected from the fetus in a pregnant woman. Ultrasound examination is also used to detect fatigue and stressed areas in metals. Sonar, an acronym that stands for **so**und **na**vigation **r**anging, uses ultrasound to locate underwater obstacles, objects, and the depth of a water body. A sonicator is a device that uses ultrasonic waves to clean objects and break down compounds for laboratory analyses. The process is referred to as sonication. Ultrasound is also used in surgery to break apart harmful growths such as kidney stones and brain tumors.

Large animals, such as elephants, hippopotamuses, and whales, use infrasonic sound to communicate. Large animals produce lower-pitched sounds compared to small animals, similar to the manner in which large musical instruments produce lower sounds compared to small instruments. For example, a tuba produces a deep, low-pitch sound as compared to a trumpet, and piccolos and flutes produce high-pitched sounds.

Another important characteristic of sound is its speed. The speed of sound in air depends on several factors including density, humidity, and temperature. The speed of sound in air was first determined accurately at the start of the seventeenth century. The time it took for the sound of a cannon to reach observers a known distance away was used to determine the speed of sound. The time was determined by measuring the interval from the firing, as determined by the cannon's flash, until the arrival of the "boom" of the cannon. In order to account for wind speed, the experiment was repeated in the opposite direction and the times averaged. In this manner, the speed of sound was calculated to be approximately 340 m/s (760 mph). The currently accepted speed of sound in air is given by the equation $(331 + 0.6T)$ m/s, where T is the temperature in degrees Celsius. Thus, the speed of sound in air at room temperature is about 343 m/s (768 mph).

Sound travels at different speeds in different substances. The high-pitched voice produced by someone who has inhaled helium is caused by the speed of sound in low-density helium being approximately 1,000 m/s (2,240 mph). Sound travels faster in both liquids and solids than it does in gases. The speed of sound in water at room temperature is 1,480 m/s (3,315 mph). In glass, the speed of sound is 5,600 m/s (12,540 mph), and in steel it is 6,000 m/s (13,440 mph).

Aircraft with the ability to travel faster than sound have been around since the middle of the twentieth century. The sound barrier was first broken on October 14, 1947, by Captain Chuck Yeager (1923–) of the United States Air Force in a Bell X-1 rocket aircraft. Traveling at faster than the speed of sound is termed supersonic speed. A measure of supersonic speed is given by the Mach number, named in honor of Ernst Mach (1838–1916).

The Mach number is the ratio of the speed of an aircraft to the speed of sound in the surrounding air. Supersonic flight has the ability to produce sonic booms, heard by observers on Earth. A sonic boom signifies the arrival at the Earth's surface of the shock wave produced by a supersonic aircraft. The shock wave, similar to that produced by a boat cutting through the water, originates from the nose and tail ends of the aircraft. The shock wave forms a three-dimensional cone spreading out behind the aircraft, with the aircraft at the apex of the cone (Figure 8.7). Because the aircraft is traveling faster than the speed of sound, the sound waves generated by the aircraft overlap and intensify. When the shock wave reaches the ground, the sudden rise in pressure creates an explosive boom. This is immediately followed by a steady drop in pressure until the low-pressure pulse from the tail hits, creating a second boom. The overall process takes place very quickly and is generally over in a fraction of a second. The nature of the booms depends on a number of factors such as the dimensions of the aircraft and its altitude and speed, and atmospheric conditions.

Music

Much of our knowledge of sound is a result of human history's obsession with

Figure 8.7
Sonic boom shock wave

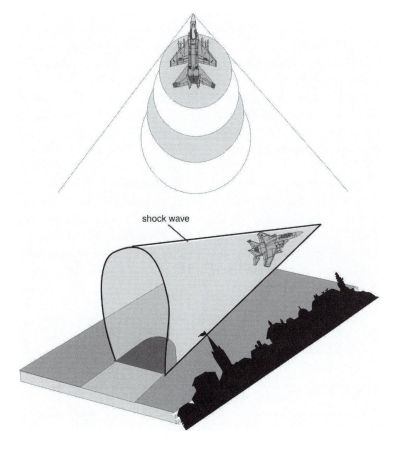

shock wave

music. Music, from a scientific perspective, can be considered sound that possesses aesthetic characteristics. Of course, beauty is in the eye of the beholder, and a cultural aspect is embodied in all types of music. Traditional western music has evolved around a systematic adherence to mathematical patterns that result in common scales, chords, and various other elements of music. In this section, a more detailed examination of music will be examined using the basic principles of sound outlined in the first part of this chapter.

As mentioned at the start of the section on sound, the Pythagoreans developed a theory of the universe based on the harmonic musical sounds produced by vibrating stings. The Pythagoreans observed that vibrating strings sounded together produced pleasant sounds when the ratio of the lengths of the strings matched specific whole-number ratios. Because the frequency of a plucked string depends on its length, certain frequencies when sounded together produced a pleasant sound one might call music, while other strings produced noise. The distinction between music and noise depends on how sound waves combine according to the superposition principle to produce an overall sound wave with specific qualities. Noise tends to give a random pattern, while most Western music tends to be repetitive, following distinct patterns.

When a string is plucked or struck, the string vibrates in a number of modes as standing waves. The frequency of the wave determines the tone as described by the note sounded. The tone is sounded with a definite pitch or frequency. For example, a standard A-tone used to tune instruments vibrates with a frequency of 440 Hz. Multiples of 440 Hz also produce the A notes lower or higher in pitch. A frequency of 220 Hz is an octave lower, while a frequency of 880 Hz is one octave higher. Specific frequencies can be associated with specific notes. Table 8.2 gives the frequency of notes A through G for two full octaves. An octave is the distance between two corresponding notes, including the notes themselves. Starting from A (220 Hz) through A (440 Hz), there are eight notes, hence the term octave. Table 8.2 demonstrates that a corresponding note one octave higher vibrates at twice the frequency (the numbers in Table 8.2 are not exactly twice because they were rounded to the nearest whole number). The whole-number ratios observed by Pythagoras are also reflected in the ratio of the frequencies. For instance, the ratio G:A is 392:220 or approximately 6:5, D:C is 294:262 (9:8), and G:C is 392:262 (3:2). The ratio can be extended to three notes together, such as G:E:C, which is 392:330:262 (6:5:4). Notes sounded together give rise to chords. When three notes are sounded that harmonize, a triad or three-note chord is produced. The combination of the G, C, E tones produces a G chord.

Another concept related to the frequencies is that of musical scales. Musical scales consist of a series of tones (notes) arranged

Table 8.2
Notes and Frequencies

Note	A	B	C	D	E	F	G
Frequency (Hz)	220	247	262	294	330	349	392
Frequency (Hz)	440	494	523	587	659	698	784

in order of increasing (decreasing) pitch. The familiar do-re-mi-fa-so-la-ti-do is called the diatonic scale and has been the most common scale employed in Western music for the last 500 years. In the diatonic scale, five full steps and two half steps, called semitones, separate the notes in making the octave. For instance, the C (major) scale starts with middle C at a frequency of 262 Hz. Proceeding from C, it is seen that the difference in frequencies between successive tones or notes varies. From middle C to D the difference in frequency is 32 Hz, and from D to E is 36 Hz. When moving from E to F, the difference is only 19 Hz, and from F to G is 43 Hz. The remaining steps are 48 Hz (G to A), 54 Hz (A to B), and 29 Hz (B to C). The steps between both and F and B and C are only about half that of the distance between surrounding steps. This scale is played by starting at middle C on the piano and playing the white keys through the next C. The chromatic scale consists of half steps. The chromatic C major scale consists of the notes C, C# (C sharp), D, D#, E, F, F#, G, G#, A, A#, B, returning to C. A chromatic C scale is played if adjacent keys, both black and white, are played starting from middle C. In the equally tempered scale, the distances between half steps are all equal. Therefore, for the chromatic scale the change in pitch between each half step is $21/12 \approx 1.0594631$ times greater. This is because the frequency doubles over the chromatic scale, and there are 12 steps. If the steps are equally spaced, then each half step must be the twelfth root of two times as great as the previous half-step.

It has been only in recent history that musicians have agreed on standard pitches associated with musical tuning. Historically, middle C has varied between 256 and 280 Hz, depending on the geographic region and local custom. In 1939 at an international conference in London, 440 Hz was established as the standard for the A above middle C (this note is referred to as A-4). This standard makes middle C itself equal to 262 Hz. There is a scientific middle C with a frequency of 256 Hz. Although not accepted by musicians, this frequency is a mathematical convenience because it equals 2^8. Consequently, the difference between octaves can be expressed in powers of two, for example, one octave above middle C equals 2^9 (512) Hz.

Most of the examples relating sound to music in this chapter have used vibrating strings. While stringed instruments comprise one substantial portion of musical instruments, music produced by vibrating air columns, in horns and woodwinds, makes up another substantial segment of musical instruments. In fact, stringed instruments also depend on vibrating columns of air since in most cases strings are coupled to a sounding box that causes a column of air to vibrate and amplify the sound. In brass instruments the vibrating lips of the musician create vibrations in a column of air contained in the horn's tube. In a woodwind instrument, air blown across a reed causes it to vibrate, and these vibrations are transferred to a column of air.

When playing a stringed instrument, the pitch is adjusted by varying it's the string's length. Moving the fingers up and down along the length of the fingerboard accomplishes this. In order to change the pitch of a brass or woodwind instrument, fingers are placed over holes or valves are pressed to vary the length of the air column. This column must be open at one end and may be open at both ends. Examples of instruments with both ends open include the flute and bassoon.

Trumpets and clarinets are examples of instruments open at only one end. Standing waves are produced in the columns of air associated with brass and woodwind instruments. An antinode exists at the open end or ends of the instrument (in reality the antinode is not exactly at the end, but near the end) because at the open ends the pressure is equal to atmospheric pressure. Because an antinode must exist at each open end, there will be two antinodes for instruments with two open ends. The fundamental wavelength will equal twice the distance between antinodes, or twice the effective length of the column of air. For example, a bassoon that is 3 meters long will produce a fundamental wavelength 6 meters long. The frequency of the note produced can be calculated by dividing the speed of sound by the wavelength: 343 m/s ÷ 6 m ≈ 60 Hz. A 0.5 m flute produces a fundamental frequency of 343 Hz (343 m/s ÷ 1 m). Varying the length of the column as it is played produces different frequencies or notes. In addition to the fundamental frequency, overtones are produced. In an instrument open at both ends, the air pressure wave is partially reflected at the open end back into the column. Reflection from an open end changes phases as it is reflected. This means that if the pressure was slightly greater than atmospheric before it is reflected, then it will be slightly lower after it is reflected. The reflected pressure waves will interact with original waves and other reflected waves in the column. Overtones will occur when the length of the tube is an integer multiple of one-half wavelength. This can be expressed as $L = n(1/2$ wavelength), where $n = 1, 2, 3$, and so forth. The overtones occur at frequencies given by the equation $n(343$ m/s$)/2L$. for the 0.5 m flute, with $n = 1$, the fundamental frequency is 343 Hz. The first overtone has a frequency of 686 Hz, which can be found by setting n equal to 2.

For an instrument with one open end, there is an antinode at the open end and a node at the closed end. The distance between a node and an adjacent antinode is one-fourth of a wavelength. In this case, overtones will occur when the length of the tube is an odd number of multiple of one-fourth wavelengths: $L = n(1/4$ wavelength), where $n = 1, 3, 5$, and so on. The overtones occur at frequencies given by the equation: $n(343$ m/s$)/4L$. So for a clarinet that is 0.70 m long, the fundamental frequency ($n = 1$) will be 122.5 Hz, and the first overtone ($n = 2$) is 245 Hz. An organ is an instrument that utilizes both types of air columns, with pipes open at both ends and at only one end.

Another term associated with music is beat. In music, beat refers to the rhythm in the sense that there are a number of beats to a measure. Beats as related to sound refers to a pulsating intensity of sound caused by the interference of sound waves of slightly different frequency. Two waves of different frequencies will combine to produce a sound that periodically increases and decreases. This can be understood by considering a simple example. Consider two waves, one with a frequency of 6 Hz and another with a frequency of 4 Hz. Although these frequencies are below the audible range of humans, we will use them to illustrate how beats are produced. If a wave crest signals the beginning of each wave, there will be a wave crest every one-sixth second for the 6 Hz wave and a crest every one-fourth second for the 4 Hz wave. Table 8.3 gives the times in seconds when each wave crest occurs. Table 8.3 shows that a crest will occur simultaneously for both waves at 0.5 second and at 1 second. Every one-half second the two waves add to

Table 8.3
Beats

4 Hz	$\frac{1}{4}$	$\frac{1}{2}$	$\frac{3}{4}$	1		
6 Hz	$\frac{1}{6}$	$\frac{1}{3}$	$\frac{1}{2}$	$\frac{2}{3}$	$\frac{5}{6}$	1

produce a maximum. This is also illustrated in Figure 8.8. In general, the interference of two sound waves will produce a beat frequency that equals the difference in the two sound frequencies. Stringed instruments can be tuned to themselves by eliminating the beats between two strings. For example, if one string is tuned to a D, the fret on another string can be held to give a D, and the two strings can be sounded together. By eliminating the beat frequency, the second string can be tuned to a D.

Acoustics

The general study of sound is called acoustics. Acoustics also refers to the specific study of sound as related to materials, structures, and rooms. This is often referred to as architectural acoustics, and this is how the term acoustics in this section is used. The acoustics of a room, auditorium, or concert hall depends on the nature of sound waves and factors related to the design of the structure, such as its size, shape, and the type of construction materials. Important sound wave characteristics are reflection, refraction, and diffraction. Of these, reflection of sound waves is the most important. An echo is caused by the reflection of a sound wave. An echo is generally associated with large outdoor locations such as canyons. Sounds within enclosed structures, such as a room or auditorium, reach an observer both directly and indirectly. Direct sound waves reach the listener first, followed by sounds reflected off the walls, ceiling, floor, and objects in the room. Rather than a distinct echo, the

Figure 8.8
Beats occur when waves combine to produce alternating regions of constructive and destructive interference. If each vertical line represents a wave crest, wave crests occur at 1/2 second and 1 second.

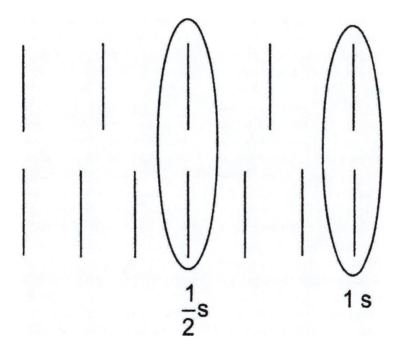

listener is more likely to experience **reverberations.** Reverberations are the multiple reflections that tend to reach the ear in a relatively short period. The time over which reverberations occur is less than 0.1 second. This is because the brain cannot distinguish between sound waves less than 0.1 second apart. Waves reaching the ears less than 0.1 second apart blend together, prolonging the original sound. This blending means the listener's brain interprets the closely spaced waves as a prolonged sound. Reverberations give fuller and richer sounds. This is demonstrated by singing in a shower stall, speaking through a tube, or shouting in a tunnel. In a shower stall, reverberations off the enclosure surfaces produce a deep rich sound, as opposed to the same sound in an open area.

Hard dense surfaces reflect more sound per unit area that than soft surfaces. This is observed in uncarpeted rooms with hard walls, where sounds tend to be louder and sharper. Designers of concert halls and auditoriums use different materials to selectively reflect or absorb sound in different areas of the structure. Smooth and rough surfaces differ in their ability to reflect waves at sharp angles. When a sound wave reflects off a smooth surface, the reflected angle is distinct and sharp. This is similar to light reflecting off a smooth flat mirror (discussed in the next chapter). The reflected angle equals the incident angle. This means a person sitting in a room will tend to receive reflected sound waves from a limited number of locations on the surrounding walls, ceiling, and floor (Figure 8.9). Conversely, a rough surface will produce sound waves that are diffusely reflected. This means the listener will experience reflected sound waves from many more points in the room.

The size and shape of a room are also important factors in determining how sound is perceived in a room. A large room can produce echoes as reflected waves reach the listener more than 0.1 second apart. Straight surfaces will reflect sound at distinct angles, while a room with projections and a number of angles gives a greater number of reflections. Curved walls can focus sounds like a lens. Interestingly, several of the greatest concert halls as rated by musicians were built in the 1800s. These halls tend to be narrow and long, shaped like a shoebox. This seems to be related to how the reflected waves reach the listener. In a wide hall, the first reflected waves come from the ceiling, which is generally closer to most listeners than the side walls, and reach the ears simultaneously. In a narrow hall, the first reflected waves generally come from the two side walls and reach the ears at slightly different times. This produces reverberations leading to a fuller, richer sound, as discussed above.

Transmittance, diffraction, and refraction are other characteristics of sound that impact acoustics. Sounds are transmitted through walls and other solid objects. This is observed each time we sit in enclosed buildings and hear sounds from outside. Air is a poor conductor of sound. The greater density of liquids and solids leads to greater conductivity of sound in these phases. The transmission of sound involves how the sound wave changes as it moves from one medium to another. The most common form of transmission involves sound waves in air being transmitted through a solid material such as a wall. When a sound wave in air encounters a solid, part of the sound wave is reflected and part is transmitted. Because the wave is moving from a region of less density to a region of greater density, the reflected wave is inverted. The transmitted wave is not inverted. Both the reflected and inverted waves have lower amplitude. The frequency of the transmitted wave and reflected wave are the same. The reflected wave may interact with the other waves from the source,

Figure 8.9
Sound is reflected from smooth surfaces (left) at a distinct angle, while rough surfaces (right) give a diffuse reflection. For reflection, the angle of reflection equals the incidence angle.

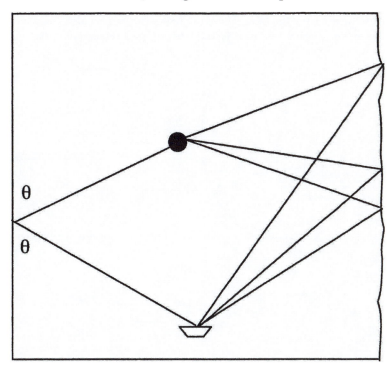

and this explains how sounds in air columns in musical instruments are produced.

Diffraction refers to the bending of sound waves around obstacles or the spreading out of waves past a barrier. Diffraction, along with transmission, allows sound to be heard around corners. The amount of diffraction (spreading out of a sound wave) depends on its wavelength and the size of an opening through which it passes. Greater diffraction occurs as the size of the sound's wavelength increases compared to the size of an opening through which it passes. High-frequency sounds, with shorter wavelengths, will have less diffraction than low-frequency sounds, with longer wavelengths. Therefore, high-pitched sounds will spread out over a smaller area than low-pitched sounds. Cone-shaped loud speakers, megaphones, and sirens have

small openings where the sound is introduced in order to enhance diffraction and, therefore, the dispersion of sound coming from them.

While diffraction involves the bending of waves around obstacles and openings, refraction refers to the bending of a wave as it moves from one medium to another or through regions of different densities. Refraction occurs because one part of the wave is slowed down (or sped up) as it travels across regions where the speed of the sound wave varies. For example, consider sound moving over an area where the temperature increases with height. Such a situation might be above a lake on a hot day. The part of the sound wave near the lake moves slower, where the air is cooler. This results in the sound wave refracting or bending toward the lake (Figure 8.10). Figure 8.10

Figure 8.10

As the sound wave moves over the lake, where the temperature is cooler, the part of the wave closer to the lake moves slower. This results in the bending or refraction of the wave toward the lake.

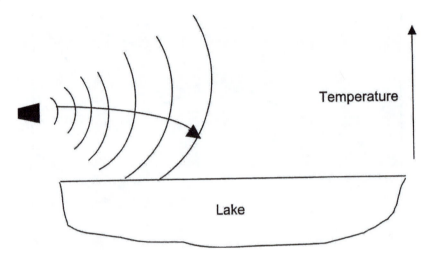

Figure 8.11
Diagram of the human ear

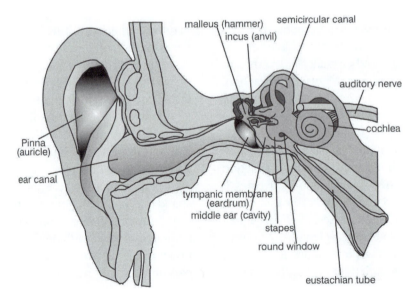

shows that sound waves bend downward where the atmospheric conditions are stable. In an unstable atmosphere, where cool, more dense air aloft exists over warm, less dense air near the ground, sound is refracted away from the ground.

The Ear and Hearing

Everything in this chapter would be impossible if it were not for our ears and our ability to hear. The human ear consists of three main parts called the outer, middle, and inner ear. The most conspicuous portion of our ears is the external ear. The external ear is called the pinna and, along with the auditory canal, comprises the outer ear (Figure 8.11). The auditory canal funnels sound into the interior of the ear. The end of the auditory canal leads to the middle ear and the tympanic membrane or eardrum. Sound is channeled toward the eardrum, causing it to vibrate. These vibrations are transferred to air-filled cavities in the middle ear and to a set of three small bones called the hammer, anvil, and stirrup. These bones transfer sound vibrations to the inner ear.

The inner ear is where the sensory perception of sound originates. It consists of fluid-filled semicircular canals and the cochlea. The cochlea is a coiled tube containing a membrane covered with several hundred thousand minute hairs. These hairs are stimulated by the movement of fluid in the cochlea and transfer signals to the brain via the auditory nerve. Once the signals reach the brain they are interpreted as sound.

Summary

Sounds are ever present in daily life; it is rare to experience complete silence. This chapter discussed how sound is a wave that requires a medium through which to travel. Because sound is a wave, it experiences those characteristics common to other waves, including interference, reflection, refraction, and diffraction. The wave properties of sound dictate how sound behaves and explain a wide-range of phenomena from sonic booms to music. Many of the wave properties introduced in this chapter will be revisited in the next chapter, where the subject is light.

9

Light and Optics

Introduction

Light has intrigued humans throughout history. Prehistoric humans were undoubtedly aware that the Sun and heavenly bodies provided light and that fire was a source of light that could ward off nightly dangers. It is not surprising that creation stories, such as those found in Genesis, focused on light. The Bible starts with the creation of light and the phrase, "God said, Let there be light, and there was light." The study of light has been the focus of the greatest scientific minds over time. Attempts to explain the behavior of light has led to significant advances in our understanding of physics. This in turn has forced a frequent reshaping of standard scientific theories. This chapter will trace the history of our understanding of light and explore light's basic behavior. We will also examine the optical applications that range from simple mirrors and lenses to lasers and fiber optics.

While we all have a familiar understanding of light, light can best be understood scientifically as a type of electromagnetic wave. A thorough understanding of electromagnetic waves will be developed in this chapter. An electromagnetic wave consists of two parts: an electrical part and a magnetic part (Figure 9.1). The electrical and magnetic parts of the wave travel perpendicular to each other at a speed of approximately 3.0×10^8 m/s (in a vacuum) and influence each other. Because light can be considered a wave (its particle nature will be the considered in the next chapter), it can be characterized by its wavelength and frequency. Electromagnetic waves with wavelengths ranging from 400 nm (violet) to 700 nm (red) comprise visible light (Figure 6.7). The abbreviation nm stands for nanometer. "Nano-" is the SI prefix representing 10^{-9}, therefore, 1 nanometer is 10^{-9} meter. The frequency of visible light ranges from 4×10^{14} Hz (red) to 8×10^{14} Hz (violet). Many sources use light in a more generic sense to describe not only visible light, but also electromagnetic waves outside the visible range, for example ultraviolet light. Throughout this chapter the emphasis will be on visible light, and the term "light" will be synonymous with visible light. When electromagnetic waves outside the visible range are the subject, specific reference will be made to the classes of electromagnetic

Figure 9.1
Electromagnetic wave. The electric and magnetic fields propagate as waves perpendicular to each other.

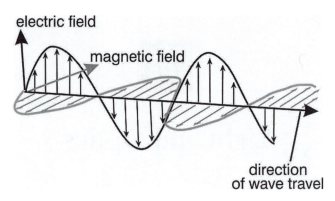

waves shown in Figure 6.7, for example, x-rays, radio waves, infrared, and so forth.

A Brief History of Light

Light has always occupied a significant place throughout history. Greek science examined light as it related to vision. The Pythagoreans believed that objects emitted a steady stream of light particles. According to the Pythagoreans, vision was the result of the eyes emitting a stream of particles that illuminated objects. According to the Pythagoreans, the source of vision was light rays emitted from the eye. Plato expanded on the Pythagorean ideas of light by postulating that vision was the result of light rays emitted from the eyes that mixed with luminous rays from the Sun. Rays from the eyes and Sun mixed and reflected off objects back to the eyes.

Euclid (325–265 B.C.E.) wrote *Optica* and *Catoptics* around 300 B.C.E. The latter was a work on mirrors. In *Optica,* Euclid expanded on Plato's work and noted that light rays emitted from the eyes traveled in straight lines. Euclid formulated optical laws using geometry. His work was the first definitive work on perspective and was

highly valued by artists during the Renaissance. Euclid questioned whether vision was dependent on light beams emitted from the eyes because he noted that when the eyes were shut and then opened that distant objects, such as stars, appear instantaneously. His argument was countered by speculating that the speed of light was incredibly fast.

Ptolemy (85–165 B.C.E.) also wrote a work on optics around 150 B.C.E. Unfortunately, of Ptolemy's five volumes relating to optics only one has survived. He formulated early ideas on the refraction of light and studied atmospheric refraction of light from planets and stars in making his astronomical observations. In addition to refraction, Ptolemy studied color, mirrors, and reflection and conducted physical experiments to formulate his ideas. Ptolemy, like Euclid and other Greek philosophers, held that vision originated in the eyes of the observer.

Abu Ali al-Hasan ibn al-Haitam, known to the West as Alhazen (965–1039), was an Arabian scholar who dispelled many of the ideas on light handed down from ancient Greece. At the same time, he provided the first modern interpretation of light. Educated in Baghdad, Alhazen spent most of his adult life in Cairo. His work was so influential on later studies in optics that he is often called the father of modern optics. He studied a wide range of natural optical phenomenon such as eclipses, shadows, sunsets, and rainbows. His studies on atmospheric refraction allowed him to estimate the depth of the atmosphere at approximately 60 miles. He examined colors, refraction, and reflection and examined optical objects such as mirrors and lenses. In examining the passage of light through various media, Alhazen formulated the general laws of refraction. Alhazen experimented extensively, and one of the most notable of his experiments involved developing the camera obscura. Alhazen's camera obscura (pinhole camera) involved hanging lanterns outside a darken room through which light could only enter through a small opening. Alhazen observed that inverted images of light from the lanterns formed on the wall opposite the small opening. From his experiments, Alhazen deduced that light travels in straight lines. More importantly, Alhazen applied his camera obscura experiments to vision. He believed that the eye was analogous to the small opening in the camera obscura, and images outside the body were reflected into the eyes. This idea refuted the long-held belief that vision was the result of the eyes projecting rays. Thus, Alhazen correctly explained vision in terms of light reflecting off objects and entering the eye. In addition to the advances Alhazen made in optics, his work was very influential in advancing the scientific method later in Europe. Alhazen coupled theory and experiment, and his methods provided a blueprint for the European scientific revolution.

After Alhazen, periodic progress was made on the study of light. Roger Bacon (1214–1292) continued Alhazen's inquiries into light and experimented with lenses and mirrors. Bacon is credited with using lenses for magnification and applying this knowledge toward the invention of spectacles. He advanced concepts on reflection and refraction as well as studying the physiology of the eye.

Optics was naturally related to the science of astronomy. Galileo's application of the telescope to astronomical observations stimulated the study of lenses used for magnification. Galileo's contemporary Kepler made a number of significant findings in optics. In his study on Mars, Kepler made observations using the camera obscura, and these observations gave contradictory results when compared to direct observations. This led Kepler to study the camera obscura, and he subsequently use this work to correctly explain how the eye works by forming an inverted image on the retina. Kepler published a major work on optics in 1604. In response to Galileo's work, Kepler conducted a study of lenses, and his work published in 1611 included the design of an astronomical telescope using two lenses.

The seventeenth century, which started with the work of Kepler and Galileo, saw significant progress made in the understanding of light as the greatest scientists of the time examined various topics in optics. Willebrod Snell (1580–1626) derived a mathematical formula explaining refraction. His law, known as **Snell's law** (see the section on refraction), was originally published by René Descartes some years after he met Snell in 1625, and it was called the law of Descartes. Around 1675 the Danish astronomer Olaus Roemer (1644–1710) used varying times observed for eclipses of the moons of Jupiter to derive the speed of light. Roemer's work resulted in a speed of 226,000 km/s, approximately 75% of its modern value.

As the seventeenth century drew to a close, two contrasting views of light were put forth that would dominate views on light for the next 200 years. Christiaan Huygens in his *Traite de la lumiere (Treatise on Light),* written in 1678 and published in 1690, proposed that light was a wave phenomenon. Huygens view was that light waves were like the ripples that formed when a pebble was tossed into a pool of water. Since it was known at that time that waves required a medium through which to move, Huygens wave theory also gave rise to a mysterious substance called ether that permeated the universe. According to Huygens, the ether consisted of minute elastic particles that retransmitted light and thus served as secondary sources of light waves. In this manner, light filled space by being retransmitted from an infinite number of points. The wave theory was supported by a number of other scientists at the time, including Robert Hooke.

In contrast to Huygens' wave theory of light, Newton proposed a corpuscular or particle theory. According to Newton, light consisted of tiny particles. Newton published his work on light in 1704 in his book *Opticks. Opticks* presented work that Newton had conducted on light throughout his life. Early in his studies, around 1665, Newton had demonstrated that white light could be separated by a prism into a spectrum of colors. This indicated that white was a mixture of colors. Newton then carried his spectrum work further by adding a second prism to recombine the separated colors back into white light. Newton's scientific rival, Robert Hooke, proposed alternative theories to a number of Newton's idea on light. Hooke accepted the wave theory, and Newton did not publish *Opticks* until after Hooke's death. While Newton's ideas rested on the particle theory of light, he proposed that light particles stimulated the ether to induce light waves. In this manner, Newton was able to resolve some of his own ideas with those of Huygens and the wave theorists.

For the next 100 years the wave theory and particle theory of light were both used to explain the behavior of light. Many individuals accepted the particle theory because of Newton's great reputation as a scientist. At the same time, the wave theory seemed more suitable for explaining the observed behavior of light. At the turn of the eighteenth century, the English physicist Thomas Young (1773–1829) published a series of papers in which he discussed the interference of light. Young also demonstrated how Newton's ideas on light were more easily interpreted using wave theory. The French engineer and scientist Augustin Jean Fresnel (1788–1827) developed a wave theory of light to explain diffraction, reflection, refraction, and polarization around 1820. Young and Fresnel's work provided major support for the wave theory, and by the mid nineteenth century the wave theory was widely accepted, while interest waned in the particle theory.

Michael Faraday's work on electricity and magnetism in the first half of the nineteenth century demonstrated that electricity and magnetism were related phenomena. Around 1830, Faraday (1791–1867) and Joseph Henry (1797–1878), an American, independently demonstrated that a changing magnetic field induces an electric current in a wire. Faraday proposed that electric and magnetic lines of force surrounded magnets and wires that carry currents.

Building on Faraday's ideas, the Scottish physicist James Clerk Maxwell (1831–1879) developed a mathematical theory that described light as an electromagnetic wave. Maxwell worked on his theory of light over a number of years, but in 1864 published a paper in which he defined light as an electromagnetic wave propagated as mutually perpendicular electrical and magnetic waves. Maxwell developed mathematical equations that defined the behavior of light, known as Maxwell's equations. Maxwell's

Figure 9.2
As the wave fronts spread out from their source, they approach a parallel configuration.
This is also true for the rays represented by the arrows.

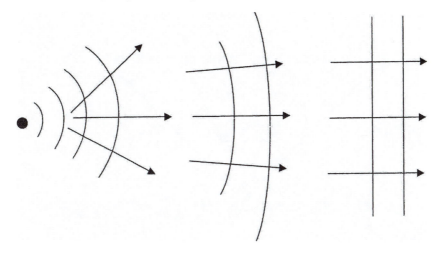

work established light as an electromagnetic wave. This brought a temporary order to the understanding of light around 1875, but this understanding was short lived. Experiments involving the speed of light, attempts to detect the ether through which light supposedly traveled, and the ability of light to create electrical current in some situations but not in others are a few of the major problems that developed at the close of the nineteenth century. This in turn ushered in a whole new interpretation of light in terms of quantum theory and relativity. Quantum theory will be examined in chapter 10 and relativity in chapter 15. The remainder of this chapter will focus on the classical theory of light.

Behavior of Light

In the previous section, a number of properties of light were introduced. Since several of these are associated with other wave phenomena, they were discussed in the last chapter with respect to sound, for example, refraction, reflection, and interference. In this section, these and several other properties unique to light are explored. The wave properties of light are often discussed by examining wave fronts and light rays. Wave fronts consist of an imaginary line drawn along all points in phase. For example, when standing on the shore and watching waves, the wave fronts are represented by crests running parallel to shore that break regularly. A light ray is a line drawn perpendicular to wave fronts. If light waves are considered analogous to ripples created by dropping a pebble into water, the wave fronts spread out from the source as shown in Figure 9.2. As these ripples spread out from their source, the wave fronts become flatter and approach a parallel configuration. Rays are imaginary lines from the source perpendicular to the wave fronts. As the waves move farther away, the wave fronts move from a spherical to a planar shape. Therefore, at a certain distance, wave fronts can be considered parallel planes with rays that are perpendicular.

Reflection

Reflection is one of the most obvious properties of light. Mirrors are commonly

Figure 9.3
Reflection off a surface is described using the incident and reflected rays and angles

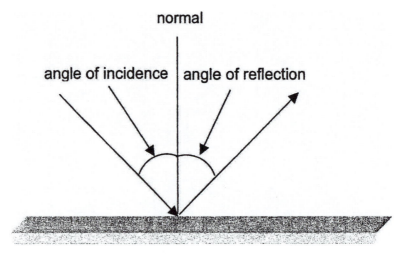

used in a variety of instruments to apply this property for human use, but reflection from other surfaces also takes place. Reflection is referenced by referring to the incident rays and reflected rays. Incident rays are the rays before they strike the surface, and reflected rays are the rays after striking the surface (Figure 9.3). The angle of incidence is the angle between an incident ray and a normal drawn perpendicular to the reflecting surface. The angle of reflection is the angle between the reflected ray and the normal. The reflection of light off a mirror or other smooth surface is called specular reflection. **Specular reflection** results in parallel rays after reflection. This is in contrast to **diffuse reflection,** where rays are not reflected parallel to each other (Figure 9.4). Specular reflection is characteristic of smooth surfaces, while diffuse reflection occurs off rough surfaces, similar to sound waves. The difference can be notice when comparing a waxed, glossy surface to a surface that is dull. The waxed surface appears brighter because the parallel reflected rays are aligned, as compared to diffuse rays, which are reflected in many directions from a dull

surface. The mathematical law describing reflection of light comes from **Fermat's principle** of least time, proposed by Pierre de Fermat (1601–1665) in 1657. Fermat's principle states that a ray of light moves from one point to another along the path that requires the least time. From this principle, the law of reflection states that the angle of reflection equals the angle of incidence and that the reflected and incident rays will lie in the same plane.

The most common application of reflection in daily life is the use of mirrors. Mirrors are generally one of first items used in the morning and the last used at night. The most common mirror in general use is a flat or plane mirror. The image formed when looking into a plane mirror has the following characteristics: it is the same size as the object, it is upright, it appears as far behind the mirror as the object is in front of the mirror, and it is reversed. The last characteristic is the reason the word "ambulance" is spelled backward on emergency vehicles. A person observes the mirror image of an object when some of the light from points on the object reflects off the mirror and

Figure 9.4
In specular reflection the reflected rays are parallel to each other, and in diffuse reflection the reflected rays are reflected at various angles

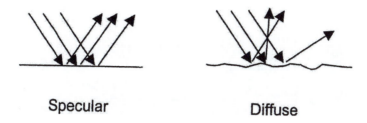

Specular   Diffuse

enters the person's eyes. The light rays from each point on the object create an image that seems to originate from a specific point behind the mirror (Figure 9.5). This type of image is referred to as a **virtual image** because the light rays do not come from behind the mirror but only appear to come from behind the mirror. Images in which the light actually comes from where the image forms are called **real images.**

Most of us have been entertained by viewing ourselves with curved mirrors that create odd images. These mirrors are called spherical mirrors because they represent part of the surface of a sphere. A section of the outside of a sphere (viewed from the outside) forms a **convex mirror,** and a section of the inside of a sphere forms a **concave mirror.** Convex mirrors are also known as diverging mirrors because the light reflected from them will diverge from a point located behind the mirror. Concave mirrors are converging mirrors because these mirrors cause reflected light to converge through a point in front of the mirror (Figure 9.6). Spherical mirrors have a center of curvature, radius of curvature, and principal axis. The principal axis passes through the center of curvature and the center of the mirror. All reflected rays diverge from or seem to diverge from or converge through a single point called the focal point. The distance between the

Figure 9.5
Reflection from a plane mirror. Ray tracing shows how the image is produced behind the mirror.

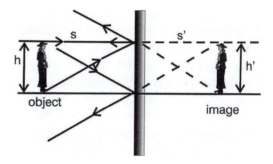

focal point and mirror is the focal length. By using geometry, it can be shown that the focal length is equal to one-half the radius of curvature. The focal point is located in front of concave mirrors and behind convex mirrors. Figure 9.6 illustrates the path of light rays drawn parallel to the principal axis. In reality, rays strike the mirrors at all angles and at various distances from the principal axis. Rays that strike the mirror close to the parallel axis have a small angle of incidence. These are called paraxial rays and will be reflected through the focal point. Those that strike the mirror farther away from the principal axis do not converge exactly at the focal point. Therefore, the focal length is one-half the radius of curvature only for

Figure 9.6
In spherical mirrors, the center of curvature, C, and focal point, F, lie on the principal axis. The radius is the distance between C and the mirror. Reflected rays in a convex mirror diverge as though they originate from the focal point, F.

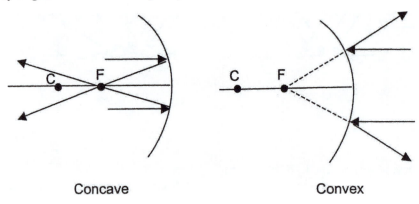

Concave   Convex

paraxial rays. The fact that rays farther from the parallel axis do not pass through the focal point gives rise to **spherical aberration.** Spherical aberration results in a blurred image. To reduce spherical aberration, mirrors should be made with a large radius of curvature. This restricts the distance that rays vary from the principal axis. Spherical aberration can also be eliminated by constructing a mirror with a parabolic shape. The Hubble telescope contained a parabolic mirror, but manufacturing imperfections necessitated a special mission to outfit the telescope with special optics to compensate for its imperfections. Parabolic mirrors find widespread use in manufacturing high-quality telescopes and other devices used to concentrate electromagnetic waves, for example, antennas. Parabolic reflectors are used to produce parallel concentrated rays in flashlights, headlamps, and car lights. By placing a bulb at the focal point of the mirror, rays are reflected parallel to the principal axis in a narrow beam.

When light rays strike a concave mirror, the rays converge through a focal point in front of the mirror. When light rays strike a convex mirror, they diverge as though coming from the focal point located behind the mirror. The type of image produced from a mirror can be described in several ways: virtual or real, upright or inverted, reduced or magnified. In order to determine the type of image that is produced, rays from different points on an object can be traced. An image is produced where reflected rays intersect. Rather than trace rays randomly, several types of principal rays are used to define the image. A parallel ray is one that is parallel to the principal axis and will be reflected through the focal point, for a concave mirror. In the case of a convex mirror, a parallel ray is reflected as though it originates from the focal point behind the mirror. A ray that passes through the focal point of a concave mirror will be reflected parallel to the principal axis. For a convex mirror, a ray that extends through the focal point will be reflected parallel to the mirror. Finally, an incident ray passing through the center of curvature of a concave mirror or a ray that extends through the center of curvature for a convex mirror will be reflected back along the same path as the incident ray.

Figure 9.7
Ray tracing results in an inverted, reduced, real image when the object is beyond the radius of curvature for a concave mirror.

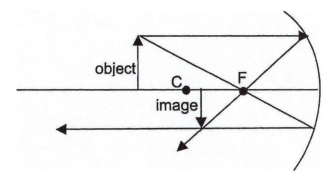

Figure 9.8
The reflected image from the convex mirror is upright and reduced. Because the image comes from behind the mirror, it is a virtual image.

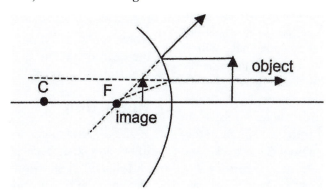

To illustrate how ray tracing works, two examples will be used. First, consider an object located beyond the radius of curvature of a concave mirror. Figure 9.7 illustrates how the image is determined by ray tracing. The parallel ray is reflected through the focal point, while the ray through the focal point is reflected parallel. This results in an image that is inverted and smaller compared to the object. The image is real because the light from comes directly from the image. Now consider the convex mirror in Figure 9.8. In this situation, the rays are extended to illustrate how to locate the image. Convex mirrors produce upright, reduced, virtual images.

Another way to determine the properties of reflected images is mathematically with the use of the mirror equation:

$$\frac{1}{d_o} + \frac{1}{d_i} = \frac{1}{f}$$

In this equation, d_o and d_i represent the distance from the mirror to the object and image, respectively. The focal length equals f. In using the mirror equation, distances in front of the mirror have positive values, and distances behind the mirror are negative. To illustrate how the mirror equation is used, it will be applied to the two ray tracing examples above. For the concave mirror example given in Figure 9.7, consider the object to be located 25 cm in front of the mirror with a focal length of 10 cm. Substituting 25 cm for d_o and 10 cm for f, then solving for d_i gives an image distance of approximately 16.7 cm. The positive value for d_o shows that the image is located 16.7 cm in front of the mirror, which corresponds to the image location shown in the figure. Magnification is the height of the image divided by the height of the object. An equation for magnification can be derived using the mirror equation and geometry: magnification $= -d_i/d_o$. For the concave mirror example, the magnification equals -16.7 cm/25 cm $= -0.7$. The negative sign indicates that the image will be inverted and only 0.7 times as large as the object, again agreeing with Figure 9.7.

In the convex mirror example, illustrated in Figure 9.8, assume the value of d_o is 10 cm and the value of f is -10 cm. The focal distance is negative because it is located behind the mirror. In fact, concave mirrors will always have a positive value of f and convex mirrors a negative value of f. Using the mirror equation, the calculated value of d_i is -5 cm. This indicates the image is virtual and located halfway between the mirror and the focal point. The magnification equals $-(-5/10) = 1/2$. The positive value for magnification means the image is upright.

Concave and convex mirrors are commonly employed in a variety of uses. Concave mirrors made so that the object is located inside the focal point serve as cosmetic mirrors. These mirrors produce a magnified, upright image. In reflecting telescopes, light from a distant source falls on a concave mirror, where it is focused on a second plane mirror and then to an eyepiece.

Table 9.1
Index of Refraction for Common Substances

Substance	Index of Refraction
Vacuum	1.0000
Air	1.0003
Water	1.33
Ice	1.31
Fused quartz glass	1.46
Sugar solution (50%)	2.11

Convex mirrors are widely employed in stores for security, on vehicles to increase the field of vision, and on roads to view around blind corners.

Refraction

Light refraction refers to the bending of light and occurs because the speed of light varies in different media. The speed of light in a vacuum to three significant figures is 3.00×10^8 m/s. The ratio of the speed of light in a vacuum to the speed of light in some medium is called the **index of refraction.** Table 9.1 gives indices of refraction for a number of common substances. Refraction can be explained by a simple analogy. Consider a wagon moving down a smooth sidewalk and then one side encounters gravel. The side encountering the gravel will be slowed considerably, while the side on the smooth sidewalk will not. This will cause the wagon to rotate around the wheels on the slow side. This rotation is similar to the bending of the wave front of light when it passes from one substance to another where its speed differs (Figure 9.9).

Figure 9.9
When light passes from one medium to another it is bent or refracted

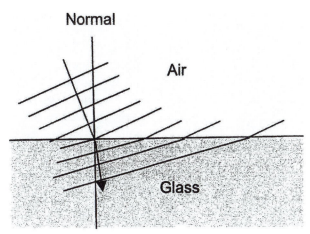

Figure 9.9 shows that as light passes into a medium where its speed decreases (to a medium with a higher refractive index) that it is bent toward the normal. The mathematical relationship describing the refraction of light is given by Snell's law, named in honor of Willebrod Snell (1591–1626). Snell's law states that when light passes from a substance with a refraction index given by n_1 into a substance with a refractive index given by n_2, that the angle of refraction, θ_2, is related to the angle of incidence, θ_1, by the following equation: $n_1 \sin \theta_1 = n_2 \sin \theta_2$. In using Snell's law, the angles of incidence and refraction are measured from the ray to the normal. The normal is the perpendicular line that runs through the line that forms the boundary between the two media. The incident and refracted ray as well as the normal all lie in the same plane.

Refraction is responsible for making an object like a pencil appear to be bent when placed in a glass of water. In this situation, the light rays leaving the pencil are refracted away from the normal upon leaving the water and entering the air before reaching the observer's eyes (Figure 9.10).

Refraction causes objects under water to appear closer to the surface and is responsible for mirages. A common mirage is the appearance of water in the distance on a hot road on a dry day. This is due to the changes in refractive index of air with temperature. On a hot day, the air directly adjacent to the pavement is significantly hotter than the air higher above the pavement. The change in temperature causes incoming rays from the Sun to bend upward, and the mixing of air makes the refracted rays seem to shimmer. The upward bending causes the rays to seem to come from the pavement. This same phenomenon occurs in the hot desert to create the appearance of an oasis in the distance.

A ray of light moving from a medium with a larger refractive index to a medium with a smaller refractive index will be bent away from the normal. In this situation, not all of the light is refracted, but part of the ray is refracted and some is reflected (Figure 9.11). As the incident angle progressively increases, the refracted angle will reach 90°. At this point, the refracted ray is parallel to the interface between the two media, and the incident angle is known

Figure 9.10

A pencil appears magnified, displaced, and bent when placed in a glass of water due to the refraction of light

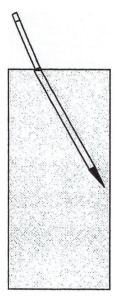

as the critical angle. The critical angle, θ_c, is given by the formula n_2/n_1, where n_2 is the medium with the smaller refractive index. Once the incident angle exceeds the critical angle, no light will be refracted, and it will all be reflected. This phenomenon is referred to as **total internal reflection.**

Total internal reflection is the basis of fiber optics. Optical fibers, which can be thinner than a human hair, consist of an inner transparent core with a relatively high refractive index surrounded by a layer with a relatively low refractive index. By projecting light into the fiber at an angle greater than the critical angle, light can be transmitted down the length of the fiber. The length of fibers may only be a an inch or hundreds of miles, depending on the application. In communications technology, a light source such as a laser is used to transmit information through fiber cables packed in bundles. In this manner, large amounts of information

are transmitted. In addition to the vast amounts of information optical fibers carry, they are also relatively free from the electrical interference associated with traditional metal wires. Another major application of optical fibers is in medicine. Endoscopes are optical fibers used to examine the interior of the body. For example, in arthroscopy, they are used to examine joints such as the knee, and surgery performed with the aid of these scopes is termed arthroscopic surgery. A colonoscopy uses optical fibers to examine and remove polyps in the colon.

Just as mirrors are used for reflection, lenses are based largely on the refractive properties of light. A lens is typically a thin piece of glass or other transparent material such as plastic that has the ability to redirect light rays to form an image. Lenses that cause incident parallel light rays to converge through a focal point are called converging or convex (also referred to as positive) lenses. Diverging or concave (negative) lenses cause incident parallel light rays to diverge from a focal point. Similar to mirrors, lenses are characterized by a focal point, focal length, principal axis, and center of curvature (Figure 9.12).

Ray tracing similar to that employed for mirrors can be used to locate and describe images produced from lenses. Parallel incident rays (those parallel to the principal axis) are refracted through the focal point on the opposite side they enter for converging lenses. For a diverging lens, parallel incident rays are refracted as though they originate from the focal point on the incident side. Rays drawn from the object through the focal point will refract from the lens parallel to the principal axis. Rays that pass through the middle of the lens are relatively unaffected and continue on their original path. Two examples of ray tracing for a converging and a diverging lens will illustrate the method of locating an image for a lens. Consider the

Figure 9.11
Total internal reflection can take place when light moves from a medium with a higher refractive index to a medium with a smaller refractive index. In the refracted case, part of the ray is refracted and part is reflected. At the critical angle, the refracted angle is 90°, and the reflected ray is also shown. Total internal reflection occurs when the incident ray is greater than the critical angle.

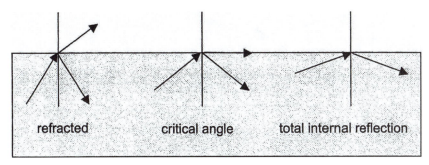

refracted   critical angle   total internal reflection

Figure 9.12
A lens is characterized by a focal point (F), a focal length, and a radius of curvature. The lens shown is a double convex lens, so it has two focal points and two radii of curvature.

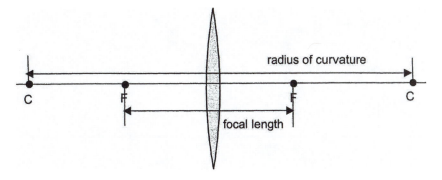

radius of curvature

C   F   F   C

focal length

double converging lens in figure 9.13. An object within the focal point of this lens will produce a virtual, magnified, upright image. The lens depicted in Figure 9.13 represents a magnifying glass. This can be compared to Figure 9.14 representing a diverging lens. In this case, the object outside the focus results in a virtual, upright, reduced image.

$$\frac{1}{d_o} + \frac{1}{d_i} = \frac{1}{f}$$

The mathematical treatment of locating the image for a lens is identical to that for the mirror. The equation is referred to as the thin-lens equation and is identical to the mirror equation: . In working with lenses, a real image is formed on the side opposite the object, and the distance to the object would be taken as positive. The focal length for a converging lens is positive and for a diverging lens is negative. For the lens shown in Figure 9.13, if the focal length is 8 cm and the object distance is 5 cm, then the value

Figure 9.13
An object place behind a converging lens will form a magnified, upright, image. The image forms where the rays or their extensions indicated by dotted lines intersect. This lens is typical of a magnifying glass, where the viewer would observe the object from the right in the diagram.

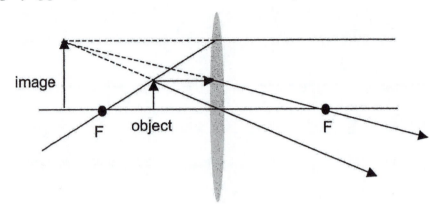

Figure 9.14
The image from this diverging lens forms a virtual, upright, reduced image. Notice that the image forms where the rays or their extensions indicated by dotted lines intersect.

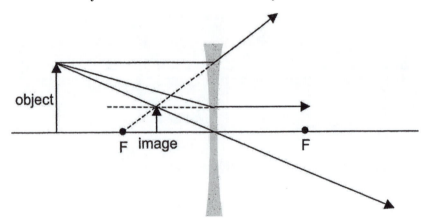

for d_i would be approximately -13 cm. The negative sign indicates that the image would be a virtual image formed on the same side as the object. When the object is located to the left of the lens and the rays move to the right, a positive d_i indicates a real image formed to the right of the lens and a negative sign signifies a virtual image formed to the left of the lens. Using the magnification equation, magnification $= -d_i/d_o$, the magnification

for the previous example is $-(-13/5) = 2.6$. The positive sign for the magnification indicates that the image is upright.

The equation for calculating the image location for lenses is called the thin-lens equation because the geometry upon which it is based assumes that the lens is a line forming one side of a triangle. Other equations exist for finding the image of a thick lens. As with mirrors, lenses produce spherical aberration because light passing through the lens farther from the principal axis is refracted at a slightly different angle. For a converging lens, this results in the ray not meeting at a precise point, but rather converging over a limited area. Chromatic aberration results from light of different colors being refracted at different angles. Therefore, the different colors would focus at slightly different points. Aberration is corrected in optical instruments by the use of diaphragms, employing compound lenses (two or more lenses acting together), special coatings, and shaping the lens to compensate for aberration.

Lenses are employed in a variety of uses such as telescopes, cameras, binoculars, and microscopes. By far, the most common use is for corrective lenses worn by people throughout the world. Over 150 million people in the United States wear some form of corrective lenses. There are reports of eyeglasses appearing as early as the eleventh century in China. The first reported use in Europe probably occurred around the end of the thirteenth century in Florence. Throughout history, advances have been made in types of corrective lenses. Corrective lenses assist the eye in forming an image on the retina. In the human eye, light enters through a lens and, when vision is normal, focuses images on the retina. Corrective lenses are used when the image does not focus on the retina. Farsightedness is the condition when people can focus on far objects, but cannot focus on nearby objects. This occurs when the image focuses behind the retina, often due to a shortened eye. In this case, converging lenses are used to focus the image on the retina. Nearsightedness is the condition when far away objects cannot be brought into focus. It occurs when the image focuses in front of the retina. Diverging lenses are used to correct this condition. In astigmatism, the thin covering over the lens, called the cornea, is distorted but can be corrected using a cylindrical-shaped lens. In addition to these standard conditions, many adaptations such as bifocal or trifocal lenses can be employed to remedy several vision problems with one pair of glasses.

Dispersion and Scattering

Dispersion of light refers to the spreading out or separation of light due to its frequency. This occurs because the speed of light through various materials depends on the frequency (color) of light. The speed of light through a medium is inversely proportional to its frequency. This means that light at the violet end of the visible spectrum (high frequency) moves slower than red light. This can also be interpreted as higher frequency light having a higher refractive index. It should be emphasized that dispersion requires a medium called a dispersive medium, and that the speed of light with different frequencies is constant in a vacuum.

Dispersion causes sunlight to be separated by a prism into its constituent colors. Violet will be slowed the most, and because violet has the highest refractive index, it will experience the greatest refraction. Red will be refracted the least. Rainbows are nature's best example of dispersion. Raindrops form the dispersive medium through which sunlight is dispersed. A rainbow begins when sunlight enters a water droplet and is dispersed into different colors. The different colors are then reflected off the back of the droplet and are refracted again as they leave the raindrop and enter the atmosphere (Figure 9.15). Due to the refractive indices of water and air, the light leaving the water droplet has its greatest intensity between 40° and 42° with respect to the incoming rays of sunlight. The 40° angle characterizes the maximum intensity for violet light and 42° for red light. The other colors of the

Figure 9.15
Rainbows occur when light is refracted and reflected in a water droplet. Red light is viewed at a greater angle, thus the red arch appears higher in the sky.

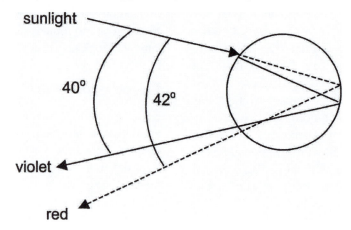

spectrum are intensified at angles between 40° and 42°. The dispersion at different angles produces the arched rainbow. Since an observer only sees color coming from a specific location, different colors are viewed from different rain droplets at different elevations. Red comes from droplets at higher elevations, and other colors come from droplets at lower elevations.

Refraction and dispersion are responsible for rainbows. The blue color of the sky is due to the differential scattering by the atmosphere of the different colors present in sunlight. The British physicist John Tyndall (1820–1893) studied the scattering of light in solutions and discovered that dissolved particles that were large enough had the ability to scatter light. This phenomenon is known as the **Tyndall effect.** Tyndall's work was more fully developed by John W. Strutt, Lord Rayleigh (1842–1919), who showed that the amount of scattering of light was inversely proportional to the fourth power of the wavelength. The scattering of sunlight off molecules in the atmosphere is referred to as **Rayleigh scattering.** Rayleigh scattering produces greater scattering

for light at the violet end of the visible spectrum by scattering it through greater angles with more intensity. Violet light is scattered approximately 10 times more than red light. Human vision is more sensitive to blue (as compared to violet), and therefore, the sky appears blue. Rayleigh scattering involves the scattering of sunlight off small particles such as molecules of nitrogen and oxygen in the atmosphere. Other types of scattering exist for larger particles. As noted, on a clear day the sky appears blue, but when pollution and large particles are present, the sky has a more whitish hue. This indicates that a greater percentage of other wavelengths are being scattered toward the Earth's surface. Rayleigh scattering also explains why the Sun appears red at sunrise and sunset. At these times the Sun is actually below the horizon and the sunlight seen is refracted light that appears before the Sun rises or after it sets. At these times, the Sun's rays travel through a longer atmospheric path. Blue light is scattered outside the Sun's rays reaching the Earth's surface. Furthermore, the greater path through which the Sun's rays travel allows longer wavelengths to be scattered, giving the Sun a red appearance. At noon, when the Sun is close to directly overhead, little light is scattered and the Sun appears almost white. At other times of the day it appears yellow, orange, or red, depending on the how different wavelengths are scattered.

Diffraction

Diffraction is the bending of light around obstacles or around the edges of an opening. Diffraction is due to the wave nature of light. In the seventeenth century, Huygens proposed that light travels out from a source as wave fronts. According to Huygens, each point on the wave front can be considered a secondary source of wave fronts. Further, each point on these secondary wave fronts can act as a source of additional wave fronts. This can be considered a continuous process, so that there are an infinite number of wave sources. The idea that each point on a wave front can serve as a source for waves is known as **Huygens' principle.**

The diffraction of light can be compared to the movement of water waves through an opening. When the size of the opening is large compared to the wavelength of the waves, only a small amount of diffraction takes place. As the opening gets smaller and approaches the size of the wavelength, diffraction becomes much more significant

(Figure 9.16). The amount of diffraction is inversely proportional to the size of the opening. Shining light through a large opening produces a sharp image about the size of the opening on a screen. When light is projected through a narrow slit, though, the image spreads out, and a distinct shadow doesn't form at the edge. Rather, the image consists of alternate bands of light and dark. The central large area of the image is bright and is called the primary maxima. The primary maxima is surrounded by dark bands referred to as minima, and moving outward there is a fading series of light and dark secondary maxima and minima (Figure 9.17). The alternating bands of light and dark (maxima and minima) are due to alternating regions of constructive and destructive interference. The dark bands represent areas of destructive interference (see chapter 8) caused by wavelets from different regions of the slit meeting at the screen. Light areas occur where constructive interference takes place. The intensity of the light decreases moving away from the centerline of the projection.

Diffraction and interference also occur when light is projected through multiple openings. Thomas Young's double-slit experiments performed at the beginning of the nineteenth century provided strong evidence for the wave theory of light (Figure 9.18). Young showed that when

Figure 9.16
Diffraction increases as the size of the opening gets smaller as compared to the sized of the wavelength

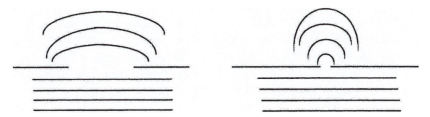

Figure 9.17
Diffraction pattern from a single, narrow slit

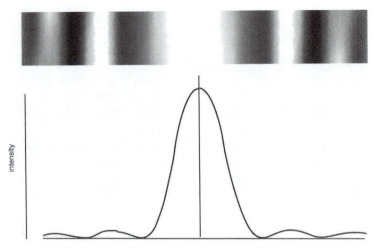

Figure 9.18
Thomas Young's double slit diffraction pattern

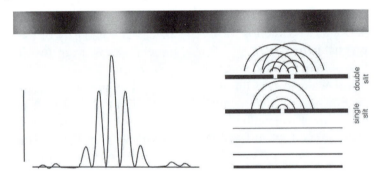

light was projected through two slits that an interference pattern of maxima and minima resulted. This would be exactly what would be expected if light were a wave. A glass slide with multiple slits is called a diffraction grating. A diffraction grating typically has thousands of openings per inch. A diffraction grating disperses color through the interference patterns produced by light passing through the grating.

Polarization

One way to describe light is as a transverse electromagnetic wave. An important aspect of transverse waves is **plane polarization.** Plane polarization means that a wave vibrates in one plane. Light from light bulbs and other common sources is unpolarized. This means the electromagnetic waves are vibrating in many directions. A radio antenna produces a polarized electromagnetic wave because the electrons in the antenna move in only an up and down direction. This can be contrasted to electrons vibrating in atoms that move in all directions in a light bulb. Certain materials have the ability to convert unpolarized light to polarized light. These polarizers work by letting light vibrating in one plane pass, while blocking out other light. An analogy of a polarizer would be a picket fence through which boards are trying to be passed. When the boards are in a vertical position, they can easily be passed between the slats in the fence. If the boards aren't nearly vertical, they won't fit through the fence. When attempting to randomly pass boards through the fence without regard for their orientation, only the vertical ones will make it through. A polarizer is like a fence that only allows light with a specific orientation to pass through.

A common polarizing material is Polaroid. Polarized sunglasses made of Polaroid have a vertical transmission axis that is most effective at blocking light vibrating in a horizontal direction. The reason for this is that light reflecting off horizontal surfaces is partially polarized in the horizontal direction. Glare is the common term used to describe this polarization. A large amount of horizontally polarized light comes from large flat surfaces such as lakes, and in these instances Polaroid sunglasses are very useful.

The polarization of light is applied in the display of a number of liquid crystal devices such as watches and calculators. A liquid crystal is a substance that has properties of both solids and liquids. The crystals can flow and transmit or block light, depending on the voltage applied to them. A liquid crystal display works by rotating polarized light so that it either passes through or is blocked by a second polarizer. Liquid crystals cause the light to be blocked by the second polarizer, and the display registers a dark image (Figure 9.19). These dark images are displayed as numbers or letters in the liquid crystal display. By mixing specific chemicals with liquid crystal molecules color displays can be made.

Polarization is one characteristic of light. Several other characteristics are important when considering light. Light that is monochromatic consists of light of a single color, that is, a single wavelength. Coherent light consists of light in phase, and incoherent light is light showing a phase shift. Coherent light might be considered equivalent to a marching band, where each person represents a wave crest. Using this analogy, each row and column is lined up and spaced out evenly. Conversely, a crowd emptying out of a stadium or arena would represent incoherent light; people are not lined up but scrabbled and random.

The diffraction interference patterns discussed previously require the light source to be monochromatic and coherent. Regular incandescent light bulbs produce incoherent light. Coherent light can be obtained by using a laser. The word "laser" is an acronym that stands for **l**ight **a**mplification by **s**timulated **e**mission of **r**adiation. To understand how a laser works, the production of light from atoms must be considered. Light is the result of electrons moving from a higher to a lower energy level in an excited atom (see chapter 10). For example, when an electric burner is turned on high, it glows red. The electrical energy supplied to the burner energizes electrons to a higher energy level in the burner coil. These energized electrons continually move back down to a lower energy level (then are reenergized) and in the process emit light predominately in the red end of the visible spectrum. Light from most sources produces incoherent light because electrons are randomly moving between numerous energy levels. An

Figure 9.19
In a liquid crystal display, unpolarized light passes through a polarizer. The polarized light is either unaffected (1) or rotated (2) by the liquid crystals, depending on the voltage applied to the liquid crystals. Light either passes through the second polarizer and is reflected by the mirror, producing a light image, or is blocked by the second polarizer, in which case a dark image appears.

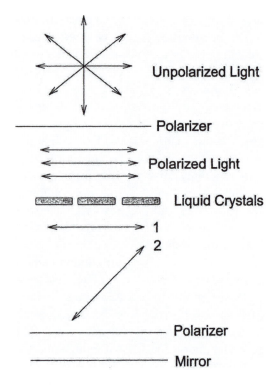

incandescent light bulb emits white light due to all colors being emitted. In a laser, a significant number of electrons exist in an excited configuration that is relatively stable. When most electrons exist in the excited state as compared to the normal ground state, the condition is known as **population inversion.** The electrons in this metastable state can be stimulated to undergo emission rather than spontaneously returning to a lower state. In a helium-neon laser, the electrons in the neon atoms are excited to a metastable state directly using high-voltage electrical energy and indirectly through collisions with excited helium atoms. The electrons in the excited neon are then stimulated to release energy and produce light with a specific wavelength. This stimulation is the result of the emitted light from one electron forcing a neighboring electron to emit at the same frequency. This process all takes place in a sealed tube. One end of the tube contains a regular mirror to reflect the coherent light back into the tube, and the other end has a half-silvered mirror. The photons move back and forth between the mirrors, but the half-silvered mirror allows some of the light to escape, and this is what is seen as the laser beam. Laser light is monochromatic and coherent. As it leaves the laser it travels as a narrow, concentrated, straight beam.

Figure 9.20
Reflection of laser light from a series of "pits" produces a binary code that carries information. Where a pit meets the flat area of the disc, destructive interference creates an "off" signal.

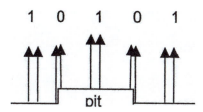

Lasers are used in numerous applications each day. They are used as simple pointers, for leveling instruments in surveying and construction, and as scanners at checkout stands to read the universal product codes (UPCs) on merchandise. In the latter application, the laser beam is reflected and absorbed as the white and dark bars of the UPC label pass through the laser beam. This is translated into a binary code of 1s and 0s that contains information about the type of item and its price. A similar coding takes place with a compact disc (CD). Whereas a UPC codes information as light and dark bars, a CD's information is coded as a series of elevated and flat areas on the disk. These bumps are called pits because when viewed from the side opposite the laser they form tiny indentations in the disc. A series of pits concentrically arranged on the disc codes the information. The depth of the pit (or height of the bump) is made so that destructive interference occurs between light reflecting off the edge of the pit and the adjacent flat area. In this manner, an "off" signal is generated everyplace where a pit meets the flat CD surface (Figure 9.20). As the disc spins, the binary coded signal is detected and converted into music, pictures, or other information. Another area where lasers have found widespread use is in surgery. Laser eye surgery such as LASIK (laser in situ keratomileusis) uses ultraviolet lasers to cut and reshape the cornea.

Summary

The nature of light has played a role in shaping physical thought throughout history. This chapter has focused on its behavior as an electromagnetic wave. The classical description of light has been discussed with primary emphasis on optics. Although the wave nature of light explains many aspects of its behavior, it does not explain everything. Early theories of light proposed light either as a wave or as a particle. Experiments in first half of the nineteenth century supported the wave nature of light. As the nineteenth century grew to a close, though, new experiments laid the groundwork for reshaping yet again the nature of light. Quantum theory developed as the twentieth century unfolded, and the fact that light behaved both as a wave and a particle established the wave-particle duality concept of light. This topic will be further discussed in the next chapter. Light also played a primary role in the development of the theory of relativity, which will be examined in chapter 15.

10

Quantum Physics

Introduction

The previous chapter examined the classical theory of light in which light was considered an electromagnetic wave. The discussion in chapter 9 focused on visible light. Many of the properties described for visible light apply to other forms of electromagnetic radiation. The electromagnetic spectrum consists of electromagnetic waves that range from gamma rays to radio waves (Figure 6.7). Gamma and x-rays have a high frequency and short wavelength and originate through nuclear processes within the atom. X-rays are also produced when electrons undergo acceleration or make energy transitions in atoms. Radiation with wavelengths immediately below visible light is called ultraviolet radiation, sometimes referred to as UV light. The Sun is a strong emitter of UV light. Atmospheric gases such as ozone are strong absorbers of UV light and help protect life on Earth from its harmful effects. Infrared radiation consists of those wavelengths longer than visible light up to approximately 1 mm. Radiation close to the red end of the visible spectrum is called **near infrared,** while infrared farther away from visible light falls into the **far infrared** range. Infrared radiation is extremely effective

in causing molecules to vibrate, and we sense this as energy in the form of heat. Beyond infrared, lie microwaves and waves typically associated with communications. The later include, in order from shorter to longer wavelengths: FM radio, television, short wave, and AM radio. The discussion of quantum mechanics in this chapter will expand the concepts of light introduced in the last chapter. In this chapter, the discussion will not be limited to visible light but include other forms of electromagnetic radiation as well.

The ideas of quantum physics were the result of the failure of classical physics to explain observations made at the atomic level. Discoveries on radiation and the atom that marked the beginning of the twentieth century could not be fully understood with nineteenth-century physics. Quantum theory resulted in a large part from the failure of classical theories to resolve new results. In the process, modern physics was born.

Blackbody Radiation

In chapter 6, radiation as a form of energy transfer was introduced. It was noted that all bodies radiate energy in the form of electromagnetic waves. The idea of a blackbody as a perfect emitter and absorber of radiation was introduced. Blackbody radiation can be described using **Wien's law** and the **Stefan-Boltzmann law.** Blackbody radiation from an object is emitted over a range of wavelengths. When the emitted wavelengths fall into the visible range, it is characterized by specific colors. For example, a heated iron rod may glow red. According to Wien's law, named in honor of the German physicist Wilhelm Wien (1864–1928), a body's peak radiation wavelength is inversely proportional to its absolute temperature: $wavelength_{peak} = 0.0029/T$, where T is the absolute temperature in kelvins of the object and the wavelength is in meters. Wien's law indicates that hotter objects have their maximum radiation at shorter wavelengths. Accordingly, the Sun, with a surface temperature of approximately 6,000 K, radiates extensively in the visible range, while most of the radiation coming from the Earth falls in the infrared range. The Stefan-Boltzmann law was discussed in chapter 6 and can be used to calculate the amount of radiation at different temperatures. Wien's law and the Stefan-Boltzmann law are illustrated in Figure 10.1, which shows blackbody radiation curves for objects at several different temperatures.

At the end of the eighteenth century, theoretical attempts to predict radiation curves such as those shown in Figure 10.1 were unsuccessful. According to classical theory, the intensity would increase quickly toward infinity as the frequency increases. The classical result did not make sense since it represented a physical impossibility. According to classical theory, the intensity continually increased with increasing frequency. This meant it approached infinity at ultraviolet wavelengths, and this dilemma was termed the **ultraviolet catastrophe.** In order to explain blackbody radiation, the German physicist Max Planck (1858–1947) derived an equation that required a radical rethinking of how energy was radiated. Rather than energy being radiated continuously through all wavelengths, Planck's mathematics required that

Figure 10.1
Radiation curves. As temperature increases, the peak wavelength at which a body radiates energy decreases.

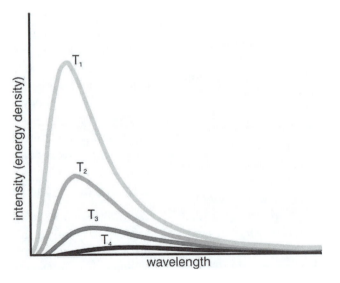

energy be radiated as a discrete unit called a **quantum** (plural is quanta). According to Planck's model, it was increasingly difficult to acquire the necessary quanta to radiate at higher frequencies, and this would explain why the radiation curves drop in the ultraviolet region rather than continue to increase. Planck's formula assumed that the energy radiated was in integral multiples of hf, where h was a constant and f was the frequency of radiation in hertz. The constant h is referred to as Planck's constant, and its accepted modern value is $6.6260755 \times 10^{-34}$ J-s. According to Planck's equation, the smallest quantum of energy radiated at a given frequency is equal to hf. The next highest quantum is $2hf$, and then $3hf$, and so forth.

The idea of quantized energy was a radical departure from classical physical thought and marked the beginning of modern physics. While today modern science accepts a quantum viewpoint, it was not readily accepted when Planck introduced the idea in 1900. Energy and matter can be considered continuous in the macroscopic everyday world, but they take on a quantum nature on the atomic level. For instance, the quantity of matter can be considered continuous and measured very precisely in fractions of grams. Although when a piece of iron is considered, it really consists of a tremendous number of individual iron quanta called iron atoms. Likewise, energy comes in discrete quantum units.

The Photoelectric Effect

Planck introduced the idea of quantized energy in 1900 to explain blackbody radiation. The next significant evidence that supported quantum theory came from interpreting the **photoelectric effect** using the new model. The photoelectric effect refers the emission of electrons from the surface of certain metals when illuminated with light. Some important observations associated with the photoelectric effect were that electrons were emitted immediately when the metal was illuminated by light, the brightness did not affect the kinetic energy of ejected electrons but the frequency did, and electrons couldn't be emitted with certain wavelengths no matter what the intensity.

Albert Einstein used quantum theory in 1905 to explain these observations. Planck assumed that blackbody radiation was emitted in discrete quanta. Einstein reason that objects absorbed radiation in a quantized fashion. Einstein knew that a specific threshold energy was required to eject an electron from a metal. This threshold energy depended on the metal. Einstein reasoned that if the light was insufficient to supply this threshold energy, then no electrons would be emitted from the metal. No matter how intense the light, it would not have sufficient energy. Light at a higher frequency had more energy, according to Planck's equation: $E = hf$. If light with sufficient energy at or above the threshold level were available, electrons would be ejected immediately. Classical theory predicted a time lag for the ejection of electrons because energy along the diffuse wave front of light would need to accumulate before an electron could be ejected. The kinetic energy of the ejected electron was directly related to the frequency of the light, while light intensity determined the number of electrons that were ejected. Einstein's interpretation of the photoelectric effect using a quantum model provided strong support for the model itself. Furthermore, Einstein's reasoning demonstrated that light in terms of the photoelectric effect behaved more like a particle rather than a wave. When light is considered as a particle, it is termed a **photon.** A photon can be considered light packaged as a discrete bundle of energy with an amount of energy given by hf.

The Compton Effect

Planck and Einstein provided initial evidence for the quantization of physical properties. Concurrently, advances were taking place on the composition of matter and radioactivity. J.J. Thomson (1856–1940), Ernest Rutherford (1871–1937), Niels Bohr (1885–1962), Erwin Schrödinger (1887–1961), Werner Heisenberg (1901–1976), and others established the quantum mechanical model of the atom, which will be examined in the next chapter. Although these advances were being made, the quantum theory was controversial and reluctantly accepted. The wave nature of light had been firmly established during the previous century, and new discoveries seemed to resurrect the 300-year-old question of light as a particle or wave. Evidence for the photon model of light was firmly provided in 1923 by Arthur Holly Compton (1892–1962) in an experiment involving the scattering of photons. Compton illuminated a graphite block with x-rays. The x-rays scattered electrons in the graphite, similar to a cue ball hitting another billiard ball. Compton found that the wavelength of scattered x-rays increased after striking the graphite. Compton was able to use conservation of energy and momentum principles to explain the scattering of x-rays using a particle interpretation of light. When an x-ray struck an electron, it transferred some of its energy, resulting in a gain in kinetic energy of the electron. The x-ray's loss of energy meant that its frequency decreased according to the equation $E = hf$. A smaller frequency is equivalent to a longer wavelength. Compton derived an equation for the change in wavelength that depended on the angle of scattering. The scattering of electromagnetic radiation off particles is known as the **Compton effect.** Compton's work provided convincing evidence that electromagnetic radiation could be viewed as photons. Compton received the 1927 Nobel Prize in physics for his work.

Wave-Particle Duality

During the first quarter of the twentieth century the particle nature of light had been reestablished. In contrast to the previous question of light as particle versus wave, the twentieth century framework was light as both particle and wave. Light could be viewed as either a wave or a particle depending on the experiment. It was perfectly acceptable to use the wave interpretation when discussing experiments involving diffraction or interference, whereas a particle or photon model made more sense in explaining the photoelectric or Compton effect. With the photoelectric and Compton effects, light interacts with matter. In the former case, light hits a metal, and in the latter case light (x-rays) illuminates a carbon block. In these examples, photons are used to explain experimental results. When light moves through space, as it does with diffraction through a slit, it is viewed as a wave. Therefore, light is most appropriately viewed as photons when it interacts with matter and as waves when it moves through space. That light manifests itself as either a particle or a wave in an experiment is known as the **complementarity principle.** This principle was put forth by Niels Bohr, who said no experiment could be designed to reveal both the particle and wave nature of light simultaneously.

That light can be viewed as both a wave and a particle, depending on the situation is readily accepted today. It took several centuries for this view to evolve, in large part because experiments that exposed the particle nature of light did not appear until the twentieth century. These were preceded by experiments biased toward light's wave properties. As an analogy, consider how money exists.

If there were no preconceived ideas about money, someone could argue that money exists as metal objects in the form of pennies, nickels, and so on. Another person might say that money exists as paper. Obviously, money can be both metal and paper. Likewise, a sheet of paper viewed directly on edge can be described as a line, but when viewed from above, it looks like a plane (Figure 10.2). If we didn't know it was sheet of paper, it would be sensible to describe it as a line or a plane, depending on our viewpoint. In reality though, it is neither a line nor a plane, but a three-dimensional object.

It was one thing to consider that light could be viewed as both a wave and a particle, but what about matter? This seems absurd. Matter as an accumulation of particles called atoms, should only be seen as particles. This seemed to be the case until Louis de Broglie (1892–1987) reasoned that if light, which was considered a wave phenomenon, could act like a particle, that a particle might exhibit wave behavior. De Broglie made his proposal as a graduate student in his doctoral thesis in 1924. De Broglie proposed that the wavelength associated with a particle was equal to h/p, where h is Planck's constant and p is the momentum of the particle. Compton used this same equation in working out the mathematics of the Compton effect. The equation h/p is called the **de Broglie wavelength** and

can be used to calculate the wavelength of a moving particle. For example, consider the wavelength of a 1,000 kilogram car moving at 10 m/s. The car's momentum would be 10,000 kg-m/s, and its wavelength would be 6.6×10^{-34} m. This length is so small it can be considered unobservable. The case for all real-world macroscopic objects is that their de Broglie's wavelengths are so small they can be considered nonexistent. At the atomic level, this is not always the case. For example, consider an electron moving at 1% of the speed of light. The electron's momentum would be equal to its mass times its speed: 9.1×10^{-31} kg $\times 3.0 \times 10^{6}$ m/s $= 2.4 \times 10^{-10}$ m. While small, this distance is of the same order of magnitude as the interatomic spacing in crystal solids. Because of this, diffraction of electrons is possible through crystals.

The wave nature of matter is illustrated in Figure 10.3. In this example, electrons behave like waves in Young's double-slit experiment. Electrons are fired at a pair of thin slits. Electrons exhibiting only particle behavior would produce two bands, corresponding to the two slits. Electrons behaving like waves would form a diffraction pattern, similar to the wave pattern in Young's double-slit experiment.

De Broglie's idea of the wave nature of particles was verified with experiments conducted by Clinton J. Davisson (1881–1958) and Lester H. Germer (1896–1971). Davisson

Figure 10.2
A sheet of paper is perceived as a one-dimensional line when viewed directly on edge. From the top it looks like a two-dimensional plane. In reality, a sheet is three dimensional.

Figure 10.3
Electrons fired at a double slit would be expected to produce a pattern of two bands as the electrons move through the slit-like particles. If they behaved like waves, an interference pattern of maxima and minima would be produced.

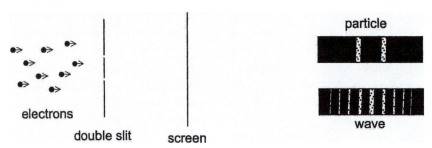

and Germer directed a beam of electrons toward a nickel crystal. The electrons scattered from the nickel produced a diffraction pattern, and they used the maxima to calculate the wavelength of the electron. This showed that a particle, in this case electrons, indeed did exhibit wave behavior. Results from the Davisson-Germer experiment verified de Broglie's hypothesis. De Broglie was awarded the Nobel Prize in physics in 1929 for his work.

De Broglie's idea of particle waves and its verification by Davisson and Germer solidified the concept of wave-particle duality. In the vast majority of cases, nature can be viewed as behaving distinctly as waves or particles. It makes complete sense to frame problems using either a particle or a wave model. At the atomic level, the demarcation between models becomes blurred and at times forces one to use both models to make sense of experimental results.

The Schrödinger Equation and the Heisenberg Uncertainty Principle

In the middle of the 1920s Edwin Schrödinger (1887–1961) and Werner Heisenberg (1901–1976) developed independent quantum mechanical models of the atom.

Schrödinger put the idea of a matter waves into the context of a wave function. Schrödinger's equation described matter waves in terms of a wave function, which described the system in terms of probabilities. The square of the wave function gave the probability of locating electrons within the atom. The quantum mechanical nature of the atom will be addressed in the next section.

Similar results for the quantum nature of the atom were found by Heisenberg, who used a branch of mathematics called matrix mathematics to draw similar conclusion about the probabilistic nature of matter. Heisenberg's work led to the **Heisenberg uncertainty principle.** This principle can be stated in several forms but boils down to the fact that anytime an observer interacts with a system by making a measurement on it, that system is changed by the observer. The uncertainty principle is another one of those concepts that can be ignored for common macroscopic phenomenon but must be considered at the atomic level. For example, consider the act of taking a person's temperature. When a thermometer at room temperature is placed in the mouth of the patient, it has a minuscule affect on the patient's temperature. This occurs because heat from the patient is transferred to the thermometer. In fact, this transfer of heat is the basis for making the temperature measurement. Now if heat is transferred to the thermometer from the patient, the question is how can the patient's temperature be determined before placing the thermometer in the patient's mouth? The answer is it can't. By placing the thermometer in the patient's mouth, the patient's temperature is altered, and a true temperature cannot be determined. It can be argued that the effect of the thermometer on the patient is so negligible that it is ridiculous to even consider it, which is true. Now consider probing an electron with light. When light interacts with electrons or other atomic particles, part of the energy of the light is transferred to the atom. In order to make more precise measurements on atoms, a smaller wavelength is required. A smaller wavelength has greater energy and changes the position of the electron. According to the Heisenberg uncertainty principle, the uncertainty in an atomic particle's position and momentum are related according to the equation $\Delta \pi \Delta x \geq h/2\pi$. This equation states that the product of the particle's uncertainty in momentum and uncertainty in position is greater than Planck's constant divided by 2π.

A Brief History of Atomic Thought

The study of the behavior of motion using physics developed up to the twentieth century (such as Newton's laws) comprises classical mechanics. The development of mechanics based on the ideas of quantization is called quantum mechanics. The first part of this chapter focused on the general ideas that revolutionized physics at the start of the twentieth century. As these ideas developed, a new picture of matter unfolded. Elementary particles that were thought to form the basic building blocks of atoms were discovered, and a quantum mechanical atomic model emerged. The second part of this chapter will trace the development of the atom from ancient thought

through the first part of the twentieth century. In the next chapter the quantum mechanical nature of the atom will be expanded based on discoveries of modern physics.

As early as 600 B.C.E., Greek philosophers sought a more basic understanding of matter. Anaximander of Miletus believed air was the primary substance of matter and that all other substances came from air. For example, fire was a form of thin air, and earth was thick air. Heraclitus of Ephesus believed fire was the primary substance and viewed reality in terms of change. Matter was always in a state of flux, and nature was always in a process of dissolving and reforming. His ideas on change are summarized in his saying: "You cannot step twice into the same river." Heraclitus' philosophy also considered reality in terms of opposites. The way up and the way down were one and the same, depending on one's perspective.

The Eleatic school of Greek philosophy, centered in the southern Italy city-state of Elea, flourished in the sixth and fifth centuries B.C.E. The Eleatics directly opposed the ideas of Heraclitus and believed the universe did not change. They felt knowledge acquired through sensory perception was faulty because our senses distorted it. One of the leading Eleatics was Parmenides, who viewed the world as a unity and believed change was an illusion. In order to make their ideas about the absence of change conform to observations, the Eleatics expanded upon Anaximander's idea of one primary substance. Rather than consider air as the primary substance, the Eleatic philosopher Empedocles of Acragas proposed four primary elements: air, earth, fire, and water.

In response to the Eleatics, the atomist school, led by Leucippus and his student Democritus, developed. The atomists refuted the idea that change was an illusion and developed a philosophy that supported the sensory observations of the physical world. The atomist school proposed that the world consisted of an infinite void (vacuum) filled with atoms. According to Democritus, "nothing exists except atoms and empty space, everything else is opinion." Atoms were eternal, indivisible, invisible, minute solids that comprised the physical world; they were homogeneous but existed in different sizes and shapes. The idea of atoms developed on logical grounds. If a piece of gold were continually divided into smaller and smaller units, the gold would eventually reach a point where it was indivisible. The ultimate unit of gold, or any other substance, would be an atom.

According to Democritus, an infinite number of atoms moved through the void and interacted with each other. The observable physical world was made of aggregates of atoms that had collided and coalesced. The properties of matter depended on the size and shape of the atoms. Change was explained in terms of the motion of the atoms through the void or how atoms arranged themselves.

Aristotle rejected the idea of atoms, but Democritus' atomist philosophy was later adopted by Epicurus and the Roman poet Lucretius (circa 95–55 B.C.E.). Aristotle had the greatest influence on scientific thought of all ancient philosophers and scholars through the Middle Ages. Aristotle's writings and teaching established a unified set of principles that could be used to explain natural phenomena. Although today we would consider his science foolish, it was based on sound logical arguments that incorporated much of the philosophical thought of his time. The teleological nature of Aristotle's explanations led to their subsequent acceptance by the Catholic Church in western Europe. This contributed to Aristotle's 2,000-year influence on Western thought. Aristotle's ideas on matter are presented in his *Meterologica* and are discussed in chapter 1.

Almost 2,000 years passed until the atomic concept was resurrected by John Dalton (1766–1844). Dalton was a Quaker and largely self-educated schoolteacher who developed an interest in science as a result of his work in meteorology. His early scientific work included essays on meteorological observations, equipment, and color blindness (Daltonism). Because of his interest in meteorology and the atmosphere, Dalton turned his attention to a study of gases. One question that Dalton considered and that guided his formulation of the atomic theory was why the gases in air formed a homogenous mixture and didn't separate according to density. Dalton knew that water vapor was lighter than nitrogen, and nitrogen was lighter than oxygen. Why then didn't these gases form stratified, or separate, layers in the atmosphere? The reason for this according to Dalton was that gases diffuse in each other. According to Dalton, the forces between different gas particles were responsible for the mixing of gases in the atmosphere. Similar gas particles repelled each other, and unlike gas particles were neutral with respect to each other.

In attempting to explain the behavior of gases, Dalton experimented on the solubility of gases in water. Dalton's work on solubility was done with his close colleague William Henry (1774–1836). Dalton was convinced that the different solubility of gases in water was due to the weight of the ultimate particle of each gas, heavier particles being more soluble than lighter particles. Dalton needed to know the relative weights of the different gases to support his ideas on gas solubility. This led him to develop a table of comparative weights of ultimate gas particles, which he presented in

1803. Over the next five years, he continued to develop his ideas on atomism and work on atomic weights. In 1808 Dalton published his table of relative atomic weights, along with his ideas on atomism in *A New System of Chemical Philosophy*. Here is a brief summary of Dalton's ideas:

1. Elements are made of tiny indivisible particles called atoms.
2. Elements are characterized by a unique mass specific to that element.
3. Atoms combine in simple whole-number ratios to make compounds.
4. During chemical reactions, atoms rearrange themselves to form new substances.
5. Atoms maintain their identity during the course of a chemical reaction.

Dalton's work on relative weights and the atomic theory didn't have an immediate effect on other scientists, but his ideas did provide a framework for determining the empirical formula of compounds. His table of relative weights was not accurate enough to give consistent results. Many scientists still debated the existence of atoms in the second half of the nineteenth century. Still, little by little, the atomic theory was adopted as a valid model for the basic structure of matter.

The Divisible Atom

In the early 1800s, Dalton proposed that atoms are the basic building blocks of matter. It is one thing to claim the existence of atoms, and another to develop the concept until it is readily accepted. Nearly a century after Dalton's work, the true nature of the atom began to unfold. Evidence accumulated during the 1800s indicated that Dalton's atoms might not be indivisible particles. As the twentieth century unfolded, so did the mystery of the atom.

One area of research that raised questions about the atom involved experiments conducted with gas discharge tubes. Gas discharged tubes are sealed glass tubes containing two metal electrodes at opposite ends of the enclosed tube (similar to small fluorescent lights). During operation, the metal electrodes are connected to a high-voltage power source. The negatively charged electrode is called the cathode, and the positively charged electrode is the anode. Before a gas discharge tube is sealed, most of the air is pumped out of it. Alternately, the tube may contain a gas such as hydrogen or nitrogen under very low pressure. During the last half of the eighteenth century, numerous improvements were made in gas discharge tubes.

William Crookes (1832–1919) observed a faint greenish glow and beam using his own discharged tubes called Crookes tubes. Crookes noted that the beam in his tubes originated from the cathode, and hence the beams became known as cathode rays. The tubes themselves became known as cathode ray tubes. Crookes conducted numerous experiments with cathode ray tubes. He observed that cathode rays moved in straight lines but could be deflected by a magnet. From this work, Crookes concluded cathode rays were some sort of particle, but other researchers believed cathode rays were a form of light.

Part of the problem involved with interpreting cathode rays resulted from the ambiguous results obtained by various researchers. For example, cathode rays were bent in a magnetic field, supporting the particle view of cathode rays, yet when subjected to an electric field, cathode rays were not deflected. The fact that cathode rays were not deflected in an electric field supported the view that cathode rays were a form of electromagnetic radiation or light. The true nature of cathode rays was ultimately determined by a group of physicists led by J.J. Thomson (1856–1940) at the Cavendish Laboratory at Cambridge University.

Joseph John Thomson was appointed head of the prestigious Cavendish Laboratory at the age of 28. Thomson was a brilliant mathematician and experimenter who tackled the problem of cathode rays at the end of the nineteenth century. Thomson's initial experiments led him to believe cathode rays were actually negatively charged particles. Using cathode ray tubes (Figure 10.4) in which the air had been evacuated until a very low pressure was attained ($\approx 1/200$ atm), Thomson demonstrated cathode rays could indeed be deflected by an electric field. The failure of previous researchers to evacuate their discharge tubes to such low pressure had prevented the cathode ray beams from being deflected in an electric field. Thomson was convinced that cathode rays were particles, or in his terms "corpuscles," but he still had to determine the magnitude of the charge and mass of his corpuscle to solidify his arguments.

Thomson and his colleagues at Cavendish conducted numerous experiments on cathode rays. They used different gases in their tubes, varied the voltage, and employed different metals as electrodes. As the years went by and Thomson accumulated more and more data, he was able to calculate a charge to mass ratio for the corpuscle. Thomson's results startled the scientific community. His results showed that the charge to mass ratio (e/m) was 1,000 times greater than the accepted value for hydrogen ions (hydrogen ions are just protons).

Thomson's Corpuscle	Hydrogen Ion
$\dfrac{\text{charge}}{\text{mass}} = \dfrac{1000e}{m}$	$\dfrac{e}{m}$

At the time, hydrogen ions were the smallest particle known to exist. Thomson's results

Figure 10.4
Cathode ray tube

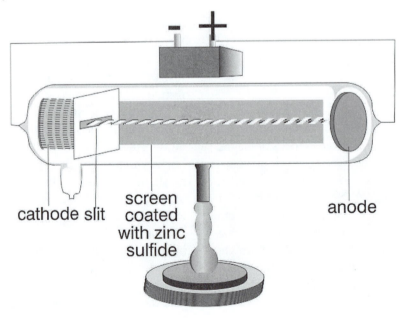

cathode slit   screen coated with zinc sulfide   anode

could be interpreted as either the charge of the corpuscle being 1,000 times greater than the hydrogen ion, or the mass of the corpuscle being 1/1,000 that of the hydrogen ion. Thomson suspected the charge of the corpuscle was equal and opposite to that of the hydrogen ion and, therefore, assumed the corpuscle had a mass of 1/1,000 of a hydrogen ion. Thomson published his results in 1897. Thomson's corpuscles eventually came to be called electrons, based on a term coined by George Stoney (1826–1911) to designate the fundamental unit of electricity.

While Thomson's work proved the existence of electrons, his studies still left open the question of the exact charge and mass of the electron. Robert Andrew Millikan (1868–1953) tackled this problem at the University of Chicago by constructing an instrument to measure the mass and charge of the electron. Millikan's instrument consisted of two parallel brass plates separated by about 1 cm (Figure 10.5). A pinhole-size opening drilled into the top plate allowed oil droplets sprayed from an atomizer to fall between the two plates. The opening between the two plates was brightly illuminated, and a microscopic

eyepiece allowed Millikan to observe individual oil droplets between the two plates. Millikan calculated the mass of oil droplets by measuring how long it took an oil droplet to move between the two uncharged plates. Next, Millikan charged the top brass plate positive and the bottom plate negative. He also used an x-ray source to negatively charge the oil droplets when they entered the space between the plates. By varying the charge, Millikan could cause the oil droplets to rise or stay suspended between the plates. Using his instrument and knowing the mass of the oil droplet enabled Millikan to calculate the charge of the electron. With Millikan's value for charge, the electron's mass could be determined. The value was even smaller than given by Thomson and found to be 1/1,835 that of the hydrogen atom. Millikan's work conducted around 1910 gave definition to Thomson's electron. The electron was the first subatomic particle to be discovered, and that opened the door to the search for other subatomic particles.

In their studies with cathode rays, researchers observed different rays traveling in the opposite direction of cathode rays. In 1907, Thomson confirmed the

Figure 10.5
Millikan's instrument used to perform oil drop experiments

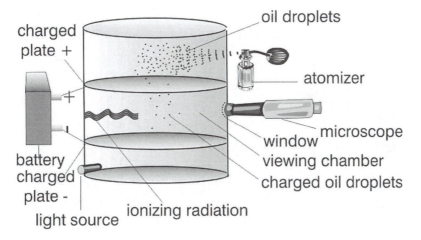

rays carried a positive charge and had variable mass depending on the gas present in the cathode ray tube. Thomson and others found the positive rays were as heavy or heavier than hydrogen atoms. In 1914, Ernest Rutherford (1871–1937) proposed that the positive rays were composed of a particle of positive charge as massive as the hydrogen atom. Subsequent studies on the interaction of alpha particles with matter demonstrated that the fundamental positive particle was the proton. By 1919 Rutherford was credited with identifying the proton as the second fundamental particle.

The last of the three fundamental particles is the neutron. Experimenters in the early 1930s bombarded elements with alpha particles. One type of particle produced had the same mass as the proton but carried no charge. James Chadwick (1891–1974), in collaboration with Rutherford, conducted experiments to confirm the existence of the neutral particle, called the neutron. Chadwick is credited with the discovery of the neutron in 1932.

Atomic Structure

With the discovery of the neutron, a basic atomic model consisting of three fundamental subatomic particles was complete. While discoveries were being made on the composition of atoms, models were advanced concerning the structure of the atom. John Dalton considered atoms to be solid spheres. In the mid-1800s one theory considered atoms to consist of vortices in an ethereal continuous fluid. J. J. Thomson put forth the idea that the negative electrons of atoms were embedded in a positive sphere. Thomson's model was likened to raisins, representing electrons, in a positive blob of pudding; the model was termed the plum-pudding model, as shown in Figure 10.6. The negative and positive parts of the atom

in Thomson's model canceled each other, resulting in a neutral atom.

A clearer picture of the atom began to emerge toward the end of the first decade of the twentieth century. This picture was greatly aided by several significant discoveries in physics. In 1895, while observing the glow produced by cathode rays from a sealed Crookes tube, Wilhelm Conrad Roentgen (1845–1923) noticed that nearby crystals of barium platinocyanide glowed. Barium platinocyanide was a material used to detect cathode rays. Roentgen was puzzled because the cathode rays should have been blocked by the glass and another enclosure containing the Crookes tube. Roentgen also discovered that a photographic plate inside a closed desk drawer developed an image of a key resting on top of the desk. Evidently whatever had caused the barium platinocyanide to glow could pass through the wooden desk, but not through the metal key. Roentgen coined the term x-rays for his newly discovered rays. He was awarded the first Nobel Prize in physics for his discovery in 1901.

Figure 10.6

Thomson proposed that the atom consisted of negative electrons embedded in a positive pool, like raisins in plum pudding

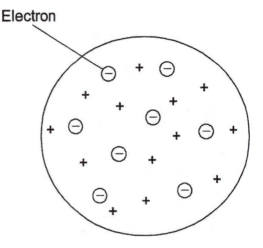

Electron

Roentgen's discovery of x-rays stimulated great worldwide interest in this new form of radiation. Antoine Henri Becquerel (1852–1908) accidentally discovered the process of radioactivity while he was studying x-rays. Radioactivity involves the spontaneous disintegration of unstable atomic nuclei. Becquerel had stored uranium salts on top of photographic plates in a dark drawer. When Becquerel retrieved the plates, he noticed the plates contained images made by the uranium salts. Becquerel's initial discovery in 1896 was further developed by Marie Curie (1867–1934) and Pierre Curie (1859–1906). Marie Curie coined the word "radioactive" to describe the emissions from uranium.

Three main forms of radioactive decay involve the emission of **alpha particles, beta particles, and gamma rays.** An alpha particle is equivalent to the nucleus of a helium atom. Beta particles are nothing more than electrons. Gamma rays are a form of electromagnetic radiation.

Physicists doing pioneer work on atomic structure employed **radioactive** substances to probe matter. By examining how radiation interacted with matter, these researchers developed atomic models to explain their observations. Rutherford, along with Johannes Wilhelm Geiger (1882–1945), creator of the first Geiger counter in 1908 to measure radiation, and Ernest Marsden (1889–1970), carried out their famous gold foil experiment that greatly advanced our concept of the atom. Rutherford's experimental setup is shown in Figure 10.7. Rutherford used foils of gold, platinum, tin, and copper and employed polonium as a source of alpha particles. According to Thomson's plum-pudding model, alpha particles would pass through the gold with little or no deflection. The uniform distribution of charge in Thomson's model would tend to balance out the total deflection of a positive alpha particle as it passed through the atom. Rutherford, Geiger, and Marsden found a significant number of alpha particles that experienced large deflections of greater than $90°$. This result surprised Rutherford, who years later described discovering the phenomenon of widely scattered alpha particles "as if you had fired a 15-inch shell at a piece of tissue paper and it came back and hit you." To account for these experimental results, Rutherford proposed a new atomic model in which the atom's mass was concentrated in a small positively charged nucleus. The electrons hovered around the nucleus at a relatively great distance. To place the size of Rutherford's atom in perspective, consider the nucleus to be the size of the period at the end of this sentence, then the electrons would circle the nucleus at distances of several meters. If the nucleus was the size of a ping pong ball, then the elec-

Table 10.1
Forms of Radiation

Type	Symbol	Mass	Charge	Equiv
Alpha	α	4	$+2$	$^4_2\mathrm{He}$
Beta	β	~0	-1	$^0_{-1}e$
Gamma	γ	0	0	

Figure 10.7
Rutherford's experimental setup that led to planetary model of the atom

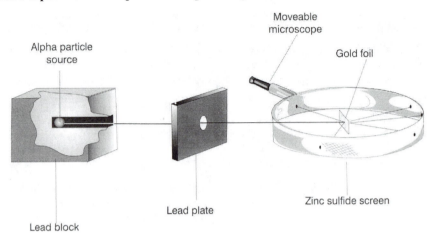

trons would be on the order of 1 kilometer away. Rutherford's planetary model represented a miniature solar system, with the electrons rotating around the nucleus like the planets around the Sun. Most of the atom's volume in Rutherford's model consisted of empty space, and this explained the results from the gold foil experiment. Most alpha particles passed through the foil with little or no deflection. This occurred because the net force on the alpha particle was close to zero as it passed through the foil. The widely scattered alpha particles resulted when the positively charged alpha particles approached the nucleus of a gold atom. Since like charges repel each other, the alpha particle would be deflected or even backscattered from the foil.

Niels Bohr advanced Rutherford's model by stipulating that an atom's electrons did not occupy just any position around the nucleus, but occupied specific orbitals to give a stable configuration. Bohr based his ideas on a study of the spectrum for hydrogen. A spectrum, as noted in chapter 9, results when light is separated into its component colors by a prism. When visible light passes through a prism, a continuous spectrum

similar to a rainbow results. Light from a hydrogen discharge tube does not produce a continuous rainbow, but gives a discontinuous pattern of lines with broad black areas separating thin lines of color. Bohr's proposed specific electron energy levels, or orbitals, to account for the lines produced in hydrogen's emission spectrum. In Bohr's planetary model, when the hydrogen in the tube became energized by applying a voltage, electrons jumped from a lower energy level to a higher level. As they moved back down, energy was released in the form of visible light to produce the characteristic lines of the hydrogen spectrum. Figure 10.8 displays this graphically. Bohr reasoned that specific orbitals were stable and while in these orbitals electrons did not radiate energy. According to classical theory, orbiting electrons would continually emit energy and lose energy as they accelerated around their nuclei. This in turn would cause the negative electrons to quickly spiral into the positive nucleus. Bohr's model explained the stability of atoms and explained how radiation was emitted when electrons jumped between energy levels. While Bohr's model successfully explained hydrogen's spectrum,

Figure 10.8
Electrons change energy levels in the Bohr atom. Level 1 is closest to the nucleus.

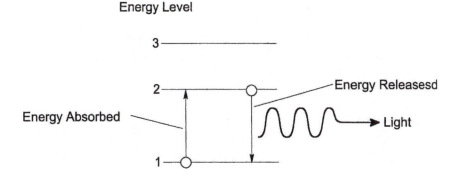

it could not account for the spectra obtained for gases with more than one electron, such as that for helium. Nonetheless, Bohr had shown that matter was discontinuous, providing further evidence for the quantum nature of matter.

Quantum Mechanics

World War I delayed advances in basic research on the atom for several years as scientists turned their attention to national security. The end of World War I brought renewed interest and further development of atomic theory in the 1920s. The discoveries regarding the atom were integrated with those of de Broglie, Schrödinger, Heisenberg, and others to the formulate quantum mechanics. De Broglie's idea of the wave nature of matter was used to explain the preferred orbitals of electrons in the Bohr model of the atom. According to de Broglie, the stable orbits occurred where there were an integral number of electron wavelengths, and the resulting standing electron waves create a form of electron resonance. De Broglie's integral wavelengths meant that electrons would be found in orbitals with discrete (quantized) energy levels.

In quantum mechanics, the position and energy of electrons are described in terms of probability. Electrons are not located in discrete orbitals, but rather occupy a region in space surrounding the nucleus. Bohr's planetary model, while still popularly used to described the atom, is an oversimplification. In quantum mechanics, quantum numbers are used to describe each electron in an atom. The numbers give the probability of finding an electron in some region surrounding the nucleus as well as other information about the electron. The quantum mechanical model is a highly mathematical and nonintuitive picture of the atom. Rather than thinking of electrons orbiting the nucleus, the quantum mechanical picture places electrons in clouds of different shapes and sizes around the nucleus. These clouds are referred to as orbitals. Several orbitals are shown in Figure 10.9.

In the quantum mechanical model, the atom's electrons surround the nucleus in orbitals in regions called shells. Shells can be thought of as a group of orbitals, although the first shell contains only one orbital. The number of electrons surrounding the nucleus equals the number of protons in the nucleus forming a neutral atom. An atomic orbital is a volume of space surrounding the nucleus where there is a certain probability of finding the electron. The electrons surrounding the nucleus of a stable atom are located in the orbitals closest to the nucleus. The orbital closest to the nucleus has the lowest energy level, and the energy levels increase the farther the orbital is from the nucleus. The location of electrons in atoms should always be thought of in terms of electrons occupying the lowest energy level available. An analogy can be drawn to the gravitational potential energy of an object. If you hold an object such as this book above the floor, the book has a certain amount of gravitational potential energy with respect to the floor. When you let go of the book, it falls to the floor, where its potential energy is zero. The book rests on the floor in a stable position. If we pick the book up, we used work to increase its potential energy, and it no longer rests in a stable position. In a similar fashion, we can think of electrons as resting in stable shells close to the nucleus, where the energy is minimized.

Orbitals are characterized by quantum numbers. A way to think of quantum numbers is as addresses of electrons surrounding the nucleus. Just as an address allows us to know where we are likely to find a person, quantum numbers give us information about the location of electrons. Just as we can never be sure a person will be at his or her home address, we can never know where

an electron will be located within an orbital. The address of electrons in an atom is given by using quantum numbers, and is referred to as the electron configuration.

Four quantum numbers characterize the electrons in an atom:

1. The principal quantum number symbolized by the letter n. The principal quantum number tells which shell the electron is in and can take on integral values starting with 1. The higher the principal quantum number, the farther the orbital is from the nucleus. A higher principal quantum number also indicates a higher energy level. Letters may also be used to designate the shell. Orbitals within the same shell have the same principal quantum number.

   Principal Quantum Numbers

n	1	2	3	4	5	6
Shell	K	L	M	N	O	P

   $\longrightarrow$ Energy $\longrightarrow$

2. The angular momentum quantum number symbolized by the letter l. The angular momentum quantum numbers gives the shape of an orbital. Values for l depend on the principal quantum number and can assume values from 0 to $n - 1$. When $l = 0$ the shape of the orbital is spherical, and when $l = 1$ the shape is a three-dimensional figure eight (see representations in Figure 10.9). As the value of l increases,

Figure 10.9
Representation of several different orbitals. Note the two different d orbitals represented.

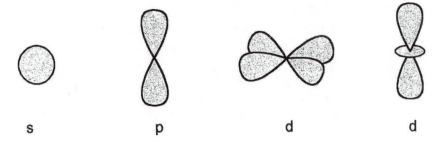

s   p   d   d

more complex orbital shapes result. Letter values are traditionally used to designate l values.

l value	0	1	2	3	4
Letter	s	p	d	f	g

$\xrightarrow{\quad\text{Energy}\quad}$

The letter designation originated from spectral lines obtained from the elements. Lines were characterized as sharp, principal, or diffuse (s, p, d). Lines after d were labeled alphabetically starting from the letter f.

3. The magnetic quantum number symbolized by m_l. The magnetic quantum number gives the orientation of the orbital in space. Allowable values for m_l are integer values from -1 to $+1$, including 0. Hence, when $l = 0$, then m_l can only be 0, and when $m_l = 1$, m_l can be -1, 0, or $+1$.

4. The spin quantum number symbolized by m_s. The spin quantum number tells how the electron spins around its own axis. Its value can be $+1/2$ or $-1/2$, indicating a spin in either a clockwise or counterclockwise direction.

Using the above definitions for the four quantum numbers, the possible quantum numbers for electrons in an atom can be determined. A basic rule when working with quantum numbers is that no two electrons in the same atom can have an identical set of quantum numbers. This rule is known as the **Pauli exclusion principle,** named after Wolfgang Pauli (1900–1958). For example, when $n = 1$, l and m_l can only be 0, and m_s can be $+1/2$ or $-1/2$. This means the K shell can hold a maximum of two electrons. The two electrons would have quantum numbers of 1, 0, 0, $+1/2$ and 1, 0, 0, $-1/2$, respectively. We see that the opposite spin of the two electrons in the K orbital means the electrons do not violate the Pauli exclusion

principle. Possible values for quantum numbers and the maximum number of electrons each orbital can hold are given in Table 10.2 and shown in Figure 10.10.

Using quantum numbers, electron configurations can be written for the elements. To write electron configurations it is assumed that electrons occupy the lowest energy levels, since this is the most stable configuration. It must be remembered that no two electrons can have the same four quantum numbers. Starting with the simplest element hydrogen, several configurations will be written to demonstrate the process. Hydrogen has only one electron. This electron will most likely be found in the K ($n = 1$) shell. The l value must also be 0. The m_l and m_s quantum numbers don't need to be considered, since the former is 0 and the latter can arbitrarily be assigned either plus or minus 1/2. The electron configuration for hydrogen is written as shown in Figure 10.11. Helium has two electrons. Both electrons have n values of 1, l values of 0, and m_l values of 0, but they must have opposite spins to conform to the Pauli exclusion principle. The electron configuration of He is, therefore, $1s^2$. Lithium, with an **atomic number** of 3, has three electrons. The first two electrons occupy the 1s orbital, just like helium. Since the 1s orbital in the K shell can only hold two electrons, the third electron must go into the next lowest energy level available. The next available energy level is in the L shell, where the n value is 2 and the l value is 0. The electron configuration of lithium is, therefore, $1s^2$ $2s^1$. As the atomic number increases moving through the periodic table, each element has one more electron than the previous element. Each additional electron goes into the orbital with the lowest energy. The electron configuration for the first 20 elements is given in Table 10.3.

An examination of the electron configuration of the first 18 elements indicates that the addition of each electron follows

Table 10.2
Quantum Numbers

Shell	n	Possible l	Possible m_l	m_s	Maximum Electrons
K	1	0	0	$+\frac{1}{2}, -\frac{1}{2}$	2
L	2	0, 1	−1, 0, 1	$+\frac{1}{2}, -\frac{1}{2}$	8
M	3	0, 1, 2	−2, −1, 0, 1, 2	$+\frac{1}{2}, -\frac{1}{2}$	18
N	4	0, 1, 2, 3	−3, −2, −1, 0, 1, 1, 2	$+\frac{1}{2}, -\frac{1}{2}$	32
O	5	0, 1, 2, 3, 4	−4, −3, −2, −1, 0, 1, 2, 3, 4	$+\frac{1}{2}, -\frac{1}{2}$	50

Figure 10.10
Shells and orbitals showing the maximum number of electrons possible in each

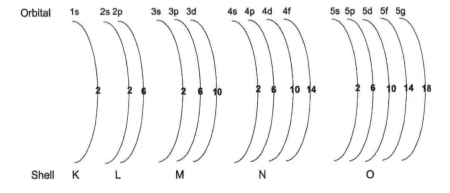

a regular pattern. One would expect that the electron configuration of element 19, potassium, would be $1s^2 2s^2 2p^6 3s^2 3p^6 3d^1$, but Table 10.3 indicates the nineteenth electron goes into the 4s orbital rather than 3d. The same is true for calcium, with the twentieth electron also falling into the 4s orbital rather than 3d. It is not until element 21, scandium, that 3d begins to fill up. The reason for this and other irregularities has to do with electron repulsion. It is easier for the electrons to go into the fourth shell rather than the already partially filled third shell.

Figure 10.11
The electron configuration of hydrogen

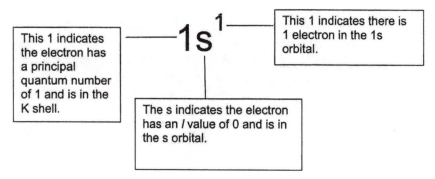

Table 10.3
Electron configurations of the first 20 elements

1 H	$1s^1$	11 Na	$1s^2\ 2s^22p^6\ 3s^1$
2 He	$1s^2$	12 Mg	$1s^2\ 2s^22p^6\ 3s^2$
3 Li	$1s^1\ 2s^1$	13 Al	$1s^2\ 2s^22p^6\ 3s^23p^1$
4 Be	$1s^2\ 2s^2$	14 Si	$1s^2\ 2s^22p^6\ 3s^23p^2$
5 B	$1s^2\ 2s^22p^1$	15 P	$1s^2\ 2s^22p^6\ 3s^23p^3$
6 C	$1s^2\ 2s^22p^2$	16 S	$1s^2\ 2s^22p^6\ 3s^23p^4$
7 N	$1s^2\ 2s^22p^3$	17 Cl	$1s^2\ 2s^22p^6\ 3s^23p^5$
8 O	$1s^2\ 2s^22p^4$	18 Ar	$1s^2\ 2s^22p^6\ 3s^23p^6$
9 F	$1s^2\ 2s^22p^5$	19 K	$1s^2\ 2s^22p^6\ 3s^23p^6\ 4s^1$
10 Ne	$1s^2\ 2s^22p^6$	20 Ca	$1s^2\ 2s^22p^6\ 3s^23p^6\ 4s^2$

The quantum mechanical model and the electron configurations of the elements provide the basis for explaining many aspects of chemistry. Particularly important are the electrons in the outermost orbital of an element. These electrons, known as the valence electrons, are responsible for the chemical properties elements display and chemical bonding. Free valence electrons are responsible for the electrical conductivity of substances and will be discussed in this role in the chapter on electricity.

Summary

In the relatively brief period of 30 years at the end of the twentieth century, physics moved from the classical to the modern era.

A quantum picture of nature developed, and with it a new understanding of how the light behaved. The development of quantum theory accompanied discoveries revealing atomic structure and resulted in a quantum mechanical model of the atom. This laid the foundation for other theories based on the quantum model, such as quantum electrodynamics and quantum chromodynamics. These will be discussed in the next chapter as the model of the atom is expanded to account for the plethora of discoveries brought about with new technology in the areas of high-energy and particle physics.

11

A Modern View of the Atom

Introduction

By 1935, it was believed the internal structure of the atom had been determined. The proton, electron, and neutron had been discovered, and although these were considered fundamental particles, still more particles were being theorized and discovered. In 1930 Wolfgang Pauli proposed a particle to account for the violation of conservation of energy and momentum during beta decay. Pauli's ghost particle was dubbed the **neutrino** (little neutron) by Enrico Fermi (1901–1954). In 1932 Carl David Anderson (1905–1991) discovered two subatomic particles while studying cosmic rays. These were the **positron** (a positively charged electron) and **muon.** Positrons are a form of antimatter. Antimatter consists of particles that have the same mass as ordinary matter, but differ in charge or some other property. Antineutrons have different magnetic properties compared to regular neutrons. By the 1960s, several hundred different subatomic particles had been identified, and the question of what was a fundamental particle was open to debate. This led to the development of the standard model of fundamental particles and interactions, generally referred to as the standard model. This chapter will focus on the standard model and discusses the fundamental particles given by this model.

Particle Detectors and Accelerators

To observe subatomic particles that are very small and can travel at speeds approaching the speed of light requires special instruments. Particle detectors are used for observing at the subatomic level and provide indirect visible evidence of particles. Several different types of detectors exist that operate by recording the tracks made by particles that interact with them. A detector can be compared to examining tracks made on a muddy road. A person would be able to distinguish whether a person, bicycle, or vehicle passed along the road. A person trained in reading tracks would be able to distinguish whether the vehicle was a compact car, SUV, or truck. In a similar manner, physicists can identify subatomic particles by the tracks produced in detectors. Early detectors consisted of photographic plates, cloud chambers, bubble chambers, and spark chambers. Photographic plates record the path of particles as they moved across the plates and expose the plates. Cloud, bubble, and spark chambers produce tracks when charged particles ionize atoms in the detector, which in turn produce tracks. The tracks themselves reveal important information about the particle. For example, when particles are subjected to a magnetic field in a detector, positive and negative particles bend in opposite directions. Furthermore, the track of heavier particles will tend to be straight, while very light particles produce tight spiral tracks (Figure 11.1).

The cloud chamber was invented by Charles T. R. Wilson (1869–1959) in 1911. Wilson's cloud chamber consisted of a container holding a supersaturated vapor of water, alcohol, or some other liquid. As charged particles moved through the chamber, they ionized molecules in their path. Vapor condensed on the ionized molecules, producing tracks of condensed droplets. These ephemeral tracks were photographed and analyzed to identify particles. Wilson received the 1927 Nobel Prize in physics for his invention of the cloud chamber. Bubble chambers produce bubble tracks in a superheated liquid. In a bubble chamber, a liquid is heated under pressure to above its boiling point. The pressure is then reduced. When a particle disturbs the liquid, it cause immediate boiling and produces bubbles revealing the track of the particle. The bubble chamber was invented in 1952 by Donald Glaser (1926–), who was awarded the 1960 Nobel Prize in physics for his work. A spark chamber operates as a particle ionizes a noble gas contained in a box with a series of thin metal plates. The ionization creates a path of sparks as it moves through the box.

A multiwire tracking chamber operates by producing an electrical signal to identify a particle rather than producing a visible track. In this type of tracking chamber, ions collect on a wire or surface, and the collected charge creates an electrical signal. The first

Figure 11.1
Bubble chamber and tracks (opposite page). Bubble chamber courtesy of Brookhaven National Laboratory. Particle tracks courtesy of CERN

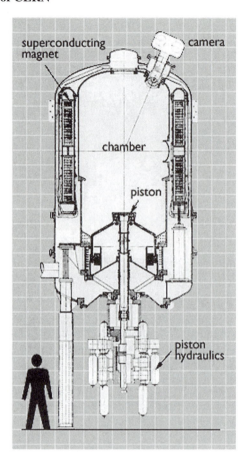

modern tracking chamber was invented by Georges Charpak (1924–), who received the 1992 Nobel Prize in physics for his work. Charpak's tracking chamber consisted of a box containing a system of wires at high voltage and filled with pressurized xenon and carbon dioxide gas. As particles move through the chamber, the ions produced migrate to the wires, producing an electrical signal. The signals are then interpreted electronically by a computer. In this manner, thousands of particles can be recorded per second.

Figure 11.1 (continued)

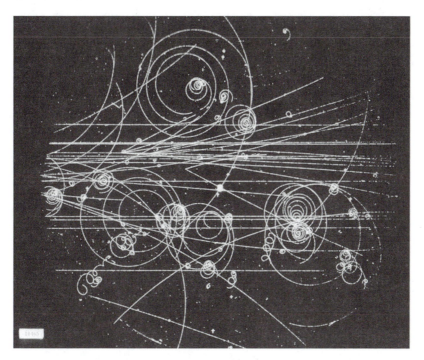

Scintillators are detectors that operate when the ions created by particles produce light through a photochemical reaction. The light photons activate electrons in a cascading effect in a photomultiplier tube. These electrons produce an electrical signal that enables precise timing and determination of the energy loss of the particle. Calorimeters are detection devices that absorb the energy of particles and through scintillation can determine the total energy of the particle.

The myriad of subatomic particles that have been produced in the last half-century is due to the development of particle accelerators. Modern accelerators are capable of accelerating subatomic particles close to the speed of light. Once accelerated to an appropriate speed, the accelerated particle collides with a target particle (or two accelerated particles traveling in opposite directions collide with each other). The colloquial term for a particle accelerator is "atom smasher." The collisions produced in accelerators reveal the internal structure of the particles being smashed. In this sense, accelerators serve as the clubs and particles the piñatas in high-energy physics experiments. In order to reveal increasing detail and detect smaller particles, larger accelerators have evolved over the years. The largest of these are world renowned, with staffs numbering in the thousands and conducting advance research. Accelerators are rated according to the energy they can impart to particles. This energy is typically given in **electron volts,** abbreviated eV. An electron volt is the energy acquired when an electron moves through a potential difference of 1 V. An electron volt is a very small unit of energy equivalent to 1.6×10^{-19} J. It is a convenient unit of energy for calculations at the atomic level. The most powerful accelerator is the Fermi National Accelerator Laboratory's Tevatron. It accelerates and collides protons and antiprotons in a 4 mile

Figure 11.2

In a linear accelerator a positive proton is accelerated through a series of tubes by using alternating current. The proton is pushed by the positive charge on first tube and pulled by the negative charge on the second tube, causing it to accelerate. By timing the change in voltage applied to the tubes, protons continues to accelerate down the length of the tube.

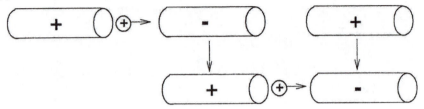

underground ring, achieving collisions at energies of 1.6 trillion eV.

Particle accelerators were first developed around 1930. Several basic types of particle accelerators are used, but are all based on the principle that charged particles can be accelerated in an electromagnetic field. The first accelerators built in the 1930s were linear accelerators, or linacs. These consisted of a series of hollow tubes separated by gaps in which the alternating current regulated down the length of the tube was used to accelerate the particle (Figure 11.2). The speed a particle can obtain using a linear accelerator is directly related to the length of the accelerator. The largest linear accelerator is located at Stanford University and is slightly longer than 2 miles.

Another design for particle accelerators is based on a circular arrangement. A cyclotron is similar to a linear accelerator wound into a spiral. Two D-shaped chambers called "dees" house electromagnets. The electromagnets accelerate particles in an outward spiral as they pass through an electric field (Figure 11.3). The invention of the cyclotron is credited to Ernest O. Lawrence (1901–1958), who received the 1939 Nobel Prize in physics for his work. According to Einstein's theory of relativity, an object's mass increases as it accelerates. In particle accelerators this is a problem

because as the mass increases, the particle slows down and becomes out of sync with the changing electric field. A synchrotron is a synchronous cyclotron in which the electric field increases to compensate for the change in mass of the particle due to relativistic effects. Synchrotrons are the most common type of accelerators. The increasing magnetic field in synchrotrons allows particles to maintain a circular path, which in turn allows them to reach high energies.

Particle accelerators are some of the largest scientific instruments. The largest particle accelerator is located at CERN (European Organization for Nuclear Research), outside of Geneva on the border of France and Switzerland. It has a circumference of 27 km and is buried 100 m below the surface of the Earth. This accelerator is known as LEP, for large electron-positron collider. Fermi's Tevatron has a 4 mile circumference and is used to accelerate protons and antiprotons in opposite directions to 99.9999% of the speed of light.

Advancing the frontiers of science and learning more about the nature of matter requires building more powerful and larger particle accelerators. The United States started construction in 1983 on what would have been the largest particle accelerator. Two billion dollars was spent on the design and initial construction of the

**Figure 11.3
Cyclotron**

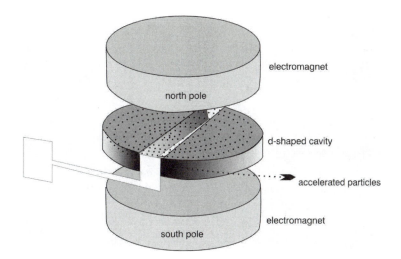

Superconducting Super Collider. This giant accelerator was started in Texas and would have used thousands of superconducting magnets to give protons 20 times the energy of modern accelerators. It would have accelerated protons and antiprotons around an oval track 54 miles in circumference. In 1993, Congress cancelled funding of this project. The most powerful accelerator in the world is currently under construction at CERN and is called the Large Hadron Collider. It is scheduled to be completed in 2007 and will share the same tunnel as CERN's LEP (circumference of 27 km), contain superconducting magnets, and be capable of accelerating protons to near the speed of light. This collider will enable physicists to observe at even smaller levels than currently probed and should provide evidence for theorized particles.

Fundamental Particles, Forces, and the Standard Model

By the early 1930s, three fundamental particles had been identified: electron, proton, and neutron. Within the next decade, two areas of research started to reveal that the universe contained a multitude of subatomic particles. Cosmic ray research involved exposing photographic plates to radiation from outer space. The photographic plates were placed at mountain top locations and subsequently analyzed for signature tracks produced by various particles. Concurrently, the development of accelerators led to the identification of hundreds of other subatomic particles. Particles are characterized by their mass, charge, and spin. Some of the subatomic particles identified in the middle of the twentieth century seemed to exhibit properties that were peculiar and were called strange particles. Throughout the last half of the twentieth century, more particles were identified in collider experiments. At the same time, physicists advanced theories that predicted the existence of new particles, which led them to reconsider what the fundamental constituents of matter are. This work developed into what is called the standard model of fundamental particles and interactions or, in abbreviated language, the standard model. The standard model proposes a set of fundamental particles for matter and accounts for three of the fundamental forces of nature: electromagnetic, weak, and strong. The fourth force is gravity and is not included in the standard model. Before examining how forces relate to the standard model, a brief examination of the fundamental particles is necessary.

According to the standard model, the fundamental particles of nature are **quarks, leptons,** their **antiparticles,** and **bosons.** Murray Gell-Mann (1929–) and George Zweig (1937–) independently proposed the existence of quarks in 1963. The term quark was coined by Gell-Mann, who obtained it from a line in James Joyce's 1939 novel *Finnegans Wake:* "Three quarks for Muster Mark!" According to the quark theory, small particles called quarks combine to produce subatomic particles. Quarks carry a fractional charge of $\pm 1/3$ or $\pm 2/3$. Six flavors, or types, of quarks exist: up, down, strange, charm, bottom, and top. Sometimes beauty and truth are used instead of bottom and top, respectively. Quark flavors are given symbols such as u for the up quark and d for the down quark. Each of these flavors can further be classified as having one of three primary colors: red, green, or blue. The flavors and colors should not be confused with our normal notion of flavor and color, but interpreted as part of a convenient classification scheme used to explain how matter behaves. Quark flavors and charges are given in Table 11.1. The quarks are listed in order of mass from the lightest to heaviest. It should be remembered that for each quark in Table 11.1 there is an antiquark, so there are a total of 12 quark/antiquark particles. Antiquarks are symbolized using the symbol for a quark with a bar over it, for example, the up antiquark is $\bar{u}$.

Table 11.1
Flavor and Charge of Quarks

Quark Flavor	Symbol	Charge	Relative Mass
Up	u	$\frac{2}{3}$	1
Down	d	$-\frac{1}{3}$	2
Strange	s	$-\frac{1}{3}$	40
Charm	c	$\frac{2}{3}$	300
Bottom (beauty)	b	$-\frac{1}{3}$	940
Top (truth)	t	$\frac{2}{3}$	34,000

Quarks never exist individually, but in groups called **hadrons.** The principle that quarks only exist in combination and not individually is known as **quark confinement.** Hadrons can be further classified as three-quark or two-quark groups. A three-quark group is referred to as a **baryon.** The word "baryon" is derived from the word "barus," meaning heavy. A quark and anti-quark combination produces a **meson.** The proton and neutron are baryons that form ordinary matter. A proton consists of two up quarks and a down quark, and a neutron consists of two down quarks and an up quark (Figure 11.4). Quarks combine to produce many other baryons, but these are short-lived unstable particles. For example, an up, down, and strange quark combine to produce a lambda particle (Λ) with a lifetime of 10^{-10} s. An up quark and two strange quarks produce a cascade particle (Ξ), lifetime also 10^{-10} s. Quarks were able to account for the known hadrons in the early 1960s, and this provided strong evidence for their existence.

The other type of hadrons are mesons and consist of a quark and an antiquark. For example, a down antiquark and an up quark produce a $\pi+$ meson called a pion. This particle has a very short lifetime of only 2.6×10^{-8} second. Mesons have a very short lifetime and have a baryon number of zero. The baryon number is the number of baryons minus the number of antibaryons comprising the particle.

Hadrons are color neutral, which means they have a color charge of zero. This means that the combination of the colors in baryons and mesons is neutral or white. For example, a proton consisting of two up quarks and a down quark would contain all three primary colors. When quarks with three primary colors combine, a neutral, or white, baryon is produced. A meson would contain one of the primary colors and its complementary colors. The complementary colors for red, green, and blue are cyan, magenta, and yellow, respectively. A meson might, therefore, consist of a red quark and a cyan antiquark to give a neutral meson.

Figure 11.4
Baryons consist of three quarks. Protons and neutrons are stable baryons made from up and down quarks.

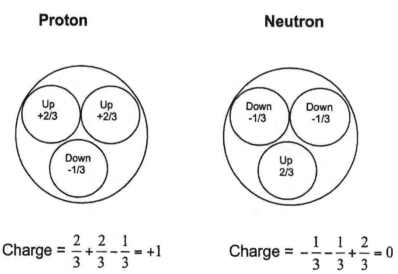

The original quark hypothesis proposed only up, down, and strange quarks. By 1970, experimental evidence of the up and down quarks was found. This evidence came through the indirect means of scattering experiments, since individual quarks could not be observed. These scattering experiments were similar to those that Rutherford performed in his gold foil experiment (see chapter 10). Data gathered on electrons scattered from matter gave results consistent with the quark model. At the same time, the theoretical models were incomplete and required more quarks. The charm, bottom, and top quarks were introduced to account for the plethora of subatomic particles found in collision experiments and to complete the model. Evidence for the charm quark was found in 1974, and the bottom quark was confirmed in 1977. Experimental evidence for the top quark was finally confirmed in 1995 at Fermi Labs. It took an additional 18 years to find experimental evidence for the top quark because the top quark was so massive that it required a more powerful accelerator to create it. Furthermore, in collision experiments, a top quark forms in only one out of billions of collisions. Therefore, it takes an enormous number of collisions to produce a single quark.

Leptons are a second large class of subatomic particles. The word "lepton" comes from the Greek *leptos,* meaning small. Leptons are much smaller than quarks and can exist as individual particles. Just as there are six quarks in the standard model, there are six leptons. The most common lepton is the electron. Two other leptons similar to the electron but more massive are the muon and the tau. The muon is about 200 times more massive than the electron and the tau particle is about 3,500 times more massive. The three other leptons are neutrinos, and each neutrino corresponds to one of the electronlike particles. Hence, there is an electron neutrino, a muon neutrino, and a tau neutrino. Neutrinos were first proposed by Wolfgang Pauli to explain the missing energy in beta decay. Neutrinos don't carry a charge and have little if any mass. The universe contains countless neutrinos, and a flux of neutrinos constantly pours out of the Sun. Neutrinos are so small that trillions of them pass through our bodies each second. The elusive character of neutrinos has made their detection one of the most challenging problems of modern phys-

Table 11.2
Leptons

Lepton	Symbol	Charge	Relative Mass
Electron	e	-1	1
Muon	μ	-1	207
Tau	τ	-1	3,478
Electron neutrino	v_e	0	?
Muon neutrino	v_μ	0	?
Tau neutrino	v_τ	0	?

ics. They were first detected in 1956. Large underground caverns filled with water serve as detectors for neutrinos. The detectors are built underground in order to shield the detector from all the other types of particles that could produce a signal at the Earth's surface. A typical neutrino detector consists of a water-filled cavern lined with light detectors to monitor scintillation when a neutrino reacts with a water molecule. The largest neutrino detector is Super-Kamiokande, located in an abandoned zinc mine in Japan. It contains over 10,000,000 gallons of water and 10,000 sensors. This detector is capable of detecting several neutrinos per day. An important question for astronomy has been whether neutrinos have mass, and if so what it is. By knowing if neutrinos have mass, the effect of gravitational forces on the expansion of the universe can be estimated. This in turn will help determine if the universe will expand forever or reach a point at which it will start to collapse upon itself. Current evidence seems to indicate that neutrinos are not massless. Table 11.2 lists the six different leptons, their symbols, and their charge.

The third class of fundamental particles consists of bosons. These are force-mediating particles. That is, they are responsible for the four fundamental forces, or interactions. The term "interaction" is often used instead of "force" to indicate the true nature of how matter interacts. In this discussion, the term "force" will be used, but be aware that many sources call it interaction instead. The four fundamental forces are gravity, electromagnetism, the weak force, and the strong force. Each of these forces can be characterized by its strength and the range over which it acts. Gravity is the weakest of the four forces but operates over an infinite range (Table 11.3). The range over which a force operates can be considered the distance over which it is felt. Gravity operates over an infinite range because everything in the universe is attracted by gravity to everything else. Although gravity operates over an infinite range, it follows the inverse square law, so that as the separation between objects in the universe increases, the gravitational force decreases. Although this force is weak, the collective mass of objects gives structure to the universe on a large scale. Gravity is not included in the standard model because it becomes inconsequential at the atomic level, and physicists have not determined how it fits in the model.

Electromagnetism is the force that causes unlike charges to attract one another and like charges to repel. Electrons and protons in atoms are attracted to each other by the electromagnetic force, and this force is

Table 11.3
Fundamental Forces

Force	Relative Strength	Range	Boson
Gravity	6×10^{-89}	infinite	Gravitron
Weak	0.00001	10^{-17} m	W^+, W^-, Z^0
Electromagnetic	0.007	infinite	Photon
Strong	1	10^{-15} m	Gluon

responsible for bonding in chemical compounds. Like the force of gravity, it operates over an infinite range and obeys the inverse square law. In contrast to gravity, the electromagnetic force can either be attractive or repulsive, depending on the sign of the charges interacting.

The weak force occurs in many different types of radioactive decay. This force is more correctly called an interaction. The weak force is important in the transmutation of quarks and leptons, for example, when a quark changes its flavor. A common example of where the weak force presents itself is the case of beta decay. In negative beta decay, a neutron transforms into a proton and emits a beta particle (electron) and an electron antineutrino. In positive beta decay, a proton transforms into a neutron and emits a positron (antielectron) and an electron neutrino. The electromagnetic and weak forces are actually manifestations of the same force, referred to as the **electroweak force.** According to the work of Sheldon Glashow (1932–), Steven Weinberg (1933–), and Abdus Salam (1926–1996), the electromagnetic and weak forces were indistinguishable in the early stages of the universe. As the universe expanded from a hot, dense state where particles were highly energized, the electromagnetic and weak forces segregated. In the early stages of the universe, the weak force carriers were identical to photons. Glashow, Weinberg, and Salam shared the 1979 Nobel Prize in physics for their work on the electroweak theory.

The coupling of two or more of the fundamental forces into a single theory is known as a **unified theory.** The attempt to unify the strong force (see below) with the electroweak force would unite three of the four forces into a grand unified theory (GUT). Physicists are currently making progress in developing a GUT. Ultimately, the holy grail of physics is to unite all four forces into one theory. This is the so-called theory of everything. Attempts to arrive at a theory of everything may require a fundamentally different approach such as the superstring theory, which is based on one fundamental particle called a string. Particles and interactions are the result of different vibrations and rotations of strings.

The final of the four fundamental forces is the strong force (or strong nuclear force). The strong force is responsible for keeping quarks together. It also manifests itself as a residual force that holds nucleons together in the nucleii of atoms. Neutrons in the nucleus are stable, but as individual particles they decay in an average time of around 15 minutes. The massive strength of the strong force is the reason that quarks are always observed in combinations and not as individual particles, that is, they obey the confinement principle. Due to the strong binding force, energy is liberated in fission reactions when heavy atoms are split or in fusion reactions where light atoms combine.

According to the standard model, the force (or interaction) between two objects is due to an exchange of a particle. When the particle is exchanged, energy and momentum are transferred between the objects. The exchange can be compared to a person on ice skates tossing a medicine ball to another person on ice skates. When the ball is thrown, energy and momentum are transferred between the two individuals. The medicine ball acts as a particle to carry or mediate the force between the two people. In the standard model, each of the forces is mediated by a particle. Gravity is thought to be mediated by **gravitrons,** but currently this is conjecture and there is no definitive proof of gravitrons. The other three forces are mediated by particles collectively called bosons or gauge particles. Bosons responsible for each of the four fundamental forces are listed in Table 11.3. The electromagnetic force is mediated by virtual photons. Virtual photons are not photons as discussed in chapters 9 and 10. Virtual photons are unobservable entities that have no definite mass and exist only for an instant. Other virtual particles are theorized to form out of the consequences of quantum theory and the uncertainty principle. They are an invention of physicists to help explain processes. For example, when two electrons approach each other, the resultant repulsive force can be explained in terms of electrons exchanging a virtual photon. It would be similar to two ice skaters throwing the medicine ball to each other. The skaters move away from each other as momentum and energy are transferred between them. Similarly, the two electrons repel each other because of the exchange of the photon.

Processes involving subatomic particles can be represented with displays called Feynman diagrams. Richard Feynman (1918–1988) was one of the leading physicists in the last half of the twentieth century who contributed to the development of the standard model. He was a leader in resolving some of the questions involving the electromagnetic and weak force through a theory called **quantum electrodynamics.** In Feynman diagrams, time is viewed from left to right. Straight lines represent observable subatomic particles, and wavy lines represent virtual particles such as the photon. Interaction between particles takes place at the vertices of the diagram. A simple Feynman diagram representing the interaction of two electrons approaching each other is shown in Figure 11.5.

The strong force holding quarks together is mediated by gluons. Gluons, like quarks, possess color, and the strong force takes place between quarks possessing different color. This can be compared to an electromagnetic interaction occurring between charged particles. In order to have an electromagnetic

Figure 11.5
Feynman diagram for electron-electron repulsion. As electrons approach, the exchange of a photon produces a repulsive force.

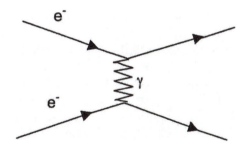

Figure 11.6
The strong force holding quarks together, as gluons are continually transferred between quarks of different colors

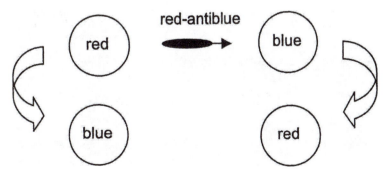

force between two objects, they must contain a charge. Similarly, two quarks must contain a color to experience the strong force. The part of the standard model dealing with the strong force in terms of color is called **quantum chromodynamics.** According to quantum chromodynamics, quarks are held together by an exchange of gluons between quarks of two different colors. An exchange of a color-anticolor gluon between the two quarks effectively switches the quark colors. For example, consider a red quark and a blue quark. If the red quark emits a red-antiblue gluon (antiblue is the same as yellow, so red-antiblue is the same as red-yellow) to the blue quark, then the red quark becomes blue and the blue quark becomes red. Another way of viewing this is that the blue quark's color is neutralized by the antiblue, and because it also accepts the red gluon, it turns red (Figure 11.6). In protons and neutrons this process occurs constantly between the three different-colored quarks. Therefore, at any instant, the quarks possess three different colors, continually changing from one color to another. The exchange of gluons keeps quarks bound to make baryons. In another process, quarks, antiquarks, and gluons produce pions (π mesons) that hold the nucleons together in the nucleus.

Figure 11.7
In negative beta decay, a down quark is converted into an up quark. In the process a W$^-$ boson decays into an electron and an electron antineutrino. The process results in a neutron being transformed into a proton.

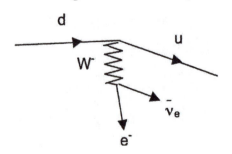

The weak force is mediated by three bosons called W$^+$, W$^-$, and Z^0. These bosons are very heavy. Photons and gluons have no mass, but weak force bosons have masses that are approximately 100 atomic mass units (about 100 times the mass of a hydrogen atom). The mass of bosons is inversely related to the range of the force. Gravity and electromagnetism have an infinite range, but their bosons have zero mass. Conversely, the weak force operates over a very small range, but the weak force bosons are heavy. As mentioned previously, the weak force occurs during beta decay. W$^+$ and W$^-$ bosons mediate the transformation of protons into neutrons and neutrons into protons, respectively. In the reactions p $\rightarrow$ n + e$^+$ + ν_e (positive beta decay) and n $\rightarrow$ p + e$^-$ + ν_e (negative beta decay), a W boson decays into a positron and electron neutrino for positive beta decay and an electron and electron antineutrino for negative beta decay. The Feynman diagram for negative beta decay is shown in Figure 11.7.

W and Z particles were detected at CERN in 1983. An interesting aspect of the electroweak theory is that another particle predicted from the weak interactions is the Higgs boson. This boson is responsible for the mass of particles. The particle was proposed by Peter Higgs (1929–) in 1964. According to Higgs' theory, the universe is permeated by a field. As objects and particles travel through this field, they interact with the field and acquire mass via the Higgs boson. It is hoped that the Higgs boson will be discovered through new experiments, possibly with the new Large Hadron Collider scheduled to be completed at CERN in 2007.

The Standard Cosmological Model

Some of the most interesting aspects of particle physics research involve the insights it has provided with respect to the history of the universe. Particle colliders have given scientists a glimpse of what conditions may have been like at different stages following the Big Bang. The Big Bang model postulates that our current universe developed from a colossal explosion originating from a singularity of infinite mass. As the universe expanded and cooled, the different forces and particles observed today separated from the original primordial soup. As time progressed after the Big Bang, it is hypothesized that different events characterized the universe. These have been given names such as the quark-lepton era or radiation era. The separation process is known as **decoupling** and can be compared to what happens when one stops shaking a mixture of water and dirt. As long as the mixture is shaken, everything stays mixed. Once the shaking stops, the dirt settles out on the bottom of the container. Heavier dirt particles settle first, while lighter particles remain suspended and settle later.

The Planck era covers the first 10^{-43} s immediately following the explosion,

Table 11.4
The Standard Cosmological Model

Time	Era	Major Events
	Planck	
10^{-43} s		Gravity decouples
10^{-35} s	GUT	Strong force decouples
10^{-33} s	Inflationary	
10^{-12} s		Electromagnetic and weak
		Forces decouple
	Electroweak	
	Hadron	Protons and neutrons form
10^{-5} s		
10^{-4} s		Electrons form
3 min		Neuclosynthesis
	Radiation	
300,000 y		Atoms form, matter and radiation decouples, galaxies form
	Matter	

which is the time limit that defines where our current laws of physics unravel. At the Planck time, the radius of the universe is the Planck length of 1.6×10^{-35} m, and the Planck temperature is 10^{32} K. At these infinitesimal scales, quantum physics must be applied to gravitation. As discussed above, gravity has not been included in the standard model, and a quantum theory of gravitation doesn't exist. In short, the Planck time sets the limit to which we can reasonably apply current physical knowledge. The Planck time, in one sense, defines the instant of creation, where an instant covers 10^{-43} s.

After 10^{-43} s, gravity decouples from the strong, electromagnetic, and weak forces; this first period is called the GUT era. In the GUT era, there is no distinction between quarks, leptons, and other particles. Matter and antimatter annihilate each other, while matter is created out of energy. The cosmic soup consists of radiation and matter in a constant flux of transition. In these transformations, matter is slightly favored over antimatter, which leads to an excess of matter to form the universe as we know it today. The three GUT forces are coupled until 10^{-35} s, at which time the strong nuclear force decouples. Quarks and leptons become distinguishable. At this stage the temperature of the universe is 10^{27} K. It is also at this stage that a rapid expansion of the universe begins and the universe is said to be in a state of false vacuum. The false vacuum results in a momentary reversal of gravity from an attractive force to a repulsive force. The inflation lasts from 10^{-35} s to 10^{-33} s and is known as the inflationary era. Following the inflationary era, the lepton-quark era is characterized by a universe filled with elementary particles. In the quark-lepton era, particles are continually annihilated by their antiparticles to form photons, and photons collide to form matter. Matter quarks are formed at a slightly greater rate than antimatter quarks. The temperature is 10^{25} K at this stage. At around 10^{-12} s, when the universe is at 10^{15} K, the electroweak decouples into the electromagnetic and weak forces. It is also around this time that the first protons and neutrons form. The formation of protons and neutrons increases as the temperature decreases, and at around 10^{-4} s electrons and positrons form. At this stage, the universe is still too hot, at about 10^{12} K, for atomic nucleii to form. The radiation era is followed by the matter era. When the temperature drops to approximately 1 billion K, several minutes after the Big Bang, helium and several other light nuclei form in a process called nucleosynthesis. At this point the universe is too cool for nuclear reactions to occur, and the universe exists in a plasma state. Plasma is a fourth state of matter consisting of ionized nuclei. The universe at this stage exists as a giant star consisting of about 75% hydrogen nuclei and 25% helium nuclei. The period starting at around 3 minutes until 300,000 years after the Big Bang is called the radiation era. Galaxies and other large structures in the universe will not form for another 500,000 years. After about 300,000 years, the temperature of the universe has cooled to about 3,000 K, and protons and electrons join to form atoms as they exist today. The size of the universe is about 1 million light years across. It is during this stage that matter starts to condense into galaxies, and the universe as known today starts to form. A summary of the standard cosmological model is given in Table 11.4.

Summary

During the last century the model of the atom has progressed from a simple planetary concept with protons, neutrons, and electrons to a model that challenges the imagination. The process of unraveling the secrets of the atom led to the creation of some of the most sophisticated and expensive scientific instruments of modern times. While these instruments have answered countless questions, in the process, countless others have been raised. The study of the smallest forms of matter has ironically provided deep insight on the galactic scale. The standard model summarizes the current state of our knowledge but falls short in that it still doesn't account for gravity. New discoveries await the completion of the Large Hadron Collider. At the same time, new theories such as superstrings seek simplicity. Ultimately, the true existence of nature will probably never be known. With each discovery, progress is made, but at the same time these discoveries continually open

12

Nuclear Physics

Introduction

Common reactions that transform matter are chemical in nature. These involve the interaction of the electrons of substances. As one substance changes into another, chemical bonds are broken and created as atoms rearrange. Atomic nuclei are not directly involved in chemical reactions but still play a critical role in the behavior of matter. Nuclear reactions involve the atom's nucleus. The nucleus contains most of the atom's mass but occupies only a small fraction of its volume. Electrons have only about 1/2,000 the mass of a **nucleon.** To put this in perspective, consider that if the nucleus were the size of a baseball, the mean distance to the nearest electrons would be over 2 miles.

At the start of the twentieth century, it was discovered that the nucleus is composed of positively charged protons and neutral neutrons (see chapter 10). These particles are collectively called nucleons. During the last half of the same century, scientists learned how to harness the power of the atom. The deployment of two atomic bombs brought a quick and dramatic end to World War II. This was followed by nuclear proliferation, the Cold War, and the current debate over developing

a nuclear defense shield. The use of nuclear power as a source of clean, efficient energy continues to be debated. While nuclear power is used by many countries to fulfill their energy needs, accidents such as Chernobyl in the Ukraine and Three Mile Island in this country have dampened the public's enthusiasm for this energy source. Because of nuclear weapons and reactor accidents, people often perceive the term "nuclear" negatively, but there are many positive aspects to nuclear science. Nuclear medicine is used extensively for both diagnostic and therapeutic procedures. Radiometric dating is an invaluable tool to scientists who use this method for applications such as dating relics, determining the age of the Earth, and studying climate change. In this chapter we will extend our examination of the nucleus started in chapter 10 and examine in detail the scientific principles behind human use of the atom.

Atomic Mass, Atomic Numbers, and Isotopes

The most basic unit of a chemical element that can undergo chemical change is an atom. Atoms of any element are identified by the number of protons and neutrons in the nucleus. The number of protons in the nucleus of an element is given by the atomic number. Hydrogen has one proton in its nucleus, so its atomic number is 1. The atomic number of carbon is 6 because each carbon atom contains six protons in its nucleus. Besides protons, the nucleus contains neutrons. The number of protons plus the number of neutrons is the **mass number** of an element. A standard method of symbolizing an element is to write the element with the mass number as a superscript and the atomic number as a subscript. Carbon-12 is written as

$$^{12}_{6}C$$

An element's atomic number is constant, but most elements have varying mass numbers. For example, all hydrogen atoms have an atomic number of 1, and almost all hydrogen atoms have a mass number of 1. This means most hydrogen atoms have no neutrons. Some hydrogen atoms have mass numbers of 2 or 3, with one and two neutrons, respectively. Different forms of the same element that have different mass numbers are known as **isotopes.** Three isotopes of hydrogen are symbolized as

$$^{1}_{1}H$$

Hydrogen

$$^{2}_{1}H$$

Deuterium

$$^{3}_{1}H$$

Tritium

Isotopes of elements are identified by their mass numbers. Hence, the isotope C-14 is the form of carbon that contains eight neutrons.

The mass number gives the total number of protons and neutrons in an atom of an element but does not convey the absolute mass of the atom. In order to work with the masses, comparative masses are used. Initially, John Dalton and the other pioneers of atomic theory used the lightest element hydrogen and compared masses of other elements to hydrogen. The modern system uses C-12 as the standard and defines 1 **atomic mass unit** (amu) as 1/12 the mass of C-12, and therefore, 1 amu is approximately 1.66×10^{-24} g. This standard means the masses of individual protons and neutrons are slightly more than 1 amu, as shown in Table 12.1.

The atomic mass of an element is actually the average mass of atoms of that element in atomic mass units. Average mass must be used because atoms of elements exist naturally as different isotopes with different masses. Thus, the atomic mass of carbon is

Table 12.1
Masses of Fundamental Atomic Particles

Particle	Mass (amu)	Mass (g)	Charge
Proton	1.007276	1.673×10^{-24}	+1
Neutron	1.008664	1.675×10^{-24}	0
Electron	5.485799×10^{-4}	9.109×10^{-28}	−1

slightly higher than 12, with a value of 12.011. This is due to the fact that most carbon (99%) exists as, C-12, but heavier forms of carbon , such as C-13 and C-14, also exist. Every element has at least one unstable or radioactive isotope, but most have several. The nuclei of unstable isotopes undergo radioactive decay. Radioactive decay is the process where particles and energy are emitted from the nuclei of unstable isotopes as they become stable.

Nuclear Stability and Radioactivity

Nuclear stability is related to the ratio of protons to neutrons. Protons packed into the atom's nucleus carry a positive charge and, therefore, exert a repulsive force on each other. In order for the nucleus to remain intact, the strong nuclear force (see chapter 11) must balance the electrostatic repulsion between protons. The strong nuclear force is the "nuclear glue" responsible for holding the nucleus together. This force is related to the ratio of neutrons to protons in the nucleus. A normal hydrogen atom's nucleus contains a single proton; therefore, no neutrons are needed because there is no repulsive force between protons. All other elements have more than one proton and require neutrons to enable the strong nuclear force to interact. Helium atoms have two protons and two neutrons in their

Figure 12.1
The ratio of the number of neutrons to protons determines the stability of atomic nuclei. As the number of protons in the nucleus increases, the number of neutrons must increase at a greater rate to be stable.

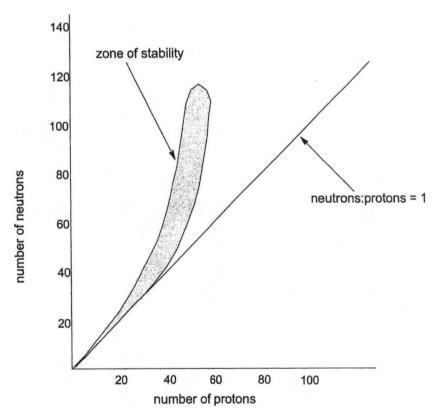

Table 12.2
Nuclear Stability and Numbers of Neutrons and Protons

Number of Neutrons	Number of Protons	Number of stable Isotopes
Even	Even	168
Even	Odd	50
Odd	Even	52
Odd	Odd	4

nuclei. The atomic number increases when moving up the periodic table because one proton is added to the nucleus with each successive element. When more protons are packed into the nucleus, more neutrons are required to overcome the resultant repulsive force.

Stable isotopes prefer certain combinations of neutrons and protons. Most stable isotopes have an even number of both protons and neutrons. A smaller number of stable isotopes have either an even number of protons and an odd number of neutrons or vice versa, and only a few have both an odd number of protons and neutrons (Table 12.2). For elements with an atomic number less than 20, stable nuclei have either an equal number of protons and neutrons or one more neutron than proton. For example, carbon has 6 protons and 6 neutrons, and fluorine has 9 protons and 10 neutrons. For elements with atomic numbers greater than 20, the ratio of neutrons to protons becomes increasingly greater than 1. The relationship between nuclear stability and the ratio of neutrons to protons is depicted in Figure 12.1. Stable nuclei tend to be found in the zone of stability. All isotopes above bismuth, atomic number 83, are radioactive. The neutron to proton ratio for isotopes outside the zone of stability characterizes unstable nuclei, and these undergo various forms of radioactive decay.

Stable nuclei are also associated with specific numbers of protons or neutrons. These islands of stability occur when the number of protons or neutrons is 2, 8, 20, 28, 50, 82, or 126. To illustrate this, considers tin with an atomic number of 50. Tin has 10 stable isotopes, but antimony with an atomic number of 51 has only 2.

Radioactive Decay and Nuclear Reactions

An unstable nucleus will emit particles or electromagnetic radiation, or both, until a stable nucleus results. The main forms of radioactive decay include **alpha (α), beta (β),** and **gamma (γ)** decay. These three forms of radiation are named for the first three letters of the Greek alphabet. An alpha particle is equivalent to the nucleus of a helium atom and consists of two protons and two neutrons. A beta particle is equivalent to an electron. Gamma radiation is electromagnetic energy with wavelengths that range from 10^{-14} to 10^{-11} m. As alpha and beta particles are emitted from an atom's nucleus, the atomic number changes. This means the atom changes from one element to another. Alpha particles have a mass number of 4, while beta particles have a mass number of zero. Therefore, the emission of an alpha

Table 12.3
Main Forms of Radioactive Emissions

Radiation	Form	Mass	Charge	Symbol
Alpha	Particle = He nucleus	4	+2	$^4_2$He or α
Beta	Particle = electron	0	−1	$^0_{-1}$e or β
Gamma	Electromagnetic radiation	0	0	$^0_0\gamma$

Figure 12.2
One possible decay series for radium. Ra-226 is transformed into stable Pb-206 through a series of alpha and beta emissions. Horizontal arrows represent beta emissions, and diagonal arrows alpha emissions.

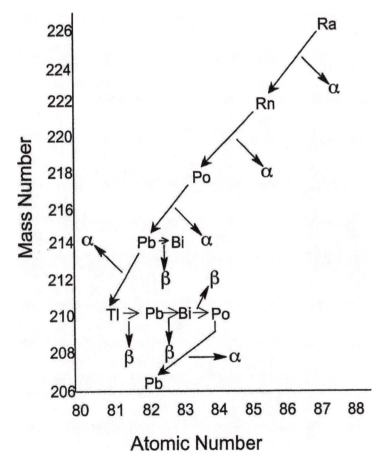

particle changes both atomic and mass numbers, but the emission of a beta particle only changes the atomic number. Since gamma radiation is a form of light, the atomic number and mass number remain constant when gamma radiation is emitted. The emission of gamma radiation does not transform an element. Table 12.3 summarizes the properties and symbols used for the three main types of radioactive emissions. Conservation of mass and charge are used when writing nuclear reactions. For example, consider what happens when uranium-238 undergoes alpha decay. Uranium-238 has 92 protons and 146 neutrons and is symbolized as $^{238}_{92}U$. After it emits an alpha particle, the nucleus has a mass number of 234 and an atomic number of 90. Since the atomic number of the nucleus is 90, the uranium becomes thorium, Th. The overall nuclear reaction can be written as

$$^{238}_{92}U \rightarrow \,^{234}_{90}Th + \,^{4}_{2}He$$

Notice that in the above equation mass is conserved, 238 = 234 + 4, and charge is conserved, 92 = 90 + 2. In a nuclear reaction, reactants and products are referred to as parent and daughter. In the U-238 example, U-238 was the parent that decayed into the Th-234 daughter.

During beta decay, a neutron is transformed into a proton. If Th-234 were to emit a beta particle, it would be transformed into protactinium-234 according to the equation

$$^{234}_{90}Th \rightarrow \,^{234}_{91}Pa + \,^{0}_{-1}e$$

Again, both mass and charge are conserved. Gamma emission often accompanies both alpha and beta decay. Gamma emission does not change the parent element.

An unstable parent nucleus may decay into either a stable or an unstable daughter. When the daughter is unstable, which is often the case, the daughter will decay. Often the journey from an unstable nucleus to a stable nucleus involves a long series of steps referred to as a radioactive decay series. One example is the decay series for radium (Figure 12.2).

Half-Life and Radiometric Dating

The process of radioactive decay for any unstable nucleus is random. In a mass of radioactive material, it is impossible to tell when any particular nucleus will decay. Although the exact time when an individual nucleus may decay is unknown, it is possible to characterize statistically when a certain proportion of the nuclei have decayed. The process can be compared to popping a bag of microwave popcorn. While it is impossible to know when any individual kernel pops, experience dictates that after a certain amount of time has elapsed, for instance 1 minute, a specific proportion of the kernels will have popped. After perhaps 3 minutes, the popping may stop, but there will still be several unpopped kernels remaining in the bag. Radioactive decay follows a similar process. At any given time a parent nucleus may decay into a daughter. Although it can't be known when any individual nucleus may decay, a time period called the **half-life** can be used as a collective measure of how long radioactive decay takes. Half-life, often symbolized using $t_{1/2}$, is the time required for half the parent nuclei in a sample to decay into daughter nuclei. An equivalent definition is the time it takes for the activity of a substance to be cut in half.

Radioactive decay follows first-order kinetics. First-order kinetics means that the decay rate depends on the amount of parent material present at any given time. In nuclear decay, the number of parent nuclei decreases in an exponential fashion (Figure 12.3). The decay rate or activity of a radioactive sub-

Figure 12.3
Exponential decay. The number of parent nuclei decreases by a factor of 1/2 for each half-life.

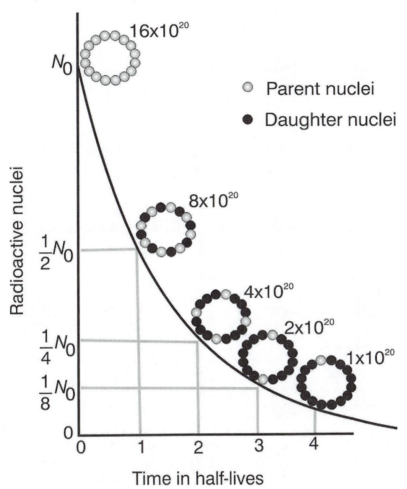

stance is equal to the decrease in parent nuclei over time:

$$\text{activity} = \frac{\Delta N}{\Delta T}$$

where ΔN equals the change in number of parent nuclei and ΔT is the change in time. The standard unit for activity is the **becquerel.** One becquerel is equal to 1 decay per second.

During each consecutive half-life the amount of parent material remaining is one-half of the amount present at the start of the half-life (Figure 12.4). The half-lives of radioactive isotopes vary over a wide range, from a fraction of a second to over billions of years. Table 12.4 lists the half-lives of some common isotopes.

The half-life of isotopes provides scientists with a nuclear clock that can be used to date objects. The concept is based on knowing the fraction of original material that is present in a sample. For instance, if half of the original isotope is present in the sample, then the sample's age is equivalent to the isotope's

Figure 12.4
The amount of a radioactive substance is cut in half after each consecutive half-life.
The amount of the original material remaining after n half-lives is $(1/2)n$.

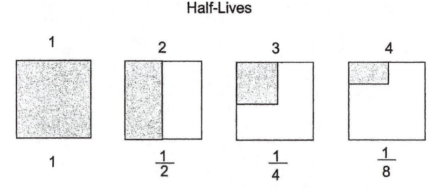

Fraction of Original Material Remaining

Table 12.4
Half-Lives of Common Isotopes

Isotope	half-life
Polonium-214	1.6×10^{-4} second
Lead-214	26.8 minutes
Radon-222	3.82 days
Strontium-90	28.1 years
Radium-226	1602 years
Carbon-14	5730 years
Uranium-238	4.5×10^9 years

half-life. If 1/4 of the original material is present, then the sample's age is two half-lives. Since the use of radiometric dating involves making accurate measurements of parent and daughter activities in the sample, it is assumed that parent material is not lost from the sample or that the sample has been contaminated.

Radiometric techniques have been applied in a variety of disciplines including archaeology, paleontology, geology, climatology, and oceanography. The isotopic method used for dating depends on the nature of the object. Geologic samples that are billions of years old would require using an isotope with a relatively long half-life, such as potassium-40 ($t_{1/2} = 1.25 \times 10^9$ years). Conversely, the movement of ocean water masses or groundwater might use tritium (tritium is hydrogen containing two neutrons, $_1^3H$), which has a half-life of 12.3 years. In general, an isotope can be used to date an object up to 10 times its half-life, for example, tritium could be used for measurement of up to 125 years.

Carbon-14, $t_{1/2} = 5,730$ years, is one of the most prevalent methods used for dating ancient artifacts. Williard Libby (1908–1980) developed the method around 1950 and was awarded the Nobel Prize in chemistry in 1960 for this work. The carbon-14 method is based on the fact that living organisms continually absorb carbon into their tissues during metabolism. Most of the carbon taken up by organisms is in stable C-12 form, but unstable C-14 (as well as other carbon isotopes) is **assimilated** along with

C-12. No distinction is made between the isotopes of carbon assimilated; so the ratio of C-14 to C-12 in the organism remains the same as that in the organism's environment, for example, the atmosphere or ocean. The natural occurrence of C-14 in the environment is about one C-14 atom for every 850 billion C-12 atoms. As long as an organism is alive, the carbon-14 to carbon-12 ratio remains constant. Once the organism dies, the C-14 stopwatch starts, and the C-14:C-12 ratio begins to decrease due to the decay of C-14. Comparing the activity of C-14 in the object to that in living material can date the sample. For example, if the activity of C-14 in an excavated bone is half that of bone from a living organism, then the bone must be approximately 5,730 years old. Carbon-14 dating is used extensively with materials such as wood, seeds, fabrics, and bone.

A popular method used to date rocks is the potassium-argon method. Potassium is abundant in rocks such as feldspars, hornblendes, and micas. The K-Ar method has been used to date the Earth and its geologic formations. It has also been applied to determine magnetic reversals that have taken place throughout the Earth's history. Another method used in geologic dating is the rubidium-strontium, Rb-Sr, method. Some of the oldest rocks on Earth have been dated with this method, providing evidence that the Earth is approximately 5 billion years old. The method has also been used to date Moon rocks and meteorites.

Nuclear Binding Energy, Fission, and Fusion

Since the nucleus consists of a collection of nucleons, it would be expected that the mass of the nucleus would equal the sum of its constituent nucleons. In fact, the mass of the nucleus is always slightly less than the sum of its parts. This decrease in mass of the nucleus is called the **mass defect**. The missing mass when the nucleus of an atom is created out of nucleons is related to the strong nuclear force that holds the nucleus together. The mass defect is converted to energy and released when the nucleons combine to form a nucleus. The amount of energy released is related to the mass defect according to Einstein's famous equation: $E = mc^2$, where m is mass and c is the speed of light. Because the speed of light is so high ($c = 3 \times 10^8$ m/s or 6.7×10^8 mph), it doesn't take much mass to produce a tremendous amount of energy. The energy calculated from Einstein's equation using the mass defect for m is called the nuclear **binding energy.** To separate the nucleus of an atom into its individual protons and neutrons would take an amount of energy equivalent to the binding energy.

The binding energy provides a measure of the stability of atomic nuclei. The greater the binding energy per nucleon of a nucleus, the greater its stability. When the amount of binding energy per nucleon is calculated for the different elements, it is found that the Fe-56 has the highest binding energy (Figure 12.5)

The fact that the binding energy curve peaks at the element iron has important consequences for obtaining nuclear energy. When a heavier nucleus splits in a process called **nuclear fission,** lighter nuclei are produced. The lighter nuclei are more stable. Whenever a process involves moving to a more stable state or configuration, energy is released. More stable nuclei can also be obtained when lighter nuclei combine to form a heavier nucleus in a process call **nuclear fusion.** The most common example of a fusion reaction takes place inside the Sun, where hydrogen nuclei fuse to form helium. The continual fusion of solar hydrogen provides the energy that makes life on Earth possible.

Figure 12.5

Binding energy curve. Elements such as iron, cobalt, and nickel have the highest binding energy per nucleon and, therefore, are the elements with the most stable nuclei.

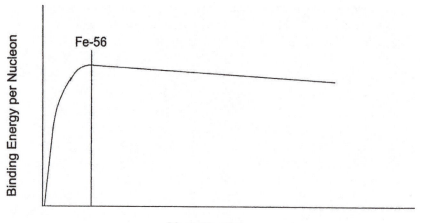

Modern nuclear power is based on harnessing the energy released in a fission reaction. The development of atomic energy started in the 1930s with the discovery that atoms could be split with neutrons. This discovery laid the foundation for building the first atomic bombs during World War II. A basic reaction representing the fission of uranium can be represented as

$$_{0}^{1}\text{n} + _{92}^{235}\text{U} \rightarrow _{56}^{141}\text{Ba} + _{36}^{92}\text{Kr} + 3\,_{0}^{1}\text{n} + \text{energy}$$

The equation shows that uranium-235 absorbs a neutron. After absorbing the neutron, the excited uranium nucleus splits and forms barium-141, krypton-92, and three neutrons. Energy is also produced in the reaction. This reaction is only one of a number of different ways that U-235 may split. Several hundred different isotopes have been identified when U-235 undergoes fission.

Three neutrons are produced when a uranium atom splits. These three neutrons have the ability to split other U-235 nuclei and start a self-sustaining chain reaction. Whether a chain reaction takes place

depends on the amount of fissionable material present. The more fissionable material that is present, the greater the probability that a neutron will interact with another U-235 nucleus. The reason for this involves the basic relationship between surface area and volume as mass increases. If a cube with a length of 1 unit is compared to a cube of 2 units, it is found that the surface area to volume ratio of the 1 unit cube is twice that of the 2 unit cube (Figure 12.6). This shows that volume increases at a greater rate than surface area as size increases. The probability that neutrons escape rather than react also depends of the surface area to volume ratio. The higher this ratio is, the more likely that neutrons escape. When a U-235 nucleus contained in a small mass of fissionable uranium is bombarded by a neutron, the probability is that less than one additional fission will result from the three product neutrons. The amount of fissionable material at this stage is called a **subcritical mass.** As the mass of fissionable material increases, a point is reached where exactly one product neutron causes another U-235 fission, which

Figure 12.6
As the size increases, volume increase at a greater rate than the surface area. This relates to the critical mass needed for a sustained nuclear reaction.

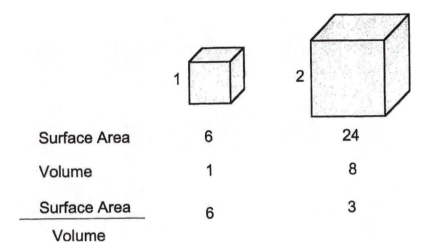

	1	2
Surface Area	6	24
Volume	1	8
$\dfrac{\text{Surface Area}}{\text{Volume}}$	6	3

in turn produces three more neutrons, and one of these splits another U-235 nucleus. The quantity of fissionable material at the point where the reaction is self-sustaining is termed the **critical mass.** For a spherical configuration of U-235, the critical mass is about 25 kilograms (55 pounds). This can be reduced to 15 kilograms (33 pounds) if the U-235 is surrounded by neutron reflector to increase the efficiency of the fission process. Once the critical mass is reach, a chain reaction can take place. If more than one product neutron reacts with other U-235 nuclei, a **supercritical mass** exists. In this case the process accelerates, creating an uncontrolled chain reaction. The development of the first atomic bombs involved bringing together simultaneously a number of subcritical masses to form a supercritical mass in a fraction of a second. Once the supercritical mass was assembled, a stream of neutrons detonated the device.

Traditional nuclear power involves using the heat generated in a controlled fission reaction to generate electricity. A schematic of a nuclear reactor is shown in Figure 12.7. The reactor core consists of a heavy-walled reaction vessel several meters thick that contains fuel elements consisting of zirconium rods containing enriched pellets of U-235 in the form of UO_2. Natural uranium is 0.7% U-235 and 99.3% U-238. Uranium-238 is nonfissionable, and therefore, naturally occurring uranium must be enriched to a concentration of approximately 3% U-235 to be used in common nuclear reactors. The reactor is filled with water. The water serves two different purposes. One purpose is to serve as a coolant to transport the heat generated from the fission reaction to a heat exchanger. In the heat exchanger, the energy can be used to generate steam to turn a turbine for the production of electricity. Water also serves as a **moderator.** A moderator slows down the neutrons, increasing the chances that the neutrons will react with the uranium. Interspersed between the hundreds of fuels rods are **control rods.** Control rods are made of cadmium and boron. These substances absorb neutrons. Raising and lowering the control rods controls the nuclear reaction.

Figure 12.7
Schematic of a nuclear power plant

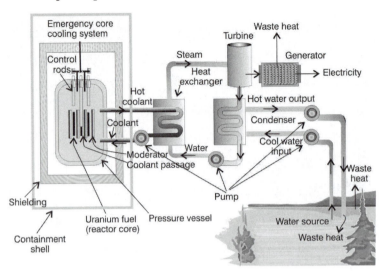

While the basic design of most nuclear reactors is similar, several types of reactors are used throughout the world. In the United States most reactors use plain water as the coolant. Reactors using ordinary water are called light-water reactors (LWRs). Light-water reactors can be pressurized to approximately 150 atm to keep the primary coolant in the liquid phase at temperatures of around 300 °C. The heat from the pressurized water is used to heat secondary water to generate steam. In a boiling-water reactor (BWR), water in the core is allowed to boil. The steam produced powers the turbines directly. Heavy-water reactors employ water in the form of D_2O as the coolant and moderator. Heavy water gets it name from the fact that the hydrogen in the water molecule is the isotope **deuterium,** $D(^2_1H)$, rather than ordinary hydrogen. The use of heavy water allows natural uranium rather than enriched uranium to be used as the fuel.

Breeder reactors were developed to utilize the 97% of natural uranium that occurs as nonfissionable U-238. The idea behind a breeder reactor is to convert U-238 into a fissionable fuel material. A reaction to breed plutonium is

$$^{238}_{92}U + {}^1_0n \rightarrow {}^{238}_{92}U \rightarrow {}^{239}_{93}Np + {}^0_{-1}e \rightarrow$$
$$^{239}_{94}Pu + {}^0_{-1}e$$

The plutonium fuel in a breeder reactor behaves differently than uranium. Fast neutrons are required to split plutonium. For this reason, water can't be used in breeder reactors, since it moderates the neutrons. Liquid sodium is typically used in breeder reactors, and the term liquid metal fast breeder reactor (LMFBR) is used to describe it. One of the controversies associated with the breeder reactor is the production of weapons-grade plutonium and consequent nuclear arms proliferation.

The world use of nuclear power to supply a nation's electricity varies widely by country. France gets around 75% of its electricity from nuclear power, and several other European countries (Lithuania, Belgium, Slovak Republic) get over half of their energy from this source. Approximately 20% of the electricity in the United States comes from 103 operating nuclear power plants. Nuclear is second only to coal, 50%, and ahead of natural gas, 15%, hydropower, 8%, and oil, 3%, as a source of electrical energy. Although once hailed by President Eisenhower in the 1950s as a safe, clean, and economical source of power, the United States nuclear industry has fallen on hard times in the last 25 years. Nuclear accidents at Three Mile Island, Pennsylvania, in 1979 and Chernobyl in the Ukraine in 1986 have raised public concerns about the safety of nuclear power. Utility companies, facing more stringent federal controls and burgeoning costs, have abandoned nuclear power in favor of traditional fossil fuel plants. The last nuclear power plant ordered in the United States was in 1978. Several nuclear plants under construction have been left abandoned, while other operating plants have been shut down and decommissioned.

The lack of emphasis on nuclear power during the last 25 years doesn't necessarily mean it will not stage at comeback in the future. Concerns about greenhouse gases and global warming from burning fossil fuels are one of the strongest arguments advanced in favor of a greater use of nuclear power. New developments in reactor design hold the promise that safe reactors can be built. Yet, concerns about the safe handling and transport of nuclear fuels, nuclear waste disposal, nuclear proliferation, and a NIMBY (not in my back yard) attitude present formidable obstacles to the resurrection of the nuclear industry in the United States.

The binding energy curve indicates that fusion of light atomic nuclei to form a heavier nucleus is an exothermic nuclear reaction. Although in theory many light elements can combine to produce energy, most practical applications of fusion technology focus on the use of hydrogen isotopes. The first application of fusion technology involved development and testing of hydrogen bombs (H-bombs) in the early 1950s. The United States and the former Soviet Union each feared that the other would develop the H-bomb, which was considered to pose an unacceptable threat to the country without the bomb. Leaders of each country knew that hydrogen bombs would be much more destructive than atomic (fission) bombs. The reason for this is related to the small size of hydrogen nuclei. Many more hydrogen nuclei can be packed into an H-bomb compared to the number of uranium nuclei in an atomic bomb. Since the energy produced in a nuclear weapon depends on the number of fission or fusion reactions that take place in a given volume, H-bombs are far more destructive. The quest to produce the first superbomb, as the H-bomb was known, initiated the arms race that characterized the relationship between the United States and the Soviet Union during most of the last half of the twentieth century. It has only been in the last decade that this race has abated with the fall of the Soviet Union, but President Bush's talk of a nuclear defense shield at the start of the twenty-first century is a reminder that the threat of mutually assured destruction remains.

Temperatures in the range of 20 to 100 million degrees Celsius are required for fusion reactions. For this reason a hydrogen bomb is triggered by a conventional fission atomic bomb. The atomic bomb produces the tremendous heat necessary to fuse hydrogen nuclei; therefore, fusion bombs are often referred to as thermonuclear. The United States exploded the first hydrogen bomb on Eniwetok Atoll in the Pacific Ocean on November 1, 1952. This bomb was based on the fusion of deuterium:

$$_1^2H + _1^2H \rightarrow _1^3H + _1^1H$$

The fusion of deuterium produces another hydrogen isotope called tritium, $^3_1$H, along with common hydrogen, $^1_1$H. The fusion of deuterium may also produce helium according to the reaction

$$^2_1H + ^2_1H \rightarrow ^3_2He + ^1_0N$$

There is hope that nuclear fusion may one day prove to be a practical energy source. The United States government has spent billions of dollars on fusion energy research, and several prototype reactors have been built. Researchers attempting to produce energy from fusion face several major obstacles. One problem is how to produce the tremendous temperatures required for fusion. Another is how to confine a fuel, such as deuterium, in a manner to produce a sustained reaction. Yet, another difficulty is obtaining more energy out of the reaction than is required to initiate fusion. One design used to produce temperatures of around 50,000,000 °C involves using a donut-shaped coiled electromagnet called a tokamak (Figure 12.8). The electromagnetic field within the cavity of the magnet acts on hydrogen isotopes, stripping them of their electrons and producing a plasma. Increasing the electric current intensifies the electromagnetic field and raises the temperature to

Figure 12.8
Tokamak design used for nuclear fusion

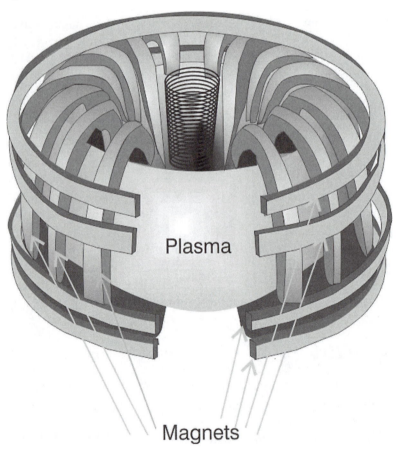

a point where the deuterium nuclei can fuse. Confining the fuel in a suspended plasma state within an electromagnetic field isolates the reactor materials from the extreme temperatures required for fusion. In another design, called inertial confinement, a series of lasers is focused on glass pellets filled with a mixture of deuterium and tritium at a pressure of several hundred atmospheres. The pellets are then injected into a reactor. At a precise point, a pellet and the laser beams intersect to create a controlled thermonuclear explosion. The heat from the explosion can then be used to produce work.

Fusion and Stellar Evolution

The fusion of hydrogen to form helium takes place throughout the universe. Stars are natural thermonuclear machines that are responsible for the formation of all naturally occurring elements. The universe is predominantly composed of hydrogen. Hydrogen is not scattered uniformly throughout the universe. Galaxies are regions of high density, while interstellar space contains little. Some regions of space contain clouds of cool hydrogen gas, called **nebula,** and it is in these regions that stars are born. The birth of a star begins when gravitational attraction causes a nebula to contract. As it contracts, the mass of hydrogen generates heat. The nebula continues to contract, and as it does its temperature rises and its interior pressure increases, forming a **protostar.** Eventually, a point is reach where the protostar's interior temperature reaches several million degrees. At this point the temperature is sufficient to initiate fusion of hydrogen nuclei. Two primary hydrogen fusion reactions are

$$_1^1\text{H} + _1^1\text{H} \rightarrow _1^2\text{H} + _1^0\text{e}$$
$$_1^1\text{H} + _1^2\text{H} \rightarrow _2^3\text{He}$$

In the first reaction a positively charged electron is formed. This particle is called a positron. In addition to the fusion of hydrogen nuclei, the helium created can also enter into fusion reactions:

$$_2^3\text{He} + _1^1\text{H} \rightarrow _2^4\text{He} + _1^0\text{e}$$
$$_2^3\text{He} + _2^3\text{He+} \rightarrow _2^4\text{He} + 2 \ _1^1\text{H}$$

Fusion reactions signify the formation of a true star. The fusion reactions balance the gravitational contraction, and equilibrium is reached. Hydrogen "burning" in the star's interior balances gravitational contraction, as long as a sufficient supply of hydrogen is available. Depending on the size of the star, hydrogen burning may last less than a million years or for billions of years. Larger stars expend their hydrogen faster than smaller stars. The Sun is considered a normal star, with an expected life of 10 billion years. Its current estimated age is 4.5 billion years, so it is about halfway through its hydrogen burning stage.

Once hydrogen burning stops, there is no longer a balance between gravitational contraction and the nuclear energy released through fusion. The star's interior at this stage is primarily helium. The core will again start to contract and increase in temperature. Eventually, a temperature is reached where helium fusion is sufficient to balance the gravitational contraction. Depending on the size of the star, several contractions may take place. It should be realized that at each stage in a star's life cycle, numerous fusion reactions take place. In addition to hydrogen and helium, many other light elements such as lithium, carbon, nitrogen, and oxygen are involved.

Once a star has expended its supply of energy, it will contract to a glowing white "ember" called a **white dwarf.** The elements produced in the interior of a star depend on the size of the star. Small stars do not have sufficient mass to produce the temperatures required to create the heaviest elements. The most massive stars, though, may go through a series of rapid contractions in their final stages. These massive stars have the ability to generate the temperatures and pressures necessary to produce the heaviest elements such as thorium and uranium. The final fate of these massive stars is a cataclysmic explosion called a **supernova.** It is in this manner that scientists believe all the naturally occurring elements in the universe were created. This would imply that all the elements in our bodies were once part of stars, and the heavier elements came from a supernova that exploded billions of years ago.

Transmutation

Transmutation is the process where one element is artificially changed into another element. Rutherford conducted the first transmutation experiment in 1919 when he bombarded nitrogen atoms with alpha particles. The nitrogen was transmuted into oxygen and hydrogen according to the reaction

$$_{7}^{14}\text{N} + _{2}^{4}\text{He} \rightarrow _{8}^{17}\text{O} + _{1}^{1}\text{H}$$

The process of transmutation produces most of the known isotopes. In fact, only about 10% of the approximately 3,000 known isotopes occur naturally. The rest are synthesized in particle accelerators. Particle accelerators were discussed in chapter 11. All elements greater than uranium, known as the transuranium elements, have been produced in particle accelerators. For example, Pu-241 is produced when U-238 collides with an alpha particle:

$$_{92}^{238}\text{U} + _{2}^{4}\text{He} \rightarrow _{94}^{241}\text{Pu} + _{0}^{1}\text{n}$$

Tritium is produced in a nuclear reaction between lithium and a neutron:

$$_{3}^{6}\text{Li} + _{0}^{1}\text{n} \rightarrow _{1}^{3}\text{H} + _{2}^{4}\text{He}$$

Radioactive isotopes produced by transmutation in particle accelerators have found wide use in nuclear medicine. Radioactive nuclei are used to diagnose and treat a wide range of conditions. **Radioactive tracers** are isotopes that can be administered to a patient and followed through the body to diagnose certain conditions. Most tracers are gamma emitters with short half-lives (on the order of hours to days) that demonstrate similar biochemical behavior as their corresponding stable isotopes. Because tracers are radioactive, radiation detectors can be used to study their behavior in the body. Several examples of common tracer studies in medicine include iodine-131 to study the thyroid gland, technetium-99 to identify brain tumors, gadolinium-153 to diagnose osteoporosis, and sodium-24 to study blood circulation.

A recent development in nuclear medicine that illustrates how advances in basic research are transformed into practical applications is positron emission tomography (PET). PET creates a three-dimensional image of a body part using positron-emitting isotopes. In PET, positron emitting isotopes such as C-11, N-13, O-15, and F-18 are used. Each of these unstable isotopes is characterized by lacking a neutron compared to its stable form, for example, C-11 needs one more neutron to become C-12. They undergo positron emission when a proton changes into a neutron:

$$_{1}^{1}\text{H} \rightarrow _{0}^{1}\text{n} + _{1}^{0}\text{e}$$

When a PET isotope is administered to tissue, it emits positrons. These positrons encounter electrons found in tissue, and the interaction

Figure 12.9

In positron emission tomography, positrons interact with elections in body tissue to produce gamma rays that are detected and converted into an image

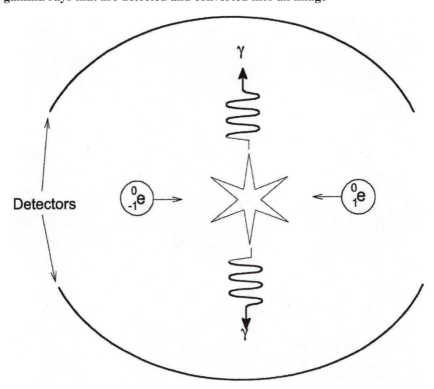

of an electron and positron results in mutual annihilation and the production of beams of gamma rays directly opposite to each other (Figure 12.9). The beams strike gamma detectors, and the two signals are translated into an image by a computer. PET has been used to diagnose neurological disorders such as Parkinson's disease, Alzheimer's disease, epilepsy, and schizophrenia. PET is also used to diagnose certain heart conditions and examine glucose metabolism in the body.

Radioisotopes are also used in radiation therapy to treat cancer. The goal in radiation therapy is to kill malignant cells while protecting healthy tissue from radiation effects. Radioisotopes such as yttrium-90, a beta emitter, may be placed directly in the tumor. Alternatively, the diseased tissue may be subjected to beams of gamma radiation.

Cobalt-60 used in radiation therapy is prepared by a series of transmutations:

$$^{58}_{26}\text{Fe} + {}^{0}_{1}\text{n} \rightarrow {}^{59}_{26}\text{Fe}$$

$$^{59}_{26}\text{Fe} \rightarrow {}^{59}_{27}\text{Co} + {}^{0}_{-1}\text{e}$$

$$^{59}_{27}\text{Co} + {}^{0}_{1}\text{n} \rightarrow {}^{60}_{27}\text{Co}$$

Radiation Units and Detection

Radiation measurements are expressed in several different units depending on what is being measured. The strength or activity of a radioactive source is the number of disintegrations that occurs per unit time in a sample of radioactive material. This is most often expressed as the number of disinte-grations per second. The SI unit for activity is the becquerel, abbreviated Bq. One bec-querel is equivalent to one disintegration per second. The **curie,** Ci, is often used when the activity is high. One curie is equal to 3.7 $\times 10^{10}$ Bq. Common prefixes such as pico- (10^{-12}), micro- (10^{-6}), and milli- (10^{-3}) are often used in conjunction with curies to express activity values. For example, the EPA considers radon levels in a home high if the measured activity in the home is greater than 4.0 pCi/L of air.

When matter absorbs radiation, a certain quantity of energy is imparted to the matter. The **dose** is the amount of radiation energy absorbed by matter. With respect to human health, the matter of concern is human tissue. The traditional unit of dose is the **rad.** One rad is defined as 100 ergs of energy absorbed per 1 g of matter. The SI unit for dose is the **gray,** Gy. One gray is equal to 1.0 J of energy absorbed per 1 kg of matter. The relationship between rads and grays is 1 Gy is equal to

100 rad. As with activity, prefixes can be attached to the gray unit.

The most important measure of radia-tion in terms of health effects is the **dose equivalent.** Dose by itself doesn't consider the type of radiation absorbed by tissue. Radiation effects on human tissue depend on the type of radiation. For instance, a dose of alpha radiation is about 20 times more harm-ful compared to equivalent doses of beta or gamma radiation. Dose equivalent combines the dose with the type of radiation to give the dose equivalent. It is found by multiply-ing the dose by a quality factor. The quality factors for gamma, beta, and alpha radiations are 1, 1, and 20, respectively. Dose equivalent is the best measure of the radiation effects on human health. The **rem** is the dose equiva-lent when dose is measured in rads. Rem is the most frequently used unit for measur-ing dose equivalent. The unit "rem" is actu-ally an acronym for **r**oentgen **e**quivalent for **m**an. Millirems (mrem, or 1/1,000 rem) are most often used to express everyday expo-

Table 12.5
Common Radiation Units

Measurement	Unit	Definition	Abbreviation
Activity	becquerel	$1\dfrac{\text{disintegration}}{\text{second}}$	Bq
	curie	$3.7\times10^{10}\dfrac{\text{disintegration}}{\text{second}}$	Ci
Dose	gray	$1.0\dfrac{\text{joule}}{\text{kilogram}}$	Gy
	rad	$100\dfrac{\text{ergs}}{\text{gram}}$	
Dose equivalent	sievert	gray $\times$ quality factor	Sv
	rem	rad $\times$ quality factor	

sures (Table 17.4). When expressing dose in grays, the unit for dose equivalent is the sievert, Sv. One rem is equal to 0.01 Sv. The various units associated with radiation measurements are summarized in Table 12.5.

Radiation comes in many different forms. Most of this chapter has dealt with nuclear radiation, those forms of radiation that come from the nuclei of atoms. Nuclear radiation originates from the Earth. When nuclear radiation comes from an extraterrestrial source, it is referred to as **cosmic radiation.** Nuclear and cosmic radiation consist of particles such as protons, electrons, alpha particles, and neutrons, as well as photons such as gamma rays and x-rays. When nuclear and cosmic radiation strike matter and free ions, the term **ionizing radiation** is used. Alpha, beta, and gamma radiation are forms of ionizing radiation. **Nonionizing** radiation includes ultraviolet radiation, responsible for sunburns, visible light, and infrared radiation. Humans sense infrared radiation as heat.

Ionizing radiation is impossible to detect using our unaided senses. It can't be felt, heard, smelled, or seen. The fact that ionizing radiation ionizes matter as it passes through it provides a basis for measuring. A common device used to measure ionizing radiation is the Geiger counter. A Geiger counter consists of a sealed metal tube filled with an inert gas, typically argon. A wire, protruding into the center of the tube, connected to the positive terminal of a power source, serves as one electrode. The negative terminal is connected to the wall of the tube and serves as the negative electrode (Figure 12.10). The voltage difference between the wire and tube wall is roughly 1,000 V. No current flows between the wire and tube wall because the inert gas is a poor conductor of electricity. One end of the sealed tube has a thin window that allows radiation to enter. When radiation enters the tube, it ionizes the atoms of the inert gas. The free electrons produced during ionization move toward the positive electrode, and as they move they ionize other gas atoms. The positively charged argon ions move toward the negative walls. The flow of electrons and ions produces a small current in the tube. This

Figure 12.10
A Geiger counter detects radiation by ionizing argon atoms. The flow of argon ions and electrons in the tube creates a current that is amplified and detected by the counter.

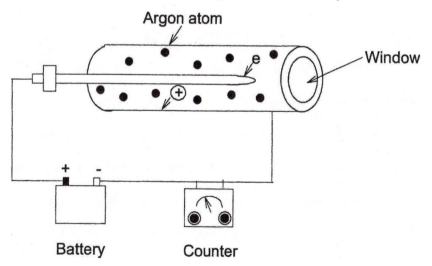

current is amplified and measured. There is a direct relationship between the current produced and the amount of radiation that enters the window, so activity can be measured.

Scintillation counters use materials that produce light when stimulated by radiation. Scintillation materials include sodium iodide crystals and special plastics. Radiation is measured by exposing the scintillation material to radiation and using a photomultiplier tube to count the number of resulting flashes. Photomultiplier tubes are photocells that convert light into electrical signals that can be amplified and measured.

Biological Effects of Radiation

Radiation is ubiquitous in the environment. Life exists in a sea of nuclear, cosmic, and electromagnetic radiation. Our bodies are constantly exposed to radiation, and radiation has several effects on our cells. Radiation may (1) pass through the cell and not do any damage, (2) damage the cell, but the cell repairs itself, (3) damage the cell, and the damaged cell does not repair itself and reproduces in the damaged state, or (4) kill the cell. Fortunately, most radiation doesn't cause cellular damage, and when it does, DNA tends to repair itself. Under certain circumstances the amount of radiation damage inflicted on tissue overwhelms the body's ability to repair itself. Additionally, certain conditions disrupt DNA's ability to repair itself. Figure 12.11 shows several different ways DNA may be damaged by ionizing radiation.

Low levels of ionizing radiation generally do not affect human health, but health effects depend on a number of factors. Among these are the types of radiation, radiation energy absorbed by the tissue, amount of exposure, type of tissue exposed to radiation, and genetic factors. Alpha, beta, and gamma radiation display different characteristics when they interact with matter. Alpha particles are large and slow with very little penetrating power. A sheet of paper or the outer layer of skin will stop alpha particles. Their energy is deposited over short distances, resulting in a relatively large amount of damage. Beta particles are smaller than alpha particles and have a hundred times their penetrating power. Beta particles can be stopped by a few millimeters of aluminum or a centimeter or two of tissue. Their energy is deposited over this distance and hence is less damaging than alpha particles. Gamma particles have a thousand times the penetrating power of alpha particles. Several centimeters of lead are needed to stop it. Gamma radiation travels significant distances through human tissue and, therefore, deposits its energy over a relatively wide area.

It is important to consider the type of radiation with respect to whether a radiation source is external or internal to the body. External alpha, and to some extent beta radiation, are not as hazardous as external gamma radiation. Clothing or the outer layer of the skin will stop alpha radiation. Beta radiation requires heavy protective clothing because it can penetrate the skin and burn tissue. The body cannot be protected from external gamma radiation using protective clothing. While external alpha particles are relatively harmless, they cause significant damage internally. When inhaled and ingested, alpha emitters may be deposited directly into tissue, where they can damage DNA directly.

As long as radiation dose equivalent exposures are low, radiation damage is nondetectable. General effects of short-term radiation exposure are summarized in Table 12.6. The dose equivalents in Table 12.6 are listed in rems. These values are several orders of magnitude greater than what humans receive in a year. Annual human exposure is normally around several

Figure 12.11
Radiation may damage DNA in several ways. It may cause breaks in the sugar-phosphate backbone or cause base cross-links. It may also result in incorrect repair.

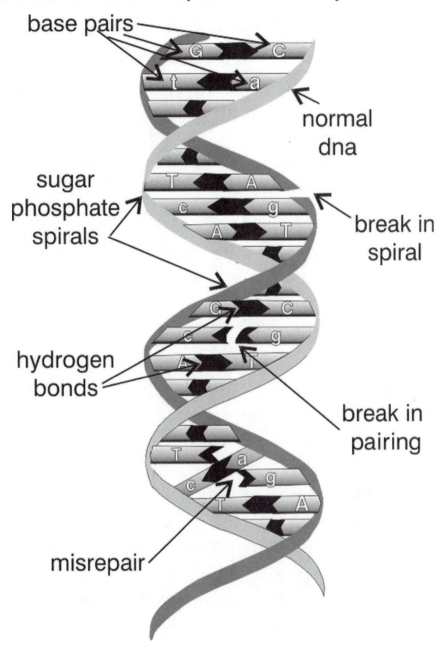

Table 12.6
Biological Effects of Short-Term, Whole-Body Radiation Exposure

Dose Equivalent In Rems	Effect
Less than 25	None
25–100	Temporary decrease in white blood cells
101–200	Permanent decrease in some white blood cells, nausea vomiting, lethargy
201–300	Nausea, vomiting, loss of appetite, diarrhea, lethargy
301–600	Initial vomiting, diarrhea, fever followed by latent period of several days to weeks; hemorrhaging, inflammation of mouth, and emaciation lead to death in approximately 50% of those receiving this dose equivalent
Greater than 600	Accelerated effects of the above, with death in nearly 100% of those receiving this dose equivalent

Figure 12.12
Annual human exposure to radiation in millirems. Note that although radon is radiation from the Earth, it is displayed as a separate category.

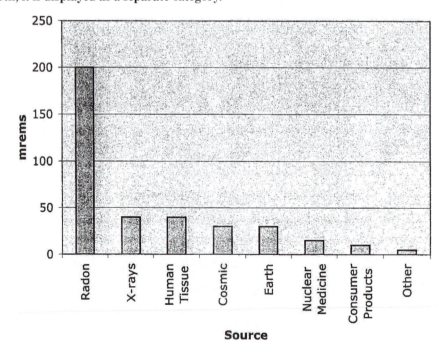

hundred milirems. Human exposure to radiation comes from both natural and human-made sources. Natural radiation is referred to as **background radiation.** It consists mainly of radon, which emanates from radioactive elements in the Earth's crust, and cosmic rays (Figure 12.12). The main human source of radiation exposure is medical x-rays. All humans carry a certain amount of radioactive material inside their bodies. Potassium, as the isotope K-40, is responsible for nearly half of the exposure humans receive from consumed food. Other sources of human exposure include C-14 and radium, which acts like calcium in the body. Radioactive isotopes are incorporated into bone and other tissues, along with stable isotopes.

Radiation exposure from both natural and human-made sources varies widely. Background radiation depends on the local geology and elevation. Areas where radioactive rocks are located close to the surface or where mining has exposed mineral deposits have higher background levels. Higher background levels also exist at higher elevations because there is less atmosphere to screen out cosmic radiation. Patients being treated with radiation therapy may have exposures of several hundred thousand milirems. Exposures of this magnitude involve radiation focused on the diseased tissue, rather than whole-body exposure.

Summary

Chemical change involves the interaction of the valence electrons in atoms as bonds are broken and formed. Even though chemical change creates new substances, atoms of elements are conserved. Nuclear changes result in the transformation of one element into another. These changes also involve a number of elementary particles and energy. Although a sea of radiation surrounds humans, the word "radiation" is often associated with negative images. Humans have used natural and synthesized radioactive isotopes for national defense, medicine, basic research, to produce energy, and in numerous commercial products. For example, smoke detectors contain a source of alpha-emitting americium, and gas lantern mantles contain radioactive thorium. Luminescent exit signs found in many schools, movie theaters, and airlines contain tubes filled with tritium. As with any technology, the benefits of nuclear applications must be weighed against their costs. This is particularly true with radioactive materials. As the use of these materials continues, scientists will also seek to control their deleterious effects.

In the future, nuclear power may be taken for granted, and our generation's burning of fossil fuels may be viewed as primitive. For now, the use of nuclear technology is in its infancy. In *Star Trek,* the starship *Enterprise* relies on two forms of nuclear power to move it through the universe. Fusion power is used for the impulse engines to achieve speeds below the speed of light. Warp speeds, which are faster than the speed of light, are obtained by using energy obtained from a reaction between matter and antimatter. Protons and antiprotons are used to power the *Enterprise*'s warp drive. While today fusion power as a practical energy source is science fiction, it may one day be a viable source of energy.

13

Electricity

Introduction

Whenever you flip a light switch, start a car, or use an appliance, you deal with some aspect of electricity. Many of our daily activities use some form of electricity. It is hard to imagine life in a world without electricity. Yet, electrical power has only been common in the United States for the past century and still is a luxury in many parts of the world. While we tend to think of electricity in terms of its practical applications, electricity also plays a primary role in the physiology of many animals including humans. Electrical impulses stimulate nerves and relay messages to the brain. Two hundred years ago it was recognized that electricity and magnetism were related properties that could be unified under a single concept called electromagnetism. Before examining electromagnetism in the next chapter, this chapter will introduce the basic principles of electricity and some of its applications in our lives.

History

The phenomenon called electricity has been observed since the earliest days of humans. Lightning struck fear in our prehistoric ancestors and led to supernatural explanations for the displays accepted today as natural atmospheric occurrences. The ancient Greeks observed that when a tree resin they called *elektron* hardened into amber, it possessed a special attractive property. When amber was rubbed with fur or hair, it attracted certain small objects. The Greek Thales observed this attractive property and noted the same property existed in certain rocks. Thales' attractive rock was lodestone (magnetite), which is an iron-bearing mineral with natural magnetic properties. The modern study of electricity commenced with the publication of William Gilbert's (1544–1603) *De Magnete* in 1600. Gilbert coined the word "electricity" from the Greek *elektron* and proposed that the Earth acted as a giant magnet.

Several individuals made significant contributions during the seventeenth and eighteenth centuries to the science of electricity. Otto von Guericke (1602–1686) created one of the first **electrostatic** generators by building a sphere of sulfur that could be rotated to accumulate and hold an electric charge. Numerous devices were built to generate static electricity through friction, and other instruments in the form of plates, jars, and probes were constructed to hold and transfer charge (Figure 13.1). One of the most popular of these was the Leyden jar, which was perfected at the University of Leyden by the Dutch physicist Pieter van Musschenbroek (1692–1761). The original Leyden jar was a globular-shaped glass container filled with water and fitted with an insulated stopper. A nail or wire extended through the stopper into the water. The Leyden jar was charged by contacting the end of the protruding nail with an electrostatic generator. Leyden jars are still used today in science laboratories.

Charles Du Fay (1698–1739) observed that charged objects both attracted and repelled objects and explained this by proposing that electricity consisted of two different kinds of fluids. Du Fay called the two fluids vitreous and resinous electricity and said these two different fluids attracted each other, but each repelled itself. Benjamin Franklin (1706–1790) considered electricity a single fluid that was either present in an object, giving it positive charge, or absent, creating negative charge. Franklin said that the electrical fluid flowed from positive to negative, thereby establishing the practice of defining electricity as flowing between positive and negative poles.

Electrostatic generators provide a brief movement of charge. The creation of continuous movement of charge, known as an electric current, began at the end of the eighteenth century and grew out of conflicting theories supported by two prominent Italian scientists: Luigi Galvani (1737–1798) and Alessandro Volta (1745–1827). Galvani and Volta held opposite views on the subject of animal electricity. Galvani, physician and professor of anatomy at the University of Bologna, devoted most of his life to the study of the physiology of muscle and nerve response to electrical stimuli. Galvani's work was a natural consequence of his profession and the period in which he lived. Electrical shock was a popular therapeutic treat-

Figure 13.1
Charge devices. Numerous devices were used to generate and hold charge: a, Guericke's generator, b, Leyden jar, c, Wimhurst generator.

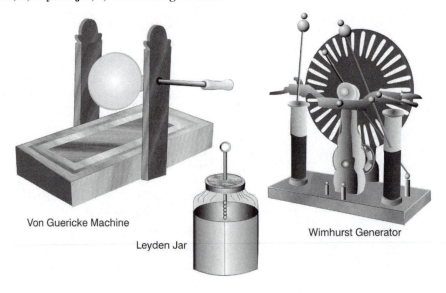

Von Guericke Machine

Leyden Jar

Wimhurst Generator

ment in the late eighteenth century. Patients were subjected to static electrical shocks as a means to treat all sorts of ailments including gout, rheumatism, infertility, toothaches, and mental disorders. The area for which this "shock therapy" seemed to hold the most promise was in treating paralysis. The hope of reversing paralysis grew out of experiments in which dissected body parts of animals responded to electrical discharge through them. Within this context, Galvani and other researchers sought a theoretical explanation for the contraction of muscles when nerves were electrically stimulated.

Galvani conducted hundreds of experiments starting in the late 1770s. These experiments consisted of using dissected body parts and subjecting them to charge held on a device such as a Leyden jar (Figure 13.1). Researchers would use various conductors and configurations to transfer the electricity to the tissue and note the response of the limb. A favorite object was frog legs and the attached spinal cord. Over the years, Galvani developed his theory of animal electricity. He explained the response of limbs to static electricity by proposing that animals contained a special electrical fluid within their muscles or nerves. According to Galvani, limbs moved when the animal electricity flowed between areas of positive and negative accumulation.

Galvani published his theory on animal electricity in 1792. Alessandro Volta, a physics professor at the University of Pavia, was a respected colleague of Galvani who initially supported Galvani's theory but raised questions on the work soon after its publication. In addition to questioning Galvani's animal electricity theory, Volta commenced his own series of experiments on the subject. Volta believed muscle contraction was not due to a flow of nervoelectrical fluid but was a result of an electrical flow between dissimilar metals used in the experiment. Volta's experiments demonstrated that the stimulation of nerves caused the contractions. A key part of Volta's experiments involved placing dissimilar metals along small segments of nerves, providing a connection between the metals, and observing that the entire limb responded when the connection was made. Volta's theory became known as the contact theory for animal electricity.

Galvani's death in 1798 left his theory to be defended by his proponents, and his theory was ultimately discredited, though modern science has revealed that Galvani was correct in believing that animal electricity could be generated from within the organism. Today it's known that electrical impulses are generated by chemical processes associated with neurons and chemicals known as neurotransmitters. Nerves can also be stimulated by external stimuli. In summary, although both Galvani's and Volta's ideas had flaws, both researchers provided fundamental knowledge on how muscles and nerves work and stimulated work on electrical currents.

The most important result of the Galvani-Volta controversy was a device that Volta created to study how dissimilar metals in contact with each other generated electricity. Volta's pile, which was actually the first electric battery, consisted of stacks of metal disks about the size of a quarter (Figure 13.2). Two different metals were used to make the disks, such as silver and zinc, and the disks were arranged so that one metal rested on another, forming a pair. Between each pair of metals Volta inserted a paper or felt disk that had been soaked in brine solution, which is a conductor. Volta's original pile contained about 20 silver-zinc pairs. Besides the pile configuration, Volta also used an arrangement called the crown of cups. The crown of cups consisted of nonconducting cups filled with a water or brine solution. The cups were connected by bimetallic conductors (Figure 13.3).

On March 20, 1800, Volta sent a letter concerning his electrical investigations to Sir Joseph Banks, president of the Royal Society of London. Included in the letter was a description of Volta's pile and his findings on how an electrical current could be generated by connecting metals. Volta observed that he received a noticeable shock when he placed two wet fingers on opposite ends of the pile. The same was true for his crown of cups, where he noted the shock progressively increased as the number of cups between his fingers increased.

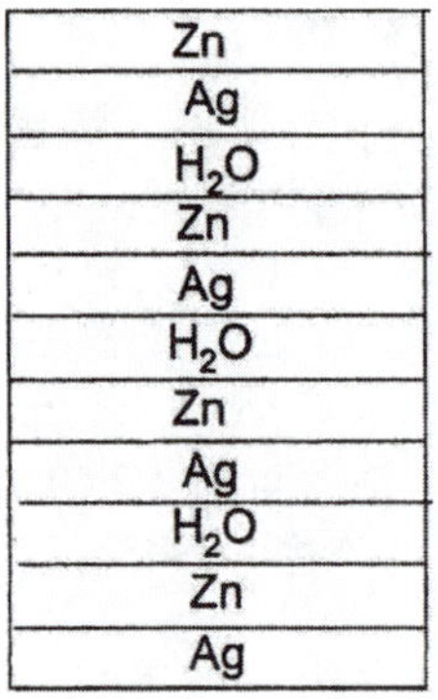

Even before its official publication, Volta's findings received considerable attention. Here was a device that could provide a steady, although small, current of electricity. Until this time, scientists used the process of generating static electricity and transferring this in pulses. Volta's results were published in English (the original letter was written in French) in *The Philosophical Magazine,* in September of 1800. Other scientists immediately constructed their own versions of Volta's pile to study this new type of electricity. One of the first experiments using a pile was done by William Nicholson (1753–1815) and Anthony Carlisle (1768–1840). They created a pile and used it to **electrolyze** water into hydrogen and oxygen.

Another important figure to emerge in the new science of electric current was Humphry Davy. Volta had attributed the current generated in his pile to the direct contact between metals. Volta's work focused on the physical relationship between the metals and, for the most part, neglected the chemical aspects of the electrical current. Davy believed a chemical reaction was the basis of the current produced in the piles or electrochemical cells. A cell is simply a container that holds chemicals needed to produce an electric current. Technically, a battery

Figure 13.2
The arrangement of disks in Volta's pile

| Zn |
| Ag |
| H_2O |
| Zn |
| Ag |
| H_2O |
| Zn |
| Ag |
| H_2O |
| Zn |
| Ag |

consists of two or more cells connected, but often the terms "battery" and "cell" are used interchangeably. Davy discovered that no current was generated in the cells if pure water was used, and better results were obtained when an acid solution was used to separate metals. Davy constructed a variety cells. He used cups of agate and gold and experimented with various metals and solutions to construct cells that were more powerful. Using a huge battery constructed by connecting over 200 cells, Davy applied current to various salts, and his work led to the discovery of a number of new elements. Potassium was discovered in 1807 when Davy ran current through potash, and several days later he isolated sodium from soda in a similar fashion. The next year he discovered magnesium, strontium, barium, and calcium.

The work of Davy was continued and expanded upon by the great English scientist Michael Faraday (1791–1867). Faraday's primary studies in electricity took place between 1833 and 1836. Faraday is responsible for providing much of our modern terminology associated with electrochemical cells. The terms **electrode, anode, cathode, electrolyte, anion, cation,** and **electrolysis** are all attributed to Faraday. Even more important than his qualitative description of

Figure 13.3
Volta's pile and crown of cups

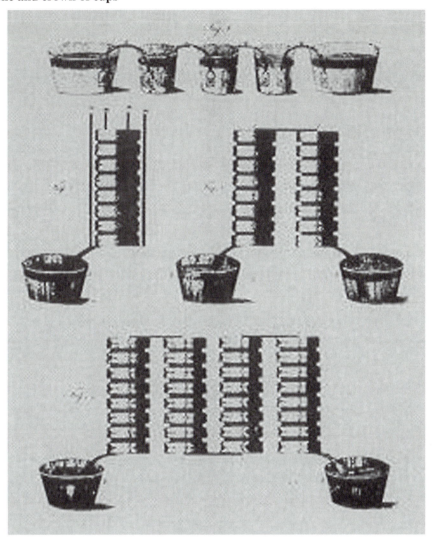

electrochemistry, Faraday did quantitative studies that led to his formulation of electrochemical laws. These laws provided a means to determine the relationship between current and the amount of materials reacting in an electrochemical reaction. Faraday's discoveries, especially those on electromagnetism, are still used today in quantifying electrical concepts and will be discussed in the next chapter.

Electrostatics

Electrostatics, or static electricity, involves charges at rest. The source of all electrical charge lies in atomic structure. The concentrated positive nucleus of an atom is surrounded by negatively charge electrons. In ordinary atoms, there are an equal number of positive protons and negative electrons resulting in a neutrally charged atom. The electrons surrounding an atom's nucleus are not fixed but can be added or removed and create ions. When electrons are added, there are more electrons than protons, creating a negatively charged atom called an **anion.** Removing electrons results in a positively charged atom called a **cation.** An object that is charged has an imbalance in the number of protons and electrons. The charge of an electron is the basic unit or quanta of charge, defined to have a value of e. The basic unit of charge, the electron, can be compared to a penny in our monetary system. Anything purchased can only be as small as 1¢, and any charge can only be as small as e. The SI unit for charge is the **coulomb,** abbreviated C, named after the French physicist Charles Augustin de Coulomb. The charge of an electron is equal to 1.60×10^{-19} C. This means a charge of 1 C is equal to the charge of 6.25×10^{18} electrons.

One way to create net charge is through friction. Charging by friction is called **triboelectricity.** Triboelectricity occurs when one object is rubbed with another object. Familiar examples are a person shuffling across a carpet and getting a shock when touching another person or piece of metal, running a comb through hair, and rubbing an amber rod with fur. In charging by friction, electrons are transferred between objects when they are rubbed together. For example, when a person wearing leather shoes shuffles across a carpet made of nylon, electrons are transferred from the person's leather shoes to the carpet. Nylon has a greater affinity for electrons, and thus, electrons will move from the shoes to the carpet. The negative charge lost by the shoes results in the shoes acquiring a net positive charge that will be distributed over the person's body. This excess charge can be transferred to a neutrally charged object, such as another person, when the person makes contact. It is important to realize that in charging by friction (and other processes

involving charge), charge is conserved. Charge is merely transferred between objects in triboelectricity. Conservation of charge is a fundamental law of physics, and in dealing with electricity it should be remembered that charge is never created or destroyed.

Substances have different affinities for electrons, and this dictates whether they will acquire a positive or negative charge when rubbed together. A tribolelectric series is a list of different substances in order of their affinity for electrons. Substances at the bottom of the series have greater affinity for electrons than those at the top. Therefore, when two objects made of substances listed in a triboelectric series are rubbed together, the object made of the substance lower in the series acquires the negative charge, and the other object a positive charge. A triboelectric series of a number of common substances is given in Table 13.1.

In addition to charging by friction, an object can be charged by contact or induction. Charging by conduction involves touching a charged object to another neutral object. Electrons are transferred between the charged object and neutral object at the point of contact. When the two objects are separated, the neutral object will have obtained a charge similar to that of the charging object. Thus, if a positively charged rod is placed in contact with a neutral object, the neutral object will acquire a positive charge.

Charging by induction involves bring a charged object close to another conducting object that is **grounded.** Grounding means providing a path for charge to flow to or from a large reservoir of charge. The Earth is such a reservoir, so grounding provides a path for charge to flow between the object and the Earth. In induction, the charged object attracts charge of the opposite sign in the conducting object and repels charge of the same sign. If the conducting object is grounded, the same sign charge will be

Table 13.1
Triboelectric Series

Substance
Human skin, dry
Leather
Rabbit fur
Glass
Hair
Nylon
Wool
Silk
Paper
Cotton
Iron
Wood
Hard rubber
Copper
Polyester
Polypropylene
PVC
Teflon

often become positively charged, while bottoms acquire a negative charge. The negative cloud bottoms induce a positive charge at the Earth's surface. As reservoirs of negative and positive charge collect, streamers of opposite charge leave the clouds and ground and form a lightning bolt. The lightning heats the air, causing it to emit light and expand rapidly. The rapid expansion creates thunder.

Conductors and Insulators

Substances through which charge readily moves are called **conductors,** and those that resist the movement of charge are called **insulators.** The ability of a substance to conduct electric charge has to do with its atomic structure. Electrons surround the nucleus of an atom in different shells (see chapter 10). Electrons in the outer shells are called **valence electrons** and take part in chemical bonding. In conductors, one to several of the valence electrons surrounding an atom are held very loosely and have the ability to move between adjacent atoms. In effect, these electrons are not uniquely associated with one atom but have the ability to move through the structure. They are referred to by several names including lose, conducting, or delocalized electrons.

Metals are excellent conductors because they are characterized by delocalized electrons in their atomic crystalline structure. For example, copper, which is commonly used in wire and other electrical devices, contains 11 valence electrons. Copper's crystalline structure consists of a copper atom bonded to 12 other copper atoms. The stable electron configuration for copper requires 18 electrons in its outer electron shell. This could be accomplished by each copper atom forming seven covalent bonds with neighboring atoms (each covalent bond consists of a pair of electrons shared by two atoms). Rather than forming seven covalent

transferred between the conducting object and the Earth. When the ground is removed, the conducting object will maintain the induced charge (Figure 13.4).

An example of where induced charge takes place on a large scale is during thunderstorms. Tops of clouds in thunderstorms

Figure 13.4
A neutral metal sphere can be charged by induction if it is grounded and a negatively charged rod is brought close to its surface. The negative rod attracts positive charge and repels negative charge through the ground. When the rod is removed and the ground disconnected, the sphere maintains a positive induced charge.

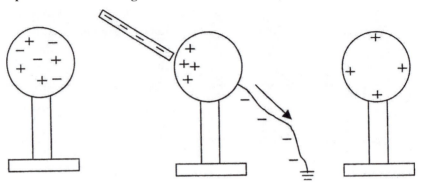

bonds, each copper atom forms five covalent bonds with five neighboring copper atoms, and the remaining valence electrons shift to form bonds with other neighboring atoms. In this manner, all 12 copper atoms are bonded using delocalized electrons. Delocalized electrons are not rigidly held by one atom but can be considered a sea of electrons with the ability to migrate through the metal. It is this ability to migrate that makes metals good conductors. The atomic structure associated with insulators doesn't contain delocalized valence electrons. The valence electrons in insulators, such as glass or wood, are strongly held by atoms and are generally not free to move through the substance.

The electrical conductivity of a material is given by its **resistance** or conductivity. Resistance is examined below in the section on electric current. One means of comparing the conductivity of different methods is with the International Annealed Copper Standard (IACS). In this system, a number precedes the abbreviation IACS and indicates the conductivity of the material compared to a standard copper wire. For example, the conductivity of tin is approximately 15 IACS, indicating it has about 15% the conductivity of copper. IACS numbers for several common metals are given in Table 13.2.

Two other classes of materials are called **semiconductors** and **superconductors.** A semiconductor is a substance that has electrical properties that fall between a conductor and insulator. Semiconductors have widespread use in the electronics industry due to their ability to conduct or insulate in different situations. Elements classified as semimetals, such as germanium and silicon, make good semiconductors. A superconductor is a material that conducts electricity without resistance below a certain temperature. In theory, absolute superconductivity would allow electricity to flow forever in a circuit. Initial studies in superconductivity required temperatures approaching absolute zero, but recent advances have been made to achieve superconductivity at higher temperatures of approximately 100 K.

Coulomb's Law and the Electric Field

Two charged objects will exert a force on each other that is attractive when the charges have opposite signs, and repulsive

Table 13.2
IACS Conductivities

Material	IACS Number
Silver	105
Copper	100
Gold	70
Aluminum	61
Nickel	22
Zinc	27
Brass	27
Iron	17
Tin	15
Lead	7

when the charges have the same sign. The relationship between two charged objects is given by Coulomb's law. Coulomb demonstrated that the force between charged objects is directly proportional to the magnitude of their charges and inversely proportional to the distance separating the charges squared. If the two objects are considered point charges (each object is considered to occupy a point in space), with charge magnitudes given by q_1 and q_2 separated by a distance r, then Coulomb's law can be stated mathematically as

$$F_{electrical} = 8.99 \times 10^9 \frac{\text{N-m}^2}{\text{C}^2} \frac{q_1 q_2}{r^2}.$$

When charges q_1 and q_2 are expressed in coulombs and the distance between charges in meters, then the force will be in newtons.

A charged object affects other charged objects in the immediate region surrounding it by exerting a force according to Coulomb's law. In order to describe how a charged object affects the region surrounding it, the region can be described in terms of an electric **field** (or electric force field). Electric fields are defined with respect to a test charge symbolized by q_o. If the force on this test charge at a point in space is F, then the electric field, designated E, is given by F/q_o. A simple example to illustrate an electric field involves the field around a point charge using a small positive test charge. If the point charge is positive, then the electric field can be represented with a set of force vectors (lines of force) pointing radially away from the positive charge (Figure 13.5). For a negative point charge, the electric field would be represented by vectors pointing radially inward (Figure 13.5). Diagrams such as Figure 13.5 are referred to as field maps. In addition to the direction of the force, the lines of force indicate the strength of the electric field. The strength is greatest where the lines are closest together. As seen in Figure 13.5, the electric field is strongest closest to the charge, as expected. Lines of force never cross each other and can be mapped for any combination of charge configurations. When mapping a dipole of opposite charges, curved lines are obtained (Figure 13.6). It should be noted that the electric filed maps are depicted in two dimensions, but the force field extends in three dimensions.

The Electric Potential

In chapter 5 it was seen that a mass has gravitational potential energy due to its position in a gravitational field, and work is equal to the change in gravitational potential energy. In order to lift a mass, work is required. Electrical potential energy exists by virtue of the position of charge in an electrical

Figure 13.5
Lines of force can be used to represent the electric field around point charges

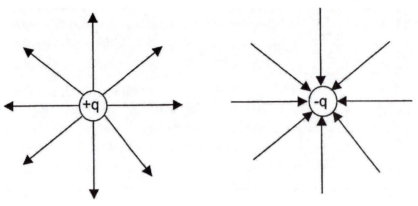

Figure 13.6
Electric field map for point charge and dipoles

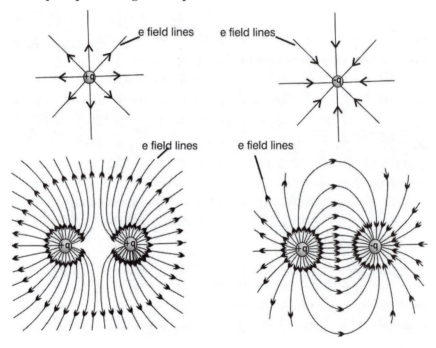

field. For example, consider a positive test charge situated between a positive plate and a negative plate. The positive test charge is attracted to the negative plate (and repelled from the positive plate), just as a mass is attracted to the Earth. Work is required to move the positive test charge away from the negative plate. Since work equals force times distance, the electrical work equals $-q_0Ed$ (E is the electric field, q_0 the test charge, and d the displacement). The negative sign indicates that the displacement and force are in

Figure 13.7
Gravitational and electrical potential energy are analogous concepts. In a gravitational field, work is done against the gravitational field to lift a mass. In an electrical field, work is done to separate a test charge, q_o, from the negative plate.

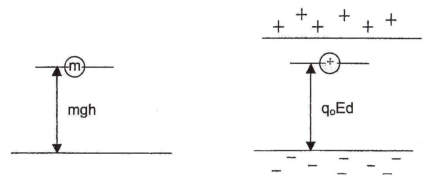

opposite directions. The increase in electrical potential energy in separating the charge from the plate is q_oEd. In this example, it is assumed that the electrical potential at the surface of the plate is zero. The equation for electrical potential energy is analogous to gravitational potential energy (Figure 13.7). In the latter case the work done against the gravitational field when a mass, m, is lifted is $-mgh$, and the mass acquires a gravitational potential energy of mgh. The electrical potential energy increases by q_oEd when a test charge is separated from a negative plate defined as having a potential energy of zero. In this example, the negative plate was defined as having an electrical potential energy of zero. In many cases, zero electrical potential energy is considered zero at the Earth's surface. It should be remembered that just as gravitational potential energy can be defined using any reference level, that electrical potential energy could use any reference. In many cases, the Earth's surface is the most logical reference level to establish zero gravitation potential energy, and ground to establish electrical potential energy.

The electrical potential energy depends on the value of the test charge. More charge will require a greater amount of work to move against the electrical field, just as it takes a greater amount of work to move a greater mass against a gravitational field. The amount of electrical potential energy divided by the charge is defined as the **electrical potential** at that point in the electric field. The unit for electrical potential energy is the joule and for charge is the coulomb. The SI unit for electrical potential is joules per coulomb and is defined as the volt, named in honor of Volta. A 12 V battery has the potential to give 12 J of energy to every coulomb of charge flowing through the battery. Another way of interpreting electrical potential is in terms of the amount of work. If it takes 30 J of work to move 5 C of charge in an electric field, then the potential at that point is -6 V (Figure 13.8).

Electric Current and Circuits

Most of our experience with electricity involves the flow of charge used to power the myriad of appliances, tools, electronics, and lights used throughout the day. Electric current involves the flow of charge. A charged object quickly discharges when connected

Figure 13.8
It takes 30 J of work to separate –5 C from the positive charge. The –5 C has an electrical potential of –6 V.

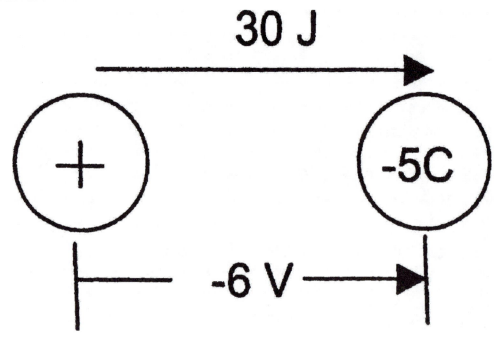

by a conductor to the ground. In this situation, there is a difference in electrical potential between the charged object and ground, and current flows for a short duration. If there were a means to maintain a net charge on the object and a difference in electrical potential, the electric current would flow continuously. Batteries, power supplies, and power plants are various means by which a difference in electrical potential is maintained to supply a continuous current.

The flow of electric current can be compared to the flow of water. The difference in electrical potential measured in volts is analogous to pressure difference in water flow. Just as a difference in water pressure in a hose causes water to flow from high to low pressure, a potential difference causes electrical charge to flow in a wire from high potential to low potential. The pressure difference for water is typically maintained

by a pump, and the potential difference for charge is often maintained by a battery. The flow of charge consists of the movement of electrons or ions through a conducting material. An ion is a charged species created when an atom or group of atoms loses or gains electrons. For example, sodium becomes a positive sodium ion, Na^+, when it loses an electron, a calcium atom becomes Ca^{2+} when it loses two electrons, and when chlorine gains an electron it becomes the negative chloride ion, Cl^-. Positive ions are called cations, and negative ions are anions. There are also **polyatomic ions** such as carbonate, CO_3^{2-}, and ammonium, NH_4^+.

Electric current is the flow of charge across a conductor per unit time. In solid conductors, the charge carriers are free electrons. Liquids that conduct electricity are called electrolytes, and those that do not conduct electricity are nonelectrolytes.

Electrolytes disassociate into positive and negative ions in solution, and these are the charge carriers. Electric current is measured in **amperes,** or simply amps, abbreviated A. This unit is named in honor of André-Marie Ampère (1775–1836). Ampère made several key discoveries in the early nineteenth century regarding electric current. One of his most important discoveries was that current flowing through a coil produces magnetism. An ampere is equal to the flow of 1 C/s. Using the definition of a coulomb, this means that 1 A equals a flow of charge equivalent to about 6.25×10^{18} electrons per second. Current can be classified as direct current when it flows in one direction, such as from a battery, or as alternating current. In alternating current, the current continually reverses direction. Standard alternating current in North America available from wall outlets has a frequency of 60 cycles per second. Current in many parts of the world has a frequency of 50 cycles per second.

The electrical current flowing through a conductor depends on the electrical potential and on the resistance in the conductor to the flow of charge. Using the water analogy, electrical resistance can be compared to friction in flow through a pipe. A smooth empty pipe offers little resistance to the flow of water, while a rough pipe containing debris will slow water moving through it. Electrical resistance is related to the interaction of conducting electrons as they move through the conductor from atom to atom. During this process the conducting electrons interact with atoms, giving up some of their energy. This process is known as dissipation of energy. Energy dissipation transfers kinetic energy from the conducting electrons to the atoms in the conductor. This in turn results in heating, similar to the process where friction results in heating.

Table 13.3
Resistivities of Common Materials

Material	Resistivity (Ω-m)
Silver	1.6×10^{-8}
Copper	1.7×10^{-8}
Gold	2.3×10^{-8}
Iron	10.5×10^{-8}
Tin	11.3×10^{-8}
Graphite	3.0×10^{-5}
Quartz	1.4×10^{16}
Rubber	10^{13} to 10^{16}

Resistance is measured in **ohms,** named in honor of Georg Simon Ohm (1789–1854). The Greek capital letter omega, Ω, is the accepted symbol for ohms. The resistance of many materials can be related to the type of material, its cross-sectional area, and its length, according to the equation $R = \rho L/A$, where R is the resistance, ρ the **resistivity,** A the cross-sectional area, and L the length. The resistivity, measured in ohms-meters, depends on the material, as compared to resistance, which depends not only on the material but its dimensions. Conductors have low resistivities and insulators high resistivities (Table 13.3). In dealing with water and other liquids, the resistance is expressed in terms of conductivity. Conductivity is the reciprocal of resistivity. The unit of conductivity is $1/\Omega$-m. The reciprocal of ohms is called mho, from ohm spelled backward. A mho is equivalent to the SI unit siemens, abbreviated S. Therefore, the proper SI unit for conductivity is

S/m. A term often confused with conductivity is conductance. **Conductance** is the reciprocal of resistance and is expressed in mhos or siemens.

The relationship between electrical potential, current, and resistance for many conductors follows **Ohm's law.** Ohm discovered in the early 1800s that the current through metal was directly proportional to the electrical potential across the metal. Ohm's discovery led to the idea of resistance in circuits. When electrical potential is plotted on the x axis, and current plotted on the y axis, a straight line with slope equal to $1/R$ results, where R equals the resistance of the circuit. Ohm's law expressed in equation form is $V = IR$, where V is the electrical potential in volts, I the current in amperes, and R the resistance in ohms. A simple example of Ohm's law takes place in a flashlight. The light bulb in the flashlight is the resistor. If the bulb has a resistance of 9 Ω and two 1.5 V batteries are used (giving a total voltage of 3 V), then 3 A flows through the circuit when the flashlight is turned on: $9 \, \Omega \div 3 \, V = 3 \, A$.

Ohm's law can be applied to many different circuits containing various configuration of resistors and power sources. In this section, several simple direct current circuits will be presented to illustrate the use of Ohm's law with parallel and series circuits. A circuit is an arrangement of components that allows electricity to flow continuously for some useful purpose. Circuits can be diagrammed using schematics that employ standard symbols for electrical components. Several standard symbols for these are given in Table 13.4.

Even though the charge carriers in solid conductors such as metal wires are electrons, conventional current refers to the flow of positive charge. Positive charge moves in the direction opposite to the flow of electrons. The movement of electrons and the

Table 13.4
Circuit Symbols

Symbol	Component
⊣⊢	Battery
⟋ᴡ⟍	Resistor
⚬	Open switch
●—○	Closed switch
⏚	Ground
Ⓟ	Lamp

movement of positive charge are equivalent descriptions for electric current. This can be seen by referring to Figure 13.9. The convention of using the movement of positive charge dates back to the 1700s when two types of fluids were thought to be responsible for electricity. Vitreous fluid (related to glass) and resinous fluid (related to resin) were based on the charge acquired when these substances were rubbed with silk and wool, respectively. Benjamin Franklin (1706–1790) suggested that vitreous flow be considered positive and that current be based on the flow of the vitreous fluid, while resinous fluid was fixed. The discovery of electrons didn't occur until the start of the twentieth century, and it was determined that the flow of electrons represented the resinous fluid. By this time, the convention of positive vitreous flow already had an 150-year tradition. The use of conventional current means that it is assumed that positive charge flows

Figure 13.9
Starting with two neutral objects, the transfer of an electron results in each object acquiring a charge. The transfer of an electron in one direction is equivalent to the transfer of positive charge in the opposite direction.

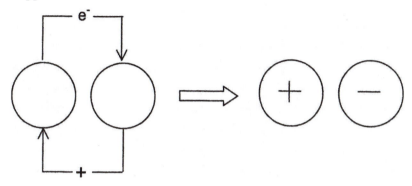

from the positive to negative terminal in a battery, when in reality electrons are flowing from the negative to positive terminal.

Using circuit symbols and conventional current, a schematic for a flashlight is shown in Figure 13.10. The two batteries are arranged so that the positive terminal of one battery is connected to the negative terminal of the next battery. Batteries connected positive to negative are said to be connected in series. In a series connection, the same amount of current passes through the components connected in series. If each battery in Figure 13.10 is a normal D-size battery, it has an electrical potential of 1.5 V. The positive terminal of the battery is represented by the thinner, taller vertical line. The voltage increase across the first battery is 1.5 V. Since the positive terminal of battery 1 is connected to the negative terminal of battery 2, they have the same potential. The overall electrical potential supplied by the two batteries is 3 V. Each battery can be compared to a pump used to raise water. Battery 1 pumps up the potential 1.5 V, and battery 2 pumps it up an additional 1.5 V. If the lamp in the circuit has a resistance of 10 Ω, then according to Ohm's law the

current equals 0.3 A. The voltage drop across the lamp is 3 V.

The previous example with one bulb can be extended to include two bulbs, as shown in Figure 13.11. The two bulbs in Figure 13.11 are connected in series. Current flows around the circuit, and voltage drops occur across each lamp, which act as resistors. Resistors in series are added to give a total resistance. If each lamp has a resistance of 10 Ω, then the total resistance of the circuit is 20 Ω. According to Ohm's law the current will be 0.15 A. The voltage drop across each lamp is 1.5 V.

One of the disadvantages of series circuits is that when one component fails, an open circuit results and the entire circuit fails. This is what happens when one bulb in a string of older type Christmas tree lights burns out. One way to alleviate this problem is by connecting components in parallel, as done when lights and outlets in a house are wired. When batteries are connected in parallel, the positive and negative terminals are connected to each other (Figure 13.12). Resistors and other components can also be connected in parallel. Parallel circuits are characterized by branches similar to a

Figure 13.10
Schematic for a direct current circuit in flashlight. Batteries have been labeled 1 and 2.

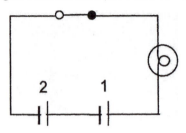

Figure 13.11
A simple direct current circuit with two batteries and two lamps connected in series

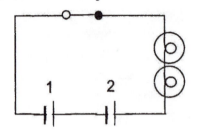

Figure 13.12
Two batteries and three resistors connected in parallel

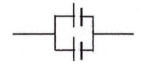
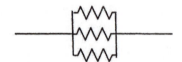

river dividing into several streams and then rejoining. The total current divides between the branches, and each branch has the same potential difference across it. The voltage drop across two identical batteries connected in parallel will be the same as the batteries, for example, 1.5 V for two standard D cells. Resistors connected in parallel can be combined to produce a total effective resistance for circuit calculations. For example, consider that the three resistors in Figure 13.12 have resistances of 2 Ω, 3 Ω, and 6 Ω and are connected to a 6 V battery. The voltage drop across each resistor will be 6 V. The amount of current flowing through each resistor can be calculated using Ohm's law. For the 2 Ω resistor the current is 3 A, for the 3 Ω resistor it is 2 A, and for the 6 Ω resistor it is 1 A. The total current through the three resistors is 6 A. The effective resistance of resistors in parallel is given by the formula: $1/R = 1/R_1 + 1/R_2 + 1/R_3 + \dots$. In this equation, R is effective resistance equal to the single resistance that can be used in place of the individual resistances. The effective resistance for a parallel resistors of 2, 3, and 6 Ω would equal 1 Ω ($1/1 = 1/2 + 1/3 + 1/6$). A circuit with a 6 V battery and resistance of 1 Ω gives a current of 6 A. This agrees with the previous calculation where the current through each resistor was determined and summed to give 6 A. It should be emphasized that throughout the previous discussion the resistance of wires and switches was considered negligible and ignored. The amounts of resistance-conducting devices added to an overall circuit can generally be ignored.

The examples on currents and circuits in this section have involved direct current. This is the type of current supplied by batteries and used to power small portable devices such as CD players and radios. Standard outlets throughout the United States use 120 V alternating current operating at 60 Hz. Many electrical devices depend on current flowing through them, and it doesn't matter if the current flows in one steady direction or

Table 13.5
Typical Power Ratings in Watts for Appliances

Clock radio	5	microwave	1,500
Clothes dryer	4,000	oven	5,000
Coffeemaker	1,000	refrigerator	500
Dishwasher	1,200	television	300
Garbage disposal	400	toaster	1,200
Hair dryer	1,200	washing machine	600
Iron	400	water heater	5,000
Room air conditioner	1,000	central air conditioner	3,000

alternates. Many devices employ a converter that plugs into a wall outlet to change alternating current to direct current. Converters employ various electrical devices such as capacitors, diodes, and transformers that allow the passage of charge in one direction while smoothing the current to produce a steady voltage.

Electric Power

Power is the rate at which energy changes or work is done. The SI unit for power is the watt. A watt is equivalent to 1 J/s or 1 A-V. Electric power, P, is given by the equation $P = IV$, where I is the current in amps and V the electrical potential in volts. Power in a resistor is dissipated in the form of heat. Using Ohm's law, the power in a resistor is given by $P = I^2R = V^2/R$. Many appliances such as toasters, irons, and incandescent light bulbs act as resistors, and the power equation for resistors can be applied to these appliances. For a 60 W light, the current would equal 60 W/120 V = 0.5 A. The resistance of the light would be 60 W/(0.5 A)^2 = 240 Ω.

The power rating on an appliance indicates the rate at which it draws electrical energy. Typical values for various appliances are given in Table 13.5. Using the power ratings, the amount of current an appliance draws can be calculated. This is important for establishing circuits in homes, which are typically wired to accommodate 20 A. If too many appliances are run off one circuit, an overload occurs, and a breaker is tripped or fuse blown. This subject is addressed more fully at the end of the chapter in the section on electrical safety.

While power is measured in watts, an electric bill is based on energy usage and is typically measured in kilowatts-hours, abbreviated kW-h. One kilowatt-hour is equivalent to the expenditure of 1,000 W for 1 h. Since 1 W is equal to 1 J/s, 1 kW-h equals 3.6×10^6 J.

Batteries

A battery consists of two or more cells connected to deliver a specific charge, although the terms "battery" and "cell" are

Figure 13.13
Dry cell battery

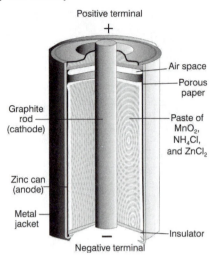

Figure 13.14
Lead storage battery

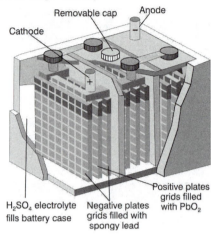

often used synonymously. Batteries are one of the most common examples of electricity associated with everyday life. Batteries, which come in all sizes and shapes, are devices that produce a difference in electrical potential from a chemical reaction. When connected in a closed circuit, the terminals of the battery are at different electrical potentials, and electrons flow from the negative terminal to the positive terminal. The negative terminal is referred to as the anode and the positive terminal the cathode. The conducting material within the battery that separates the anode and cathode is the electrolyte. Batteries differ in the materials used for the cathode, anode, and electrolyte and how these materials are arranged to make a battery. The standard dry cell battery or alkaline cell is shown in Figure 13.13. Dry cells are classified as **primary cells** because the chemicals in them can't be regenerated, in other words, primary cells can't be recharged. The dry cell battery was patented by the French engineer George Leclanché (1839–1882) in 1866. The standard dry cell consists of a zinc container that acts as the anode. The container contains an electrolyte

containing a mixture of ammonium chloride (NH_4Cl), zinc chloride ($ZnCl_2$), and starch. A carbon (graphite) rod runs down the axis of the cell. Surrounding the rod is a mixture of manganese dioxide (MnO_2) and carbon black (carbon black is carbon in the form of soot produced from burning kerosene or another hydrocarbon with insufficient oxygen). The MnO_2–carbon black mixture acts as the cathode. The MnO_2–carbon black and electrolyte exist as a moist paste, so a dry cell is not actual dry. While several reactions occur in a dry cell, the primary reactions governing its operation can be represented as

$$\text{Anode: } Zn_{(s)} \rightarrow Zn^{2+}_{(aq)} + 2e^-$$

$$\text{Cathode: } 2MnO_{2(s)} + 2NH_4^+{}_{(aq)} + 2e^- \rightarrow$$
$$Mn_2O_{3(s)} + 2NH_{3(aq)} + H_2O_{(l)}$$

$$\text{Overall: } Zn_{(s)} + 2NH_4^+{}_{(aq)} + 2MnO_{2(s)} \rightarrow$$
$$Zn^{2+}_{(aq)} + 2NH_{3(aq)} + H_2O_{(l)} + Mn_2O_{3(s)}$$

When the circuit containing the cell is closed, electrons from the oxidation of zinc flow from the negative terminal through the object being powered to the negative carbon rod that serves as an electron collector. Several reduction reactions occur at the collector; one main reaction is represented in the reactions above. The dry cell with an ammonium chloride electrolyte creates acidic conditions in the battery. The acid attacks the zinc, which accelerates its deterioration. To extend the life of a dry cell, a basic substance, typically potassium hydroxide or sodium hydroxide, can be used for the electrolyte. Alkaline cells derive their name because they contain an alkaline electrolyte.

Mercury cells, like dry cells, have a zinc anode but use a mercuric oxide (HgO) cathode. The electrolyte is potassium hydroxide, KOH. These small, flat, metallic cells are widely used in watches, calculators, cameras, hearing aids, and other applications where size is at a premium. The reactions in the mercury cell are

$$\text{Anode: } Zn_{(s)} + 2OH^-_{(aq)} \rightarrow ZnO_{(s)} + H_2O_{(l)} + 2e^-$$

$$\text{Cathode: } HgO_{(s)} + H_2O_{(l)} + 2e^- \rightarrow Hg_{(l)} + 2OH^-_{(aq)}$$

$$\text{Overall: } Zn_{(s)} + HgO_{(s)} \rightarrow ZnO_{(s)} + Hg_{(l)}$$

The voltage obtained from a mercury cell is about 1.35 V.

A common battery is the lead storage battery. This battery, composed of several cells, is used mainly in cars. The lead storage battery, the first **secondary** or rechargeable cell, was invented in 1859 by the French physicist Gaston Planté (1834–1839). A lead storage battery consists of a series of plates called grids immersed in a solution of sulfuric acid (approximately 6 M) that serves as the electrolyte (Figure 13.14). Half the

plates contain lead and serve as the anode of the cell, and the remaining plates are filled with lead dioxide (PbO_2) and serve as the cathode. The following reactions represent the **oxidation** and **reduction** reactions:

$$\text{Anode: } Pb_{(s)} + HSO_{4\,(aq)} \rightarrow PbSO_{4(s)} + H^+ + 2e^-$$

$$\text{Cathode: } PbO_{2(s)} + HSO_{4\,(aq)} + 3H^+_{(aq)} + 2e^- \rightarrow PbSO_{4(s)} + 2H_2O_{(l)}$$

$$\text{Overall: } Pb_{(s)} + PbO_{2(s)} + 2H^+_{(aq)} + 2HSO4^-_{(aq)} \rightarrow 2PbSO_{4(s)} + 2H_2O_{(l)}$$

The cell reaction generates a potential of about 2 V. A car battery is created by connecting six cells in series. The reactions show that electrons are created when lead oxidizes at the anode. The lead ions formed in oxidation combine with the sulfate ions in the electrolyte to form lead sulfate. The electrons flow through the car's electrical system back to the battery's cathode plates. Here the electrons combine with PbO_2 and hydrogen ions present in the acid and form lead sulfate. While driving, the car's alternator continually recharges a battery. The alternator carries out an electrolytic process and reverses the above reactions. In this manner the lead sulfate generated on the plates is converted back into lead and lead dioxide.

Advances in the past few decades have improved car battery technology immensely. Many lead batteries currently manufactured are labeled as maintenance free. This refers to the fact that the acid level in them does not have to be checked. The addition of water to a lead battery is necessary because the charging process causes water to undergo **electrolysis.** This process creates hydrogen and oxygen gas that escapes from the battery vents. Modern battery technology has greatly reduced and even eliminated this problem by perfecting new types of electrodes, for example, lead-calcium and lead-selenium electrodes. Maintenance-free batteries are sealed, and water does not have to be added to them. It was generally recommended that the fluid level in batteries be checked on a regular basis. Additionally, a hydrometer could be used to check the density of the fluid in the battery to indicate its condition. Many manufacturers now claim a 10- to 15-year lifetime for batteries as opposed to the typical 3- to 5-year lifetime of 20 years ago. While improved technology has produced batteries that are more reliable over the years, you still may experience a starting problem on a cold day. The inability to start a vehicle on a cold day results from charge being impeded as the electrolyte turns viscous in the cold conditions.

Another familiar secondary cell is the ni-cad, or nickel-cadmium, cell. In these cells, the anode is cadmium, and the cathode consists of a nickel compound ($NiO(OH)_{(s)}$). One of the problems with ni-cad cells is the high toxicity of cadmium. An alternative to ni-cad cells currently gaining a foothold in the market is the lithium ion cell. Lithium is much less toxic than cadmium. Additionally, lithium ion cells don't suffer from the memory effect of ni-cad cells. This effect results in the loss of efficiency when ni-cad cells are recharged before they completely lose their charge. Besides these factors, lithium has several properties that make it a highly desirable material for use in batteries. Lithium has a very low reduction potential. This is equivalent to saying it has a very high oxidation potential. Lithium's high potential means that a greater voltage can be obtained using lithium cells. Another property of lithium is that it is light, which is especially important for portable applications such as cell phones and laptop computers. A final advantage of lithium ions is that they are

Figure 13.15
Schematic of hydrogen-oxygen fuel cell

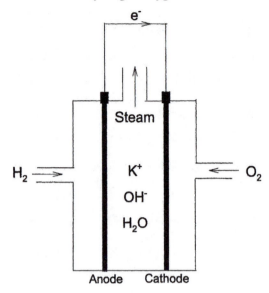

small and move more readily through materials, as opposed to larger ions.

Original lithium cells employed lithium metal as the anode, and these have been used in the military and space program for years. Lithium, though, is highly reactive, presenting fire and explosive hazards. Lithium ion cells retain the advantages of original lithium batteries but eliminate the hazards associated with them. In lithium ion cells the lithium metal anode is replace by a substance with the ability to absorb lithium ions, such as graphite. The cathode is made of a lithium compound such as $LiCoO_2$, $LiSO_2$, or $LiMnO_2$. The electrolyte separating the electrodes allows the passage of lithium ions but prevents the flow of electrons. The electrons move from the anode to cathode through the external circuit.

Numerous other types of cells exist, such as zinc-air, aluminum-air, sodium-sulfur, and nickel–metal hydride (NiMH). Companies are on a continual quest to develop cells for better batteries for a wide range of applications. Each battery must be evaluated with respect to its intended use and factors such as size, cost, safety, shelf life, charging characteristics, and voltage. As the twenty-first century unfolds, cells seem to be playing an ever-increasing role in society. Much of this is due to advances in the consumer electronics and the computer industries, but there have also been demands in numerous other areas, including battery-powered tools, remote data collection, transportation (electric vehicles), and medicine.

One area of research that will advance in years to come is the study of **fuel cells.** The first fuel cell was actually produced in 1839 by Sir William Grove (1811–1896). A fuel cell is essentially a battery in which the chemicals are continuously supplied from an external source. One simple type of fuel cell is the hydrogen-oxygen fuel cell (Figure 13.15). These cells were used in the Gemini and Apollo space programs. In the hydrogen-oxygen fuel cells, the reactions at the anode and cathode are

Anode: $2H_{2(g)} + 4OH^-_{(aq)} \rightarrow 4H_2O_{(l)} + 4e^-$

Cathode: $O_{2(g)} + 2H_2O_{(l)} + 4e^- \rightarrow 4OH^-_{(aq)}$

Overall: $2H_{2(g)} + O_{2(g)} \rightarrow H_2O_{(l)}$

The electrodes in the hydrogen-oxygen cell are porous carbon rods that contain a platinum catalyst. The electrolyte is a hot (several hundred degrees) potassium hydroxide solution. Hydrogen is oxidized at the anode where the hydrogen and hydroxide ions combine to form water. Electrons flow through the external circuit.

There are many different types of fuel cells currently under development. Many of these are named by the electrolyte or fuel used in the cell. The polymer electrolyte membrane or proton exchange membrane cell (pem) also uses hydrogen and oxygen. The electrolyte in this cell allows hydrogen ions (protons) to pass to the cathode, where they combine with oxygen to produce water. This type of fuel cell has several advantages over the alkaline fuel cell. Its lower operating temperature and relatively lightweight make it suitable for use as a power supply for vehicles. Several prototype vehicles have been built using pem cells.

Home Electricity and Electrical Safety

Electricity is a common feature of life that is taken for granted until a storm or other incident causes a power outage. Electrical power usage in the United States developed in the last decades of the nineteenth century. The demand for electrical energy grew with Thomas Edison's (1847–1931) development of the incandescent light bulb. In order to power lights sold to businesses and wealthy individuals, Edison began building central power stations to distribute electricity around 1880. Edison acquired a number of companies that produced wire, switches, generators, and various other components for his electric distribution systems. In 1892, Edison merged with a major competitor (Thomson-Houston Company) to form General Electric. Other capitalists followed Edison's lead, resulting in a rapid increase in the number of electric power companies operated by numerous entities across the country. Edison's initial power plants produced direct current of approximately 100 volts, which limited its distribution to a range of approximately one mile.

One of Edison's primary competitors in providing power was George Westinghouse (1846–1914). Westinghouse hired and employed the patents of Nikola Tesla (1856–1943). Tesla, who was originally from Croatia, had invented a number of electric systems that employed alternating current. He was hired by Edison but only worked for him for a year. Edison was convinced that electric distribution systems should be based on direct current, while Tesla's expertise was based on alternating current. The advantage of alternating current was that it could be transported long distances with minimal loss of power in conjunction with transformers. Edison and

Figure 13.16

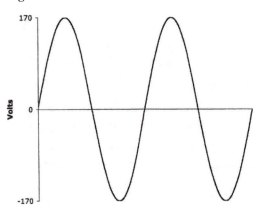

Figure 13.17

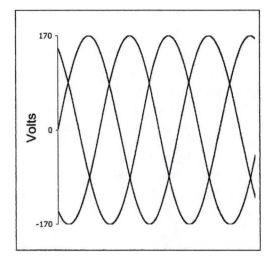

Westinghouse each waged a fierce commercial battle to convince potential customers that his system was superior. Edison argued that alternating current was inherently dangerous, and his company staged demonstrations where animals were electrocuted with alternating current to stress this point. To convince the public of the danger of alternating current, an Edison employee named Harold Brown invented the electric chair. This device was adopted by the New York State's corrections department and was first used in the United States in 1890. A major breakthrough in the adoption of alternating current took place when Westinghouse won a contract to build a power plant in Niagara Falls for industrial purposes. The plant employed the power of the falls for generators and resulted in power being transported to Buffalo, which was 20 miles from Niagara Falls. By the turn of the century, alternating current had been accepted as the standard for power generation, and Edison adopted alternating current and abandoned his direct current power system.

Modern power plants are based on Tesla's multiphase generation system. Alternating current from a wall outlet in a home is single phase. The current alternates at 60 Hz and a potential of 120 V. The voltage is not constant but varies by 170 V around a standard (zero potential) in the form of a sine curve (Figure 13.16). The 170 V is the **peak voltage.** The 120 V associated with standard voltage in the United States is the root mean square voltage. The root mean square voltage can be thought of as an average voltage. It should also be noted that the 120 V standard is often reported as 110 V because there is a 10% margin of error to account for resistance and other factors. Voltage coming into a house is single phase, while that generated at a power plant and transmitted over high voltage lines is triple phase. In a triple phase system there are three sine waves offset by 120° (Figure 13.17). This allows for a more constant voltage. High voltage power lines have three lines, each carrying one of the phases.

Electricity generated at power plants is transported over long distances along transmission lines at voltages that can be in excess of half a million volts. Through a series of electrical distribution stations and substations, voltages are stepped down using transformers (see following chapter) to several hundred volts. Houses in the United States accept split-phase 240 V. This means that a single-phase 240 V line comes into the house panel where it is divided into two 120 V lines that are 180° out of phase with each other. The two 120 V lines are used individually for most of the outlets and lights in the house. A typical three-pronged outlet will have three openings. The shorter opening on the right is "hot" and is wired to the 120 V live wire from the circuit panel. The hot wire brings in electricity from the power plant. The taller opening on the left is the neutral and completes the circuit back to ground at the power plant. The bottom semi-circular opening is the ground and is a safety device to prevent electric shock. Some large appliances in the home require 240 V. This is accomplished by wiring two live 120 V wires to the same outlet (the terminals and outlets are constructed differently for 240 V wiring). In this case, because the two 120 V lines are 180° out of phase, the difference in electrical potential between the two terminals on the outlet will be 240 V. In Europe and throughout much of the world, 240 V and 50 Hz is standard. One of the reasons for this is that European electric systems developed after those in the United States, and Tesla's ideas on alternating current using 240 V for home power were more firmly established as European systems were built.

The continuous and universal use of electricity in the United States and other technically developed countries has resulted in numerous safeguards to protect people from electric shock and electrocution. Most people associate electrical hazards with high voltages, and many warnings address solely this aspect of the electrical hazard. Shock occurs when current passes through the body. The important variables related to safety for electricity are current and the number of amps passing through the body. Since current equals voltage divided by resistance, the resistance of the human body must be considered in combination with the voltage. The resistance of the human body can vary over several orders of magnitude. For a person with wet skin at the points of contact, the resistance may be as little as a few hundred ohms. In contrast to this, the resistance of a person with dry skin may be several hundred thousand ohms.

Current passing through the human body has three potential harmful effects. It can cause muscles to contract involuntarily, result in heating effects, and interfere with the heart and nervous system. The effect of

Table 13.6
Physiological Effects of Current on the Human Body

Current (a)	Effect
0.001	Threshold of sensation
0.005	Mild sensation
0.05	Muscle paralysis, shock
0.020	Muscular contractions
0.05	Potentially lethal
0.15	Breathing stops
0.5	Severe burns

different currents on the human body is summarized in Table 13.6. Alternating current is more harmful than direct current because it produces greater muscle contractions and induces heating that results in sweating, which lowers the resistance of the human body.

In order to prevent electrical shock in the home when using powered items, a number of simple devices are employed. These include grounded outlets, polarized plugs, fuses, circuits breakers, and ground fault circuit interrupters. A three-pronged polarized receptacle is the common three-opening outlet found in homes built in recent decades. The ground, which is the semicircular bottom opening, provides a path to the ground if a short circuit occurs. The ground wire provides protection against insulation failures in the wiring. It is especially important to have where an appliance or device has an exposed metal surface. If a lose wire connects the metal surface of the object to the hot lead, then the object itself becomes hot. In this case, the ground provides a return path for current. This would typically produce a surge, since the current is not flowing through the object, and trip a circuit breaker.

Plugs for objects requiring electricity have either three prongs or two prongs. As noted, those objects with exposed metal surfaces have three-pronged plugs, with the third round prong providing a ground. Objects with two-pronged plugs, such as clocks, radios, televisions, and certain power tools, are double insulated and typically made of plastics or some other nonconducting material. The double plugs may have two flat prongs of the same size, especially if they are older, or one slot wider than the other. Plugs with a one wide prong and one narrow prong are referred to as polarized. Polarized plugs will only go into an outlet one way. A standard three-opening outlet's left opening is taller (wider) than its right opening. The narrower prong from the object fits into the shorter opening. A polarized plug prevents electricity flowing through the object when it is switched off or a malfunction occurs. It works because most all objects with polarized plugs have switches. When the switch is on, there is a connection made between the hot prong (narrower) of the object's electric cord and the hot wire in the outlet's narrower opening. When the switch is off, electricity flows from the outlet to the switch and stops; thus it never flows through the object Now assume the plug is forced in backward so that the neutral prong is inserted into the hot opening. In this case, electricity flows from the outlet through the object and stops at the switch. This generally is not a problem because objects are insulated, but if someone touches a conducting part of the object, he is in danger of shock. For this reason, a polarized plug should always be inserted into an outlet in the correct direction, and a polarized plug should not be modified or forced to fit into an outlet.

Circuit breakers and fuses are also standard protection features. These devices limit the amount of current that flow through a circuit. A typical home circuit is set at 20 A, although breakers come at numerous ratings. A circuit breaker contains a bimetallic strip that heats and bends when sufficient current passes through it. A 20 A breaker is constructed to "trip" a spring-loaded switch and open the circuit when the current is 20 A or greater. The amount of current on a particular circuit can be estimated by using the equation amps = watts/volts. By dividing the wattage of each appliance on a circuit by 120 V and summing, an estimate of the total amps going through the circuit can be made. For example, using the values in Table 13.5, a toaster and microwave on the same circuit would draw 1,200 W/120 V + 1,500 W/120 V = 22.5 A, tripping a 20 A breaker. A fuse breaks the circuit when current flowing through the fuse causes a conducting element in the fuse to melt at specific amperage.

Another common electrical safety device is the ground fault circuit interrupter (GFCI). These are commonly used in bathroom, kitchens, outdoor outlets, and anywhere water is present. A GFCI is constructed to detect the difference in current flowing through hot and neutral wires in an outlet. These should be the same, and any significant difference in the hot and neutral indicates a leak in current (possibly to the ground). When a difference of a few milliamperes between hot and neutral occurs, a breaker within the GFCI receptacle trips within a few hundredths of a second.

Summary

Electricity is an accepted necessity of modern society. While people in many countries such as the United States have come to accept electricity at the "flip of a switch," much of the world's population consider electricity a luxury. Although electric power has been available in this country for almost a century, the study of electricity continues to improve efficiency and make it more readily available. This chapter has introduced some of the basic principles of electricity, with little reference made to the related areas of magnetism and electromagnetism. A full understanding of electricity requires an understanding of magnetism and electromagnetism. Several of the topics introduced in this chapter will be explored further in the next chapter, on magnetism and electromagnetism.

14

Magnetism and Electromagnetism

Introduction

Magnets have fascinated us since we were young children and provided some of our earliest scientific experiences. These include observations that magnets are attracted to and stick to some metal objects, magnets don't stick to nonmetals, and two magnets might attract or repel each other depending how they are aligned to each other. As early as 600 B.C.E. ancient Greeks recorded similar observations on a type of iron ore called loadstone. A geographic area associated with loadstone was on the Turkish coast near the settlement of Magnesia, thus providing the source of the name used today for magnetic phenomenon. Magnetic compasses, fashioned out of natural magnetic materials such as lodestone, appeared in China around 800 C.E. William Gilbert (1540–1603), a physician in Queen Elizabeth's court, completed a major work on magnetism in 1600. Gilbert's work, entitled *De Magnete,* presented two decades of his studies on magnetism. *De Magnete* was the first comprehensive scientific examination of magnetism. It discussed aspects such as permanent magnetism, magnetic induction, the loss of magnetism upon heating, and magnetic poles. Gilbert adopted the term "electrick," from the Greek word *elektron* (*elektron* was a type of pine pitch or amber that attracted objects when rubbed with various substances), to explain the attractive and repulsive forces associated with charging by friction. Gilbert also proposed that the Earth itself was a giant magnet and introduced the idea of **magnetic inclination** or dip angle. Gilbert was a proponent of Copernicus' heliocentric model, and his ideas were very influential on other scientists. He was highly praised by Galileo and is often called the father of modern electricity and magnetism.

Until the start of the nineteenth century, magnetism was considered a distinct force separate from electricity. In 1820, Hans Christian Oersted (1777–1851), while performing a demonstration on the heating effects of electric current, observed that a nearby compass needle was deflected when current flowed through his circuit. Oersted didn't have an explanation for the compass' deflection, but continued to do experiments on the effect of current on a compass needle. Others such as Michael Faraday, André-Marie Ampère, and James Clerk Maxwell

(1831–1879) demonstrated that electricity and magnetism were related and were in fact different components of the combined phenomenon of electromagnetism.

This chapter will build on the material of chapter 13 by first examining the general concept of magnetism. We will then explore the unique relationship between electricity and magnetism and how these combine to produce electromagnetism. The technological aspects of magnetism will be addressed by examining practical uses of electromagnetism. Among the numerous common devices based on electromagnetic principles are motors, generators, and transformers. A number of the concepts of electromagnetism will be illustrated by examining these applications and other applications.

The Magnetic Field and Source of Magnetism

Oersted's initial observation and subsequent work by him and others demonstrated that an electric current creates a magnetic field. The field concept was discussed in chapter 13 on electricity. A charged object alters the space around it, and this altered space can be represented with lines of force that map the electric field. While a stationary charge creates an electric field, the source of a magnetic field is moving charge. In Oersted's experiment, the moving charge came from the current in his circuit and caused the compass needle to deflect. A magnetic field is produced whenever current flows. The magnetic field is the primary region in which the magnetic force acts.

An electric field surrounds any charged object. The movement of charge creates a magnetic field. For a magnet at rest, the source of moving charge is not readily apparent. The source of magnetism for a magnet at rest is the movement of the charged electrons in the atoms making up the material. Electrons in an atom orbit (the term "orbit" is used loosely because the quantum mechanical model of the atom does not contain electrons in well arranged orbits) the nucleus and spin on their axes. The spin of electrons is more important in determining the magnetic properties of a material. A minute magnetic field is produced from the movement of each electron in an atom. Each electron is in essence creates a subatomic-size magnet. For most materials, the electrons in its atoms are matched so that there are an equal or approximately equal number of electrons with spins in opposite directions. When an equal number of electrons spin in opposite directions, their magnetic fields cancel, and there is no net magnetic field. In some elements, such as iron, nickel, and cobalt, several valence electrons spin in the same direction, producing a net magnetic field. Materials in which the atoms have an excess of electrons with the same spin are called **ferromagnetic.** The magnetic field of each atom in ferromagnetic materials affects adjacent atoms. The magnetic fields in many atoms become aligned in magnetic "neighborhoods," called **mag-**

Figure 14.1
The degree of magnetism depends on the extent to which the magnetic domains align

not magnetic   moderate magnetic   permanently magnetic

Figure 14.2
Magnetic field maps for horseshoe and bar magnets

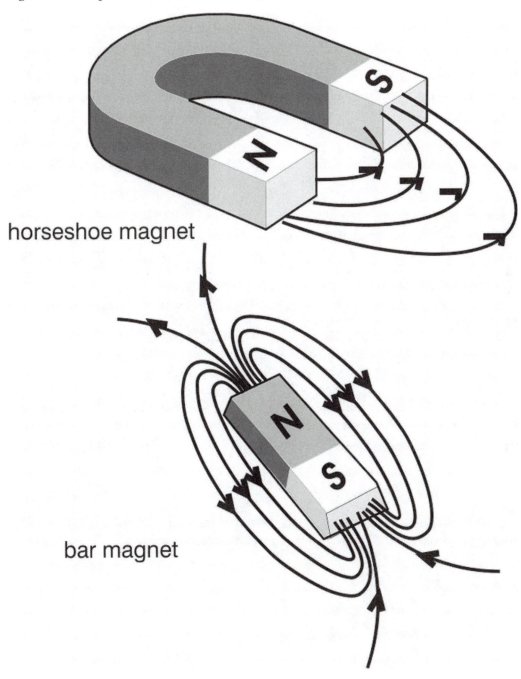

horseshoe magnet

bar magnet

netic domains. Magnetic domains consist of billions of atoms having aligned magnetic fields. Magnetic domains are about a millimeter in length. In an ordinary piece of iron, the domains are randomly oriented so that a piece of plain iron doesn't act as a magnet (Figure 14.1). However, if the piece of iron is placed in a magnetic field, the domains align with the magnetic field. A weak magnetic field will cause the domains to align in the general direction of the field. When the piece of iron is separated from the field, the iron will lose its magnetic property quickly. As the magnetic field strengthens, the domains achieve greater alignment, and the duration of magnetism increases when the field is removed. A permanent magnet results if the magnetic field is strong enough to cause the domains to remained aligned after the field is removed. Permanent magnets can lose some or all of their magnetism when heated or disturbed, for example, by hammering. The temperature at which a ferromagnetic material loses its magnetism is called the **Curie point.** Electromagnets consist of a piece of iron inside a coil of wire through which current runs. When the current is on, the magnetic domains in the iron core of the electromagnet align with the magnetic field produced, creating a strong magnetic force.

The magnetic field in space can be determined using a compass. This can be readily seen when a compass is brought near a wire conducting a current. If the compass is moved to various points around the wire, a field map can be made. The field map shows the direction and strength of the magnetic force at various points in space. The direction of the magnetic field at a point in space is defined as the direction the north pole of a compass points at that location. A magnet always has two poles: a north pole, N, and south pole, S. Magnetic field lines exit at the north pole and enter at the south pole. Where the magnetic lines of force are close together, the magnetic field is stronger. The magnetic fields surrounding two types of magnets are illustrated in Figure 14.2.

Another aspect related to the magnetic field that will be important in discussing electromagnetic induction is magnetic flux. Magnetic flux is the number of field lines that pass through an area in space. A magnetic field with strength B whose field lines passes through an area, A, at right angles has a magnetic flux of BA (Figure 14.3). As the angle between field lines and the plane of the area changes, the magnetic flux will decrease. In this case, the magnetic flux is given by the equation $BA \cos \phi$, where ϕ is the angle

Figure 14.3
Magnetic flux depends on the number of lines passing through an area. The magnetic flux is a maximum for an opening when the area is perpendicular to the field lines and zero when the field lines and area are parallel.

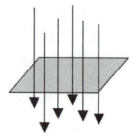
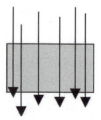

between the normal to the plane of the opening and magnetic field lines. When the area and field lines are parallel (ϕ is 90°) the magnetic flux is zero. The units for magnetic flux are teslas-meters^2 (T-m^2), and 1 T-m^2 is called a weber and abbreviated Wb. The unit is named after the German physicist Wilhelm Weber (1804–1891). Magnetic flux can be compared to filling a glass from a pitcher of water. Holding the glass upright maximizes the amount of water that goes into it when pouring from the pitcher. As the glass is rotated so that its opening is at an angle, less water goes into the glass. When the glass has been rotated 90°, the opening is parallel to the water poured out of the pitcher, and no water enters it. Magnetic flux is analogous to pouring water through an opening oriented at different angles.

Magnetic Force on a Moving Charge

A magnetic force results when a magnetic field acts on a moving charged particle. The magnetic force on a moving charged particle depends on several factors. These include the magnitude and sign of the charged particle, the speed of the particle, the strength of the magnetic field, and the angle between the velocity vector of the charged particle and the magnetic field vector. The magnitude of the magnetic force is given by the equation $F_{magnetic} = qvB \sin \theta$, where $F_{magnetic}$ is in newtons, q is the magnitude of the charge in coulombs, v is the speed of the charge in meters per second, B is the magnetic field strength in teslas, and θ is the angle between the velocity and magnetic field vectors. This equation can be rearranged to give the magnitude of the magnetic field: $B = F_{magnetic}/qv \sin \theta$. One tesla is the strength of the magnetic field when a test charge equal to 1 C moves at 1 m/s at 90° to the magnetic field and experiences a force of 1 N. Another unit used for the strength of the magnetic field is the gauss, where 1 G = 0.0001 T.

The direction of the force of a charge particle moving through a magnetic field is perpendicular to both the velocity and

Figure 14.4
Right-hand rule. When the fingers point in the direction of the magnetic field and the thumb in the direction the charge is moving, the palm is in the direction of the force.

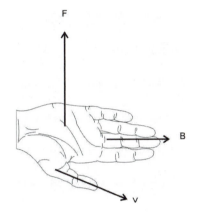

Figure 14.5
A current, i, defined in the traditional manner (as the movement of positive charge) moving through a magnetic field will cause a force, Φ, to be exerted on the wire that is out of the page

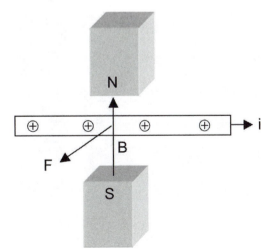

magnetic field vector and is given by the **right-hand rule.** According to this rule, the direction of the force is determined by pointing the fingers of the right hand in the direction of the magnetic field while the thumb is pointed in the direction that the charge is moving. When the right hand is positioned in this manner, the palm of the hand faces the direction of the magnetic force on a positive charge (Figure 14.4). For a negative charge, the direction is opposite, that is, it is in the direction that the top of the hand faces.

Because a charged particle experiences a force in a magnetic field, it is logical to assume that a current, which can be considered a stream of charged particles, moving through a wire will cause the wire to experience a force. If current moves through a wire as shown in Figure 14.5, there will be a force exerted on the wire according to the right-hand rule. This force will be out of the page. If the wire is formed into a loop, a torque will be exerted on the loop, and this is the basis of the electric motor. A schematic of a simple direct current motor is shown in

Figure 14.6. As current flows through the loop, a force is exerted on the wire, resulting in a torque. This torque causes the wire to rotate. The commutator consists of a split ring that rotates with the loop, while the brushes remain stationary. The brushes are the contact points through which power is supplied to the motor. As the split-ring commutator turns, it loses contact with the brushes for an instant when positioned at the opening in the ring. Although contact is momentarily lost, the loop's momentum causes it to continue its rotation. When the brushes again contact the commutator, the current in each part of the loop is reversed. This reversal ensures that the torque acting on the opposite sides of the loop produces a continuous rotation. An actual motor has many wire loops wound around a metal core. This assembly is called the armature. The power of the motor depends on the strength of the magnetic field, the number of wire windings (loops) in the armature, and the assembly of the motor. In most motors electromagnets rather than permanent magnets

Figure 14.6
Schematic of a direct current motor. When current, *i*, flows through the loop, a torque is placed on it, causing it to rotate. The opposing forces on the top and bottom of the loop cause the bottom to rotate out of the page and the top of the loop to rotate into the page.

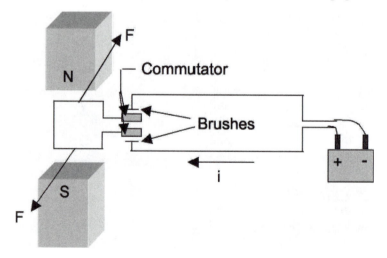

are used. Alternating current motors rely on induction to supply the current in the coils. Activity 19 in chapter 16 gives directions on how to build a simple electric motor.

Electromagnetic Induction

Oersted's experiment showed that magnetism resulted from an electrical current. A natural question immediately followed as to whether electricity could be produced from magnetism. Joseph Henry and Michael Faraday independently discovered the principle of electromagnetic induction in 1831. Although Henry discovered the idea earlier than Faraday, Faraday is given credit because he was the first to publish the work. Electromagnetic induction takes place when a changing magnetic field causes current to flow through a wire or another conducting material. A basic example of electromagnetic induction takes place when a bar magnet is passed in and out of a coil of wire (Figure 14.7). When there is no relative motion between the magnet and coil, no current flows through it. Pushing the magnet quickly into the coil creates a difference in electrical potential in the coil. This difference is the induced electromotive force (emf). The induced emf in turn induces a current if a closed circuit exists. Pulling the magnet out causes the polarity of the induced emf to reverse and current to flow in the opposite direction.

$$E_i = -N \frac{\Delta \Phi}{\Delta t}$$

The induced emf using a bar magnet and coil varies in direction depending on whether the magnet is moving in or out of the coil. The magnitude of the induced emf depends on the relative motion between the magnet and coil. When the magnet is moved in and out quickly, a greater induced emf is obtained. Another factor that influences the induced emf is the number of loops or turns in the coil. The induced emf is directly proportional to the number of loops in the coil. Faraday's law of induction relates the induced emf to the number of turns in the coil and the rate at which the magnetic flux changes with time. Faraday's law states that the induced emf is equal to -1 times the number of turns in the coil times the change in magnetic flux with time. Mathematically this is . In this equation, E_i is the induced emf, $\Delta \Phi$ the change in magnetic flux, and Δt the change in time. When Φ is in teslas-meters^2 and time in seconds, E_i will be in volts. it is important to realize that an induced emf results from a changing magnetic flux, but this does not automatically imply an induced current. The coil must be made of conducting material and connected to a closed conducting circuit in order to obtain an induced current. Of course, since electromagnetic induction is used to obtain electricity and in other applications that depend on current flow, an induced current can be assumed.

Figure 14.7
Movement of a magnet through a coil induces current through the coil

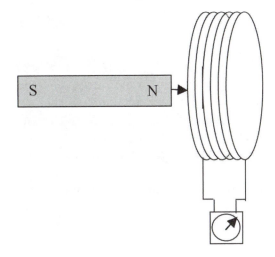

Figure 14.8
Right-hand rule for determining the direction of magnetic field produced by current flow in a wire

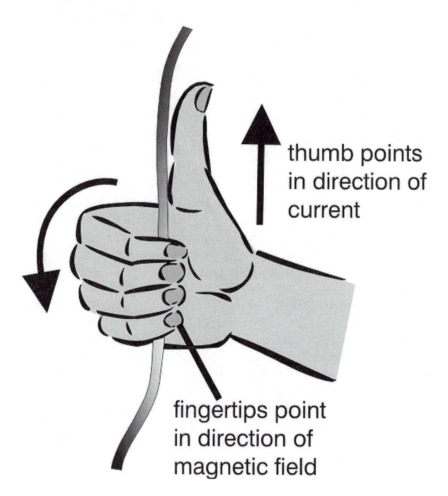

right hand

thumb points
in direction of
current

fingertips point
in direction of
magnetic field

The negative sign in Faraday's law has do with the polarity of the induced emf and how induced current flows through a circuit. In order to determine the polarity of the induced emf and the direction of the current, it is necessary to introduce another version of the right-hand rule. The second version is used to determine the direction of the magnetic field resulting from a current carrying wire. To determine the direction of the magnetic field point the thumb of the right hand in the direction of the current while curling the four fingers around the wire. The magnetic field will point in the direction of the finger tips (Figure 14.8).

An induced current will create its own magnetic field with a magnetic flux that opposes the magnetic field that created it. Because the induced magnetic field opposes the field that created it, there will be a repulsion between the two. In the bar magnet and coil example, this means there will be force

Figure 14.9
Assume a magnet is pushed into a coil. The front of the coil is to the right. The induced magnetic field resulting from the current in the coil opposes the magnetic field that created it. Magnetic field lines are indicated by the arrow. The direction of the induced current can be determined by using the right-hand rule.

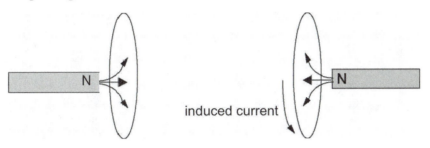

induced current

resisting the magnet as it is push in and pulled out. The fact that the induced magnetic field opposes the magnetic field that is created can be used to determine the direction of the induced current. Consider the bar magnet and coil when the magnet is pushed in north side first, as depicted in Figure 14.9. The induced magnetic field from the current in the coil can be found by assuming that a magnet is being pushed into the coil north side first in the opposite direction and using the right-hand rule. If the magnet in Figure 14.9 is pulled out of the loop, the current will flow in the opposite direction. The induced current can always be determined by assuming that it moves in a direction such that its magnetic field opposes the change that caused it. This rule is known as Lenz's law, named in honor of German physicist Heinrich Lenz (1804–1865).

Electromagnetic induction through a coil is the basis for the generator and practically all electric power produced in the world. A generator is essentially a motor in reverse (Figure 14.10). A motor uses electrical energy to create mechanical energy, while a generator uses mechanical energy to produce electrical energy. The mechanical energy for power generation is typically produced by steam, created by burning fossil fuels, expanding though a turbine or falling water, in the case

Figure 14.10
Basic generator

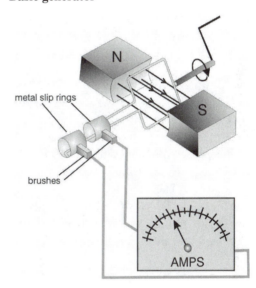

metal slip rings

brushes

AMPS

of hydropower. A direct current generator has a split-ring commutator to keep the direction of the current in the circuit constant. An alternating current generator doesn't have a commutator but uses two slip rings attached to the two ends of the loop. Split and slip rings make contact with carbon brushes that conduct electricity to the device to be powered.

The basic generator design shown in Figure 14.10 displays only one loop. An actual generator would have many loops typically wound around an iron core to form an armature. Small generators use permanent magnets, but most generators employ electromagnets to produce the magnetic field. Generators can contain numerous electromagnets and several armature coils arranged to produce current in numerous configurations. The use of three sets of armatures produces three-phase current, which was discussed in the previous chapter.

Induction motors use alternating current to convert electrical energy into mechanical energy. These motors are typically used in household appliances such as fans, clothes dryers, pumps, and power tools. An induction motor consists of a rotating part called a rotor and a stationary part (stator). The rotor is typically an iron cylinder that is slotted down it length. The slots are fitted with copper bars or fins. Surrounding the rotor is a cage of stationary conducting wire coils through which the alternating current passes. The surrounding cage, in essence, is a network of electromagnets. The passage of alternating current though stationary coils produces a rotating magnetic field, which in turn induces current and a magnetic field in the rotor. The interaction of the magnetic fields in the stator and rotor produces a torque on the rotor. This torque turns the rotor, enabling work to be performed.

Mutual Inductance and Self-Inductance

The source of induction in a loop is a changing magnetic flux through the loop. The change in magnetic flux is due to relative motion between the loop and magnet. It doesn't matter whether the magnet moves relative to a stationary loop, a loop moves relative to a magnet, or some combination. Since an electric current creates a magnetic field, induction occurs when current changes such that it changes the magnetic field. Devices that produce a changing magnetic field when

current passes through them are called **inductors** and are typically coils of wire of various shapes. Two forms of induction associated with a changing current are **mutual induction** and **self-induction.** Mutual induction takes place when the magnetic field from current flowing through a circuit creates a changing magnetic flux in a nearby circuit. Self-induction occurs because the changing magnetic field in a circuit induces an electromotive force in itself. With a direct current, the change in the magnetic field takes place when the current is switched on and off. Since this produces only a brief momentary change in the magnetic field, mutual and self-induction relate primarily to alternating current. Alternating current produces a continually changing magnetic field at the frequency of the current, for example, 60 Hz. This in turn can create an environment capable of producing a changing magnetic flux.

Mutual inductance can be illustrated by considering a coil adjacent to an independent coil through which alternating current is flowing (Figure 14.11). The coil through which the alternating current flows is called the **primary coil,** and the coil in which current is induced is the **secondary coil.** The changing magnetic field due to alternating current flowing through the primary coil creates a changing magnetic flux through the secondary coil. The changing flux in the secondary induces an emf producing a current. The induced emf in the secondary coil is related to the change in current in the primary coil with respect to time, according to the equation induced emf = $-M\Delta i/\Delta t$, where i is in amps and t in seconds. The proportionality constant in this equation, M, is the mutual inductance of the coils. Mutual inductance is expressed in the units of henries, named after Joseph Henry. One henry, abbreviated H, equals 1 V-s/A and can be interpreted as follows: when a change of 1 A/s in the primary coil takes place, a 1 V

Figure 14.11

In a transformer a voltage applied to the primary coil induces a voltage in the secondary coil

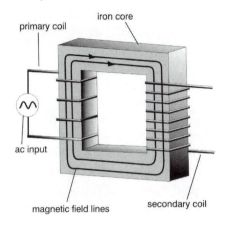

primary coil

iron core

ac input

magnetic field lines

secondary coil

electromotive force is induced in the secondary coil.

Transformers use the principle of mutual inductance to change the voltage of the alternating current produced in the primary coil to a different voltage in the secondary coil. The relation between the voltage on the primary and secondary coils is governed by the number of loops or turns in each coil. The transformer equation expresses the relationship mathematically:

$$\text{voltage}_{\text{secondary}} = \text{voltage}_{\text{primary}}$$
$$\times \frac{\text{turns secondary}}{\text{turns primary}}$$

This equation shows that when the number of turns is greater on the secondary coil, that the secondary voltage will be increased. In this case, the transformer is called a step-up transformer. A step-down transformer has fewer turns in the secondary, and therefore, the voltage induced in it is lower than the primary voltage. Although the voltage can be increased in a transformer, the average power in the primary

and secondary circuits must be the same to conserve energy. Since power equals current times electric potential, $P = IV$, the following equation holds for a transformer: $I_{\text{primary}} V_{\text{pimary}} = I_{\text{seconday}} V_{\text{secondary}}$. This equation shows that for a step-up transformer the current decreases and for a step-down transformer current decreases. The use of transformers is necessary to transmit electricity over large distances. Voltages are increased to several hundred thousand volts for transmission over high-voltage transmission lines. As the current approaches its destination, it goes through a series of step-down transformers to 120 Vs before entering a home.

$$E_{\text{induced}} = -L \frac{\Delta l}{\Delta t}$$

Self-induction occurs when alternating current flows through a coil circuit. The varying magnetic field induces an emf that opposes the emf that created it, according to Lenz's law. The induced emf for self-induction is directly related to the change in current over time, according to the equation . The negative sign indicates that the self-induced voltage opposes the original voltage. The proportionality constant L is the inductance of the coil. It is similar to the constant M in mutual inductance and is also measured in henries. The self-induced emf in a coil is often referred to as the back or counter emf because it opposes the emf that created it.

Electromagnetic Applications

Motors, generators, and transformers are three of the most universal examples that make use of electromagnetism, but there are countless other devices that employ electromagnetism in their operation. This section will discuss a few common and unique devices that employ electromagnetism and electromagnetic induction. A common class of microphones called dynamic microphones uses electromagnetic induction to generate their electrical signal. Microphones contain a thin membrane called a diaphragm that is part of transducer (a device that converts one form of energy into another). The diaphragm vibrates in response to sound. Attached to the diaphragm is a wire coil that moves with it. The coil is enclosed in a permanent magnet arrangement. The movement of the coil through the magnetic field converts the sound signal into an electrical current. This electrical signal is transferred to a speaker and amplified. This type of microphone is called a moving coil microphone. Another type of dynamic microphone has a ribbon rather than a coil attached to the diaphragm.

The firing of the spark plugs in a car depends on electromagnetic ignition. The 12 V battery provides direct current that is quickly switched on and off by a solid-state transistor (hence the name solid-state ignition). The switching function occurs through the points and distributor system in older model cars. The transistor switch is connected to the primary of the ignition coil, which is essentially a step-up transformer. The ignition coil boosts the 12 V supplied by the battery to approximately 40,000 V across the spark plug gap.

Eddy currents are a phenomenon that occurs when a flat conductor moves through a changing magnetic field, and they are applied in braking systems. If a flat conductor moves into an area where a magnetic field exists perpendicular to the flat plate, small circular currents will develop in the conductor (Figure 14.12). These currents are called eddy currents due to their rounded paths. Eddy currents retard the motion of the conductor, and this characteristic makes them useful as a nonabrasive braking system. Trains and roller coasters employ braking systems in which the wheels spin through a magnetic field. Eddy currents within the wheels retard the motion of the wheel, providing the braking action. The braking force is directly related to the speed of the wheel so that as the wheel slows down the braking force gets smaller.

A development in magnetic braking systems is the utilization of regenerative braking systems for hybrid vehicles. A hybrid electric vehicle has two power sources: a traditional gas engine and an electric motor. The gas engine and electric motor work in unison to provide greater efficiency and increased

Figure 14.12
As a conducting plate moves to the right out of the magnetic field, eddy currents are induced in the plate. The eddy currents produce forces that oppose the motion of the plate.

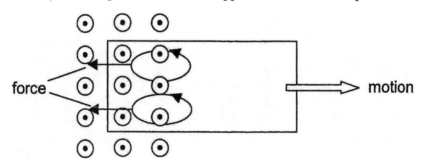

gas mileage. In a regenerative braking system, some of the energy lost through friction and braking is utilized to regenerate batteries that power the car's electric motor. When an ordinary car applies it brakes, the car's kinetic energy is converted to heat through friction between the brake and wheel and is wasted. This is called dynamic braking. In a hybrid car, the regenerative braking system recaptures the kinetic energy and uses it to recharge the batteries that power the vehicle's electric motor. The braking and electric current occur by running the motor in reverse to act as a generator. The torque induced by the magnetic field when it acts as a generator helps to slow the vehicle down and works in unison with the dynamic brake system. Regenerative braking systems are electronically controlled to act as a generator when the vehicle brakes or coasts, that is, any time the accelerator is not depressed. By utilizing the vehicle's kinetic energy during the braking, deceleration, and coasting portions of a trip to recharge the vehicle's batteries, fuel is save and pollution reduced.

Another application involving eddy currents is the use of induction hobs for cooking. The hob is the area of the stove where the cooking utensil rests. In induction cooking, alternating current flows through a coil in the hob. When a metal cooking utensil such as a pot is placed on the stove, eddy currents develop in the utensil. The currents cause only the metal in the cooking utensil to heat, and this supplies the energy needed for cooking. Induction stoves have a ceramic top that does not warm during the cooking, and this gives induction cooking an advantage from a safety aspect. The flat ceramic surface also is very convenient for cleaning spills. Another advantage of induction cooking is its 85–90% efficiency compared to approximately 50% efficiency for gas and electric stoves. Induction cooking requires the use of iron or stainless steel

cookware. Aluminum and copper cookware that are not magnetically sensitive will not work. Induction cooking has not been readily adopted in the United States, probably due to the high price of induction stoves, which cost several thousand dollars.

While trains employ magnetic braking using eddy currents, a whole new type of train has been developed in the last 30 years that makes extensive use of electromagnetism. Magnetically levitated trains, called maglevs, employ electromagnetism to suspend, guide, and propel them down a guideway. Most of the work on maglevs has been done in Japan and Germany. Although a number of maglev systems have been constructed, the technology is still under development and has not been widely adopted. Two main types of maglev systems are used. Electromagnetic suspension is based on an attractive force between electromagnets in the train's bottom and a ferromagnetic guideway that the bottom surrounds. The magnetic force is adjusted so that the train is suspended approximately an inch above the guideway by the attractive force between the magnets and the guideway. Propulsion occurs via a traveling magnetic field through interaction of magnets in the side of the train and the guideway. Electrodynamic suspension systems are based on a repulsive force between magnets in the train and an induced magnetic field in the guideway. Trains using electrodynamic systems have wheels and must obtain a critical speed (50–60 mph) before the induced forces are sufficient for levitation. In this manner they are similar to a plane taking off. Several maglev designs have employed supercooled superconducting magnetic coils in order to reduce the energy needed to propel trains, but these are very expensive and require cooling to temperatures of 4 K. Modern maglevs are capable of speeds of several hundred miles per hour. Because they ride on a cushion of air, noise, maintenance, and friction are minimized.

Geomagnetism

William Gilbert proposed that the Earth acts like a huge magnet. As a rough approximation, the Earth's geomagnetic field can be pictured as though a bar magnet extends from the north to south pole (Figure 14.13). The imaginary magnet has its south pole located at the geographic north pole. This is because opposite poles on a magnet attract. If the north pole of a compass points toward the Earth's geographic north, then the geographic north must represent the south pole of a magnet. The current locations of the magnetic poles are not at the geographic poles (true north and true south are where the Earth's axis of rotation intersects the Earth's surface) but are offset at an angle of about 12°. In 2003 the Earth's north magnetic pole was located at 78° N 104° W near Ellef Rignes Island in northern Canada, which is about 2,000 kilometers away from true north. The

Figure 14.13
The Earth's magnetic field is analogous to a bar magnet along the magnetic axis

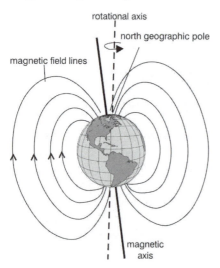

south magnetic pole is south of Australia near 65° S 140° E. The angle between true north and magnetic north on a compass is called the **magnetic declination.** Magnetic inclination, or dip angle, is the angle the Earth's magnetic field makes with the Earth's surface. The dip angle varies from 0° at the magnetic equator to 90° at the magnetic poles. Dip angle is the cue used by animals migrating between north and south to distinguish latitude.

The magnetic poles of the Earth are not stationary but wander and reverse with time. This means that the Earth's magnetic poles are normally located at high latitudes for periods of tens to hundreds to thousands of years, but these periods are punctuated by abbreviated periods lasting several thousands of years as the poles reverse. During reversals, it is believed that the Earth's magnetic field is very complex and can even display several pairs of poles spread over the Earth's surface simultaneously. The last reversal occurred 780,000 years ago. The reasons for reversals and the Earth's magnetic field itself are not clearly understood. It is speculated that convection currents in the Earth's iron and nickel outer core may be responsible for the field and its erratic behavior. This is called the dynamo effect. In addition to changing position, the strength of the magnetic field also changes. It is has declined about 10% in the last 150 years. Its current strength depends on where it is measured on the earth's surface but varies between about 0.3 and 0.6 G. It is strongest near the magnetic poles.

The magnetic field associated with the Earth itself is the internal magnetic field. The internal magnetic field is responsible for 95% of the Earth's magnetic field, while the remaining 5% comes primarily from charged particles in the ionosphere produced by solar radiation. The part of the Earth's magnetic field associated with charged particles in the atmosphere is called the external magnetic field. Because of the internal magnetic field, the geomagnetic poles, which are due to the internal field, and magnetic poles, representing the total field, are located in slightly different locations.

If the Earth's magnetic field is pictured as a simple dipole magnet, the magnetic lines of force associated with the Earth's magnetic field extend hundreds of miles into outer space. The region surrounding the Earth where its magnetic field has the greatest influence is called the **magnetosphere.** The magnetosphere deflects and traps charged particles from the **solar wind** that leaves the Sun. The solar wind is a stream of ionized gases that blows outward from the Sun at several hundred kilometers per second. As these ions approach the Earth, the magnetosphere interacts with the particles. The magnetosphere acts like the bow of a ship and deflects ions around the Earth. Concurrently, the solar wind distorts the magnetosphere so that it is compressed on the side facing the Sun and elongated on the far side. This gives the magnetosphere an asymmetrical tear-drop shape as opposed to the symmetrical lines of force associated with a bar magnet.

Not all charge particles are deflected by the magnetosphere. High-energy particles that enter the magnetosphere are trapped in regions called the Van Allen radiation belts. These belts are named for James A. Van Allen (1914–), who discovered one of the belts in 1958. The Van Allen radiation belts consist of two concentric donut-shaped regions: an outer belt and an inner belt. Van Allen discovered the inner belt, and the outer belt was found during the flights of the Pioneer spacecrafts. The outer belt extends roughly 15,000 miles out in space and traps charged particles, primarily electrons, from the solar wind by deflecting them along magnetic field lines in a spiral fashion toward the

magnetic poles. As the particles approach the poles, they are reflected back along the field lines toward the opposite pole. As a particle spirals toward the opposite pole it is again deflected back. In this manner, charged particles are trapped as they cycle back and forth along magnetic field lines. The inner belt extends roughly 4,000 miles out to space and contains primarily high-energy protons created by cosmic rays.

The movement of electrons in the outer radiation belt is responsible for the aurora displays called the aurora borealis (northern lights) and aurora australis (southern lights). These displays result when the solar wind causes electrons to be transported along magnetic field lines toward the magnetic poles where they energize gas atoms and molecules high in the upper atmosphere. Colors produced in the displays are due to different chemical species being energized. Atomic oxygen produces green, nitrogen atoms produce purple, oxygen molecules give red, and nitrogen molecules produce pink. Displays are highly visible during periods of pronounced solar activity that intensify the solar wind. During these periods of magnetic storms, the displays are brighter and their visibility extends farther toward lower latitudes.

One of the most significant discoveries involving the Earth's magnetic field has been in the area of paleomagnetism and its support of the theory of plate tectonics and continental drift. Paleomagnetism is the study the Earth's magnetic field as preserved in certain minerals found in rocks formed throughout geologic time. According to plate tectonics, the existing continents were once joined together into two large land masses. The two supercontinents broke apart and the resulting parts drifted across the face of the Earth to their present positions, where they continue to move at the rate of several centimeters per year. The theory of continental drift is attributed to Alfred Wegener (1880–1930), who proposed the theory in the early twentieth century. Wegner based his theory on the fit of the continental boundaries, fossil records, and similarities in geologic formations on different continents. It wasn't until the middle of the twentieth century that sensitive magnetometers were developed that enabled scientists to measure weak magnetic fields associated with crustal rocks. During the 1950s oceanographic surveys mapped the magnetic field in the crust at the bottom of the oceans. These surveys reveled an unexpected pattern of parallel alternating rows of relatively strong and weak magnetic fields associated with the crust when moving away from the midocean ridge system (Figure 14.14). When the Earth's field was taken into account, the pattern yielded a pattern of positive and negative magnetic anomalies. These anomalies were explained by hypothesizing that crust was being created at the midocean ridges as magma from within the Earth spewed out along the ridge. The magma became lava and when it solidified, the Earth's existing magnetic field was preserved in minerals contained in the lava. The new crust formed spread away from the ridges and carried the magnetized minerals with it. Because the Earth's magnetic field reverses, a positive anomaly would be found for crust that was created when the magnetic field pointed north, and a negative anomaly when it pointed south. For example, crust being created today would be magnetized with a north orientation. If 500,000 years from now the Earth's magnetic poles reverse, the crust created today would give a negative anomaly. Measuring the width of magnetic anomalies helps geologists determine spreading rates of plates along their boundaries. Paleomagnetic discoveries provided critical evidence that led to the universal acceptance of plate tectonics a half-century after Wegener put forth his ideas.

Maxwell's Equations

The discussion in the last two chapters has dealt with electricity and magnetism both as separate and as related phenomenon. Michael Faraday was the foremost authority on electricity and magnetism in the first half of the nineteenth century. One of Faraday's most important discoveries was that of electromagnetic induction. Faraday used lines of force to explain his findings on electromagnetism. Faraday considered tubes of force filling space and how these tubes were cut by wire loops passing through the tubes. Faraday was primarily an experimentalist, and he explained his results without developing a rigorous mathematical theory. Faraday's work was not readily accepted by the leading mathematical physicists of Europe of his time. Many of these physicists supported the concept of action at a distance. Faraday presented his findings on electromagnetism in three volumes published between 1839 and 1855 entitled *Experimental Researches on Electricity.*

Figure 14.14

Magnetic anomalies in the Earth's crust provide evidence for continental drift

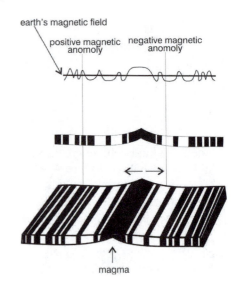

Faraday's work heavily influenced James Clerk Maxwell's monumental work on electromagnetism. Between 1864 and 1873 Maxwell built on Faraday's findings and derived a series of mathematical equations to explain Faraday's lines of force and the natural behavior of electric and magnetic fields. Maxwell's work was eventually condensed into four partial differential equations called Maxwell's equations. Maxwell's equations provided a comprehensive explanation of the properties of electricity and magnetism. Specifically, the equations showed how an electric charge produced an electric field, how currents produced magnetic fields, and how a changing magnetic field produced an electric field. Maxwell's wave equations showed that a changing electric field creates a self-propagating electromagnetic wave with electric and magnetic components. The electric and magnetic fields travel together as transverse waves perpendicular to each other, and both these waves are perpendicular to the direction of motion. The two waves are 90° out of phase (Figure 14.15).

According to Maxwell's equations, the speed of electromagnetic waves was shown to be equal to the speed of light and the same for all wavelengths in a vacuum. This demonstrated that light itself could be considered an electromagnetic wave. Maxwell's work thus provided a comprehensive theory of radiation that coupled electricity, magnetism, and light. It also accounted for other forms of electromagnetic radiation that result from different frequencies of oscillation of electric charge. Maxwell proposed that electromagnetic radiation traveled through an ether and refuted action at a distance.

Maxwell's work unified several broad areas of physics that had traditionally been interpreted as separate phenomena: electricity, magnetism, and light. According to Maxwell, a changing electric field induced a magnetic field, which in turn induced a

Figure 14.15
General diagram of an electromagnetic wave. The electric and magnetic fields are perpendicular to each other.

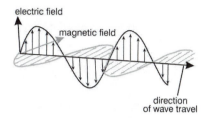

magnetic field, and this process repeated itself through space. Electric and magnetic fields always existed together, therefore, it was appropriate to speak of electricity and magnetism as electromagnetism.

Maxwell's work was experimentally verified by Heinrich Hertz (1847–1894) in 1888. Hertz used two small spheres separated by a small distances as electric oscillators to produce electromagnetic waves with a long wavelength. When Hertz charged the spheres to produce a high potential difference between them, a spark generated a wave that traveled through space and created a similar discharge across a small gap in a loop wire. Hertz went on to show that electromagnetic waves could be reflected, refracted, and diffracted, which greatly strengthened the acceptance of Maxwell's work. Hertz's work laid the groundwork for the transmission of radio waves.

Summary

Maxwell's work on electromagnetism and the work that built on it unified a body of knowledge on electricity, magnetism, light, and radiation during the last decades of the nineteenth century. Oersted's simple observation that an electric current causes a compass needle to deflect inspired Faraday, who experimentally laid the foundation on which Maxwell ultimately succeeded in developing a rigorous mathematical theory to explain the phenomena of electricity and magnetism. The basic principles of electricity and magnetism were developed into countless technologies that define modern society, such as electric power, radio, television, motors, batteries, and electronics. The speed of light in Maxwell's equations was invariant on the frame of reference, which seemed to refute Newtonian mechanics. The resolution of this dilemma took place at the beginning of the twentieth century with Einstein's work on relativity.

15

Relativity

Introduction

As the twentieth century unfolded, a revolution in physics took place. Planck's quantum model and Thomson's discovery of the electron marked the beginning of a revolutionary era in physics that reshaped how humans viewed nature. The quantum model and discoveries that redefined how matter, energy, and radiation behaved were discussed in chapters 10–12. In addition to quantum physics and the discovery of new subatomic particles, the very way humans viewed the world was reconstructed by Albert Einstein and his theory of relativity. Newton's mechanical view of the universe had served physics well for two centuries, but as the electromagnetic nature of light developed and the realm of physics expanded to the subatomic level, limitations in classical physics arose.

The world we live is readily understood using Newtonian mechanics. The theory of relativity is a more general view of nature that encompasses Newtonian mechanics. In this sense, relativity conforms to the **correspondence principle.** This principle says that any new theory must not only explain new observations but must also account adequately for everything explained by preceding theories. Thus, Newtonian mechanics is a special case of the theory of relativity that applies when speeds are negligible compared to the speed of light. Since this is the situation that defines everyday life, we develop a Newtonian worldview. This has served humans well and allowed us to develop a modern society capable of sending humans to the Moon and back and sending spacecraft to the far reaches of the solar system. This chapter will expand on the Newtonian classical mechanics by introducing relativity. In the process, some of the classical ideas that seem so logical will need to be revised. Certain aspects of relativity defy common sense and force us to stretch our imagination beyond everyday phenomenon into a world that we can explore only in our minds. This is because we lack relativistic examples in our ordinary world, and therefore, we lack an exposure to these concepts.

The Speed of Light

Because the theory of relativity is closely connected to the speed of light, it is beneficial to examine how knowledge on this fundamental constant developed historically. In ancient times, the speed of light was generally thought to be instantaneous. Empedocles of Acragas speculated that light had a finite velocity. Middle Ages scholars who studied optics, such as Alhazen and Roger Bacon, supported the view that the speed of light was finite. Kepler believed the speed of light was instantaneous, but Galileo questioned this belief. In *Two New Sciences,* Galileo has Simplicio present the ancient view on the speed of light with the statement: "Everyday experience shows that the propagation of light is instantaneous; for when we see a piece of artillery fired at great distance, the flash reaches our eyes without lapse of time; but the sound reaches the ear only after a noticeable difference." Galileo went on to suggest experiments for determining the speed of light using light sources, shutters, and timing devices. Galileo reasoned that two individuals with lanterns separated by several miles could determine the speed of light. The speed of light could be determined if each person started with a covered lantern. The first person would uncover a lantern and then wait for the second person to see its light

and uncover the second lantern. The time it took for light to make the round-trip between the two observers could be used to determine the speed of light. Galileo may have tried the experiment, but because of the incredible speed of light his results would have been inconclusive.

The first definitive speed of light was determined in 1675 by Ole Rømer. Rømer used discrepancies in eclipses of Jupiter's moon Io by Jupiter to determine the speed of light. Rømer noted that the time of the observed and predicted eclipse varied in a regular manner depending on the distance between the Earth and Jupiter. Rømer reasoned that the discrepancy between the predicted and observed eclipse was due to the longer time it took light to transverse the greater distance when the Earth and Jupiter were farther apart. Therefore, the discrepancies between the observed and predicted eclipse would be greatest when the light had to traverse the Earth's entire orbit (Figure 15.1). Using values for the distance of the Earth's orbit, Rømer determined that it took approximately 7 minutes for light from Jupiter to reach the Earth. He computed the speed of light at 225,000 km/s, which is about 75%

Figure 15.1
Rømer used the discrepancy between calculated and observed eclipses for Jupiter's moon Io when the Earth was at different position in its orbit with respect to Jupiter to estimate the speed of light. Note that the drawing is not to scale.

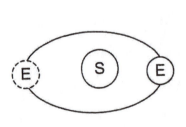
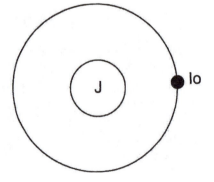

of the modern value of 300,000 km/s (more precisely 299,792 km/s).

A half-century after Rømer calculated the speed of light, James Bradley (1693–1762), an astronomy professor at Oxford, used **stellar aberration** to determine the speed of light. Bradley was working on the problem of stellar parallax when he derived his method to determine the speed of light. Stellar parallax refers to the change in apparent position of stars closer to the Earth against the background of more distant stars as the Earth orbits the Sun. The heliocentric theory of Copernicus predicted stellar parallax, but 300 years after Copernicus put forth his model, stellar parallax had yet to be observed. The reason for this failure was that the change in angular position was so small it could not be observed with the telescopes employed in the intervening 300 years. In his attempt to observe the parallax with a telescope, Bradley discovered and explained stellar aberration. Stellar aberration refers to a minute closed elliptical path a star makes when it is observed over the Earth's yearly orbit. For stars high in the sky (more directly overhead), the path of the star is open and circular and for stars low in the sky the path is flatter and closed. Stellar aberration is a consistent motion exhibited by all stars, as opposed to parallax that is greater for nearby stars.

Bradley determined the reason for stellar aberration was because stars were observed from an orbiting Earth reference frame. The phenomenon can be compared to the motion of raindrops when stationary and moving. If rain is falling from a location directly overhead in the absence of wind, then a person on the ground will need to prop an umbrella directly overhead for protection. When the person starts to walk, for example, toward the east, the rain will seem to be coming down at an angle (Figure 15.2). While walking, the person needs to point the umbrella toward the east for protection. As the person walks faster, the apparent horizontal component to the rain increases. Although the rain is still falling vertically, the person observes the rain coming at an angle. As far as the person is concerned, the rain is falling at an angle and not coming straight down. The angle depends on the speed of the observer and the speed of the falling rain.

Light from stars can be viewed as rain coming from directly overhead. As the Earth

Figure 15.2
When a person walks into a vertical rain, the umbrella must be pointed toward the direction of travel as on the right. The person perceives the rain as coming down at an angle. This illustrates the concept of stellar aberration.

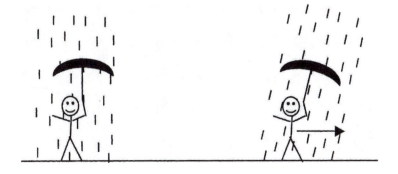

Figure 15.3
Bradley used stellar aberration and geometry to determine an accurate value for the speed of light around 1725

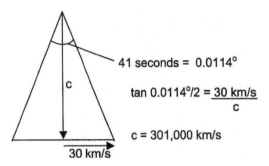

41 seconds = 0.0114°

$\tan 0.0114°/2 = \dfrac{30 \text{ km/s}}{c}$

c = 301,000 km/s

makes its annual orbit around the Sun, light will appear to come in at an angle from the direction in which the Earth moves. This results in light from the star appearing ahead of its true position. The movement of the Earth in its orbit results in stellar aberration. Bradley measured the aberration angle and used the fact that the Earth moved around the Sun at approximately 30 km/s. Bradley observed a stellar aberration of 41 seconds of arc over the year. Using geometry, he calculated the speed of light at 301,000 km/s (Figure 15.3). The speed of light is denoted in Figure 15.3 as *c*. The small letter "c" commonly used for the speed of light is derived from the Latin word *celeritas*.

More than 100 years after Bradley's work in the middle of the nineteenth century, two physicists from France born in the same year employed techniques to make the first accurate terrestrial measurements of the speed of light. Armand Hippolyte Louis Fizeau (1819–1896) and Jean Bernard Léon Foucault (1819–1868) devised methods that made it possible to use Galileo's idea of timing the speed of light as it made a round-trip over the Earth's surface. In 1849 Fizeau aimed a beam of light at a rotating wheel with 720 teeth around its outside

circumference, similar to a cogwheel. The teeth broke the beam of light into pulses. The frequency of the pulses depended on how fast the wheel rotated. The beam of light was directed so that after passing through an opening in the wheel it reflected off a mirror that was 8,600 m away. The reflected beam traversed the path back to the rotating wheel. By adjusting the speed of the wheel, Fizeau caused a light pulse to pass through the next opening in the rotating wheel. Fizeau knew the spacing of the wheel's teeth and how fast the wheel was rotating. Fizeau had a pair of telescopes arranged so that when the reflected light passed through an opening, he observed a bright image, and when the reflected light hit a tooth, the image disappeared. Fizeau determined the slowest rotation necessary to produce an image. With this information, Fizeau calculated a value for the speed of light as 313,000 km/s.

In 1875, Foucault used the same general method as Fizeau, but instead of using a rotating toothed wheel he employed a rotating mirror. Foucault aimed a beam of light at the rotating mirror. The light reflected off the mirror toward a series of stationary mirrors arranged in a ziz-zag pattern. This arrangement extended the path of light to approximately 20 m before it was reflected back toward the rotating mirror. The return beam would strike the rotating mirror at a slightly different angle on its return path. Foucault observed the displacement of the return beam, and from this he calculated a speed of light of 298,000 km/s. Foucault's technique was refined over the next 50 years by others for measuring the speed of light. In 1879 Albert Michelson (1852–1931), while teaching physics at the U.S. Naval Academy, used Foucault's method to get a value for the speed of light, which stood for 40 years. Michelson used an octagonal-shaped rotating mirror. He established a light path that was several hundred meters long and determined this distance within a few millimeters. Michelson invested in high-quality mirrors and lenses and calculated a value of *c* of 299,910 km/s. Michelson continued to improve his measurements and in 1926 got a value of 299,796 km/s, which was considered the most accurate measurement for several decades. Michelson continued to make measurements on the speed of light using other techniques until his death. He was awarded the Nobel Prize in physics in 1907 and was the first American to receive this award. Modern measurements using lasers during the last part of the twentieth century have given increasingly precise measurement for the speed of light. The modern accepted value is 299,792.458 km/s.

The Michelson-Morley Experiment

Michelson's work on the speed of light led him to one of the most important sets of experiments in the history of physics. At the end of the eighteenth century the wave nature of light was well established. Because light could be viewed as a wave phenomenon, it was natural to compare it to other familiar waves. Since waves required a medium to travel through (such as waves through water), light was also thought to require a medium. This medium was speculated to be the **luminiferous ether.** The ether was thought to permeate the universe and allowed light to travel across a room or the universe. This ether had to be very diffuse and meant there was no such thing as a true vacuum. Furthermore, the ether established an absolute frame of reference for measuring motion. Michelson sought to detect the ether with an ingenious experiment based on the experimental techniques he had developed to measure the speed of light.

Michelson reasoned that the Earth's movement through the ether should create an ether wind that would influence the speed of light. Just as an airplane moves faster with a tailwind and slower with a headwind, light would move faster or slower depending its direction relative to the ether wind as the Earth moved through it. Michelson's reasoning could be understood by looking at the analogy of two rowers racing on a river with a current, as depicted in Figure 15.4. Assume the river is 50 m wide with a current of 3 m/s and both rowers leave from the same starting point. Furthermore, both rowers can row at a speed of 5 m/s. One rower makes a round-trip to a point directly opposite its original location, while the other rower travels a distance up and down the shore equal to the direct round-trip distance, namely, 100 m. The race can be analyzed using simple geometry and equations for speed. The rower moving across the river has to aim the rowboat into the current at an angle in order to reach the shore directly opposite the starting point. Since the rower moves at 5 m/s and the current is 3 m/s, the speed across the river in both direction is 4 m/s. At this rate, the time to make the trip across the river and back is 25 s. Now consider the rower moving along the shore. Rowing against the current the speed is 2 m/s (5m/s – 3m/s) and moving with the current the speed is 8 m/s (5 m/s + 3 m/s). The time for the round-trip would be the time to go 50 m at 2 m/s, which equals 25 s, plus the time to go 50 m at 8 m/s, which is 6.25 s. Therefore, the time for rower to make the trip parallel to the bank is 31.25 s. It takes longer to make the trip along the bank as opposed to across the river. Michelson's idea was to have a race between two beams of light: one beam moving perpendicular to the ether wind and another moving parallel. Using the rowboat analogy, the light beam moving parallel to the ether wind should take longer.

Figure 15.4
The basis for Michelson's experiment to detect the luminiferous ether was based on the difference in time it takes for two rowers in a hypothetical race across a river. Both racers start at the same point. One travels to a point directly across the river, marked with a circled X, and back to the start as shown by the solid arrows. This rower must travel at an angle across the river to arrive at the circled X to account for the current. Likewise, the rower must travel at an angle to arrive back at the start. Another rower travels the same distance as the river's width but travels up- and downstream parallel to the current, as shown by the dotted arrows. As long as the rowers move faster than the current, the rower moving parallel to the current will always take longer.

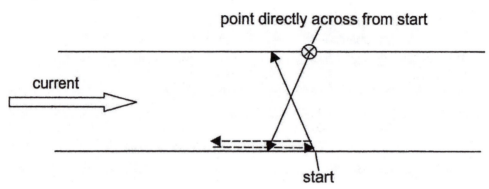

In any race, it is important to determine who crosses the finish line first. Michelson's problem was to devise a method where he could determine that one beam arrived before the other. Michelson's work started in the 1870s and continued over the next decade. In the late 1880s, while he was at Case School of Applied Science in Cleveland, he teamed with Edward William Morley (1838–1923), a chemistry professor at neighboring Western Reserve University, to conduct the now famous Michelson-Morley experiment. The experiment relied on Michelson's knowledge of measuring the speed of light. The Michelson interferometer is depicted schematically in Figure 15.5. Michelson had perfected his interferometer over the years to make very precise measurements. The actual interferometer Michelson used had a number of mirrors so that the light beams were reflected several times across the interferometer, extending the path length to 11 m

before arriving at the detector. The interferometer was mounted on a square sandstone block that measured 1.5 m on each side. The sandstone block rested on a wooden disk and floated in a pool of mercury contained in a cast iron tank. The massive block resting in mercury allowed the apparatus to be rotated through 360° with very little vibration.

In the summer of 1887, Michelson and Morley made measurements with the interferometer in the basement of Western Reserve University. The specific measurement that was made was to observe the interference pattern created by the light waves at the detector. If the light wave arrived at different times, interference fringes would be observed. Michelson and Morley observed the interference pattern with the instrument in different positions. An observation was made, and then the interferometer would be rotated to a new position and another observation made. Since the Earth orbited

Figure 15.5
In the Michelson interferometer, a light beam was directed at a half-silvered beam splitter. Part of the beam was reflected toward the adjustable mirror and back toward the detector. Another part of the beam passed through the splitter and was reflected at the fixed mirror back to the beam splitter and then to the detector. The compensating mirror was inserted so that each beam traveled through an equal amount of glass.

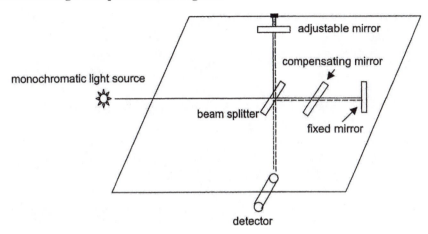

the Sun at 30 km/s, it was believed that light traveling with and against the ether wind would travel slower than light that traveled perpendicular to the ether wind. Michelson and Morley were unable to detect the ether wind no matter how the instrument was oriented or what time of the year the experiment was conducted. The inability to detect the ether puzzled both Michelson and Morley. Michelson thought there may have been an unknown flaw in the experiment. Morley likewise couldn't explain the inability to account for the ether wind and continued to make measurements using more precise interferometers for the next several decades. Several immediate explanations were presented to account for the inability to detect the wind. One was that the wind was dragged along with the Earth as it moved through space so that the Earth didn't pass through the ether wind. Another explanation was termed the **Lorentz-Fitzgerald contraction.** The idea was first proposed by the Irish physicist George Francis Fittzgerald (1851–1901) and independently expanded upon by Hendrik Antoon Lorentz (1853–1928) from The Netherlands. According to the Lorentz-Fitzgerald contraction, an object shortened in the direction of motion. This contraction could be used to explain the results of the Michelson-Morley experiment. If there was a shortening of the interferometer in the direction of motion, this meant that the light traveling in the direction of the ether wind would travel a minute distance less than the light traveling perpendicular to the wind. The distance was just enough to cause the two light beams to arrive simultaneously at the detector, accounting for a lack of a shift in the beams. While the Lorentz-Fitzgerald contraction could explain the results of the Michelson-Morley experiment, there was no experimental support for such a contraction. Lorentz himself realized it was an ad hoc means to resolve the puzzling results of Michelson and Morley.

Reference Frames

By the start of the twentieth century the world of physics was rapidly changing the view of nature. Planck had introduced quantum theory, Thomson had discovered the electron, and other discoveries were being made on light and radioactivity. Against this background a young unknown German physicist living in Switzerland burst on the scene and changed how to interpret physical reality from a Newtonian to an Einsteinian relativistic view. To appreciate relativity, one must consider relative measurements and the terms used to describe these measurements. Statements such as "that's a large baby," "that's a slow car," or "he did that job fast" imply relativity. An idea of how big a normal baby is and how long it should take to do a job provides a conscious "yardstick" for relative measurements. This conscious yardstick is what is termed our frame of reference. Motion is typically measured using the Earth as the frame of reference. The speeds of objects are measured or judged by assuming that the Earth is static and determining how long it takes them to move over the Earth. For example, a person sitting in a train car can judge the speed of the train by looking at the Earth or a fixed object attached to the Earth. Now consider a person inside a train car that happened to have its windows darkened. Also, assume that the train was moving at a constant speed on a perfectly smooth railway devoid of vibrations. In this case, the person would have no comprehension of the speed of the train. Without being able to look out the window and view the surroundings and unable to feel the vibrations, the person would have no sense of motion. For all practical purposes, the person in the train car would believe the train was at rest. A ball could be thrown toward the front of the train car and the speed of the ball judged with respect to the train, which would be the passenger's frame of reference.

Although the Earth is typically taken as an absolute frame of reference for motion measurements, the Earth is not an absolute reference frame. A person viewing motion of an object from outer space would also observe motion associated with the Earth's rotation on its axis and revolution around the Sun. If the Sun is taken as a frame of reference in order to account for rotational and orbital motions, there is the problem of the Sun's 200-year movement around the galaxy. In order to measure absolute motion in the universe, an absolute reference frame is required. Based on thinking at the end of the nineteenth century, many scientists speculated that the absolute reference frame was the luminiferous ether that permeated the universe.

Motion since the time of Galileo was interpreted in terms of Galilean-Newtonian relativity. Galileo postulated laws of motion based on a stationary frame of reference or one that was moving at a constant speed. Galileo's idea of an **inertial reference frame** and relativity was presented in his interpretation of how a person on a ship moving at a constant speed and someone on shore watching the ship interpreted motion. If a cannon ball is dropped from the top of a mast, the person on the ship would view the ball falling straight down and landing at the foot of the mast. Conversely, a person on shore would see the ball taking a parabolic path, but still striking at the foot of the mast. The person using the shore as his frame of reference would interpret the ball's path as due to the combined force of gravity and the ship's forward movement that the ball possessed when it was dropped. The ship-bound observer using the ship itself as the frame of reference would attribute the ball's flight to the force of gravity. Both observers would have valid views. Galilean transformations could be made to relate motion in one frame of reference to another. For instance, if the speed of the ship was u, the horizontal speed of the cannon ball with respect to the observer on shore would be u. If a person on the ship tossed an object forward with a speed of v with respect to the ship, then the speed with respect to the shore observer would be $u + v$.

Newton developed his laws of motion based on Galileo's assumptions and the principles formed what was known as Galilean-Newtonian relativity. According to Galilean-Newtonian relativity, the laws of motion were equally valid for observers in inertial reference frames. An inertial reference frame is one that is either stationary or moving at a constant speed with respect to the observer. The term "inertial" signifies that Newton's law of inertia (first law of motion) applies. In inertial reference frames, the laws of physics and motion were considered equivalent. This means that if one takes measurements in a stationary (with respect to the observer) reference frame and one moving at constant speed with respect to the observer, then the same results should be obtained. Another way of stating this is that observers in a stationary reference frame and inside an opaque enclosed reference frame moving at a constant speed would have no means to determine that one was stationary and one was moving. The observers in the stationary and the opaque enclosure would arrive at equivalent results when making measurements, for example, they would both find $F = ma$.

Newton extended the Galilean view of the reference frame in the sense that Newton's laws of motion were considered to be valid in any frame that was at rest or moving at a constant velocity to one at rest. All that was needed to specify a reference frame was a set of three coordinates to define space and time. When a reference frame was defined, motion in all other inertial frames could be related by the Galilean transformations. Problems arose with Maxwell's interpretation of light. According to Maxwell's equations, the speed of light was not constant in inertial reference frames, and the Galilean transformations that could be applied to motion did not hold for light. Maxwell's equations required one absolute reference frame, and this reference frame was thought to be the luminiferous ether. This meant that Newton's laws applied for inertial reference frames, but Maxwell's laws for light did not. Applying a Galilean transformation to the speed of light measured in a manner similar to that of the ball thrown forward on a moving ship meant that the speed of light would be greater than c. For example, consider a car moving down a highway at a speed of v with respect to an observer on the side of the road. Now assume that the driver of the car turns on the headlights. The speed of light from the headlights with respect to the driver of the car equals c. The speed of the light with respect to the roadside observer according to the Galilean transformation should equal the speed of light plus the speed of the car: $c + v$. The belief that the laws of physics were the same in all inertial reference frames coupled with Maxwell's result that the speed of light was constant (in a vacuum) created a serious problem for physics. The Lorentz-Fitzgerald contraction was a mathematical convenience to rectify the problem, but Lorentz himself realized that it was just an ad hoc solution with no real physical basis. The predictions of light's speed from Maxwell's equations coupled with the lack to detect the ether wind in the Michelson-Morley experiments presented a serious challenge to physics as the twentieth century unfolded. It was at this time that Einstein provided a solution to the problem with his special theory of relativity.

Special Theory of Relativity

Albert Einstein introduced his special theory of relativity in 1905. Einstein's 1905 paper on relativity would have been sufficient to establish his reputation, but he also published other profound work that caused the world to take note of this obscure physicist working in a Swiss patent office. Einstein was born in Ulm, Germany, and educated at schools in Germany and Switzerland. His education in physics and mathematics was sound, although unremarkable. He graduated from the Zurich Polytechnic Institution in 1900 and held several temporary positions until finally gaining a full-time position as an examiner at the Swiss Patent Office in 1902. His work gave him the freedom to develop his ideas independently, and these were introduced to the world in 1905 with several seminal papers. It was during this year he received his doctorate from the University of Zurich. In one of his 1905 papers, Einstein proposed light was a quantum phenomenon. This introduced the idea of light existing as photons or bundles of energy and was later used to explain the photoelectric effect. Another paper dealt with **Brownian motion** and provided evidence for the existence of molecules and their size. The paper entitled *On The Electrodynamics of Moving Bodies* introduced the special theory of relativity. Although in retrospect the year 1905 can be considered a watershed year for Einstein and the world of physics, his genius was not immediately recognized. Einstein continued to work as a patent examiner and waited four more years, until 1909, to obtain an academic position.

Einstein's special theory of relativity was based on two postulates. The first states that the laws of physics are the same in all inertial reference frames. This means that there is no absolute preferred frame of reference and, therefore, no such thing as an absolute speed. Consequently, there is no experiment that could distinguish whether someone was on the ground or in a room moving at constant speed with respect to the ground. The first postulate explained the failure to detect the ether from the Michelson-Morley experiment. In the Michelson-Morley experiment, it was thought that the speed of light measured on the Earth, which itself was moving with respect to stationary ether, would be affected by the ether. Einstein's first postulate means that in the absence of an absolute reference frame, that the speed of light would be the same in all directions since it was being measured relative to the Earth as opposed to nonexistent ether.

Einstein's second postulate was that the measured speed of light in a vacuum is constant regardless of the relative speed of the observer and light source. The second postulate defies the commonsense of the Galilean-Newtonian viewpoint that had been accepted for 250 years. For example, consider a fictitious car capable of traveling at 50% of the speed of light, $0.5c$, when it turns on its headlights. When the car is at rest, the speed of light will be measured by both a passenger in the car and an observer at rest with respect to the car as c. Now assume the car moves at $0.5c$ and its headlights are turned on. It would seem that since the car is moving at $0.5c$ and the speed of light is c that an observer at rest with respect to the car would measure the speed of light as $1.5c$. This represents the Galilean-Newtonian interpretation of the situation and seems perfectly logical. Einstein's second postulate states that the speed of light cannot be greater than c. Therefore, the observer at rest should still measure the speed of light as c and not $1.5c$. According to special relativity the resultant velocity, V, when adding two velocities together is given by the equation:

$$V = \frac{v_1 + v_2}{1 + \dfrac{v_1 v_2}{c^2}}$$

Applying this equation to the car example with v_1 equal to c and v_2 equal to $0.5c$ gives a value of c for the resultant velocity.

The equation for adding two velocities is applicable to all situations and not just those involving light. For example, consider a person riding in a car at 60 mph who fires a gun and the velocity of the bullet is 500 mph in the same direction as the car is moving. We'd expected that the velocity of the bullet relative to the road to be 560 mph. Using the relativistic equation for adding velocities, the speed with respect to the road is about 559.999999999969 mph, which for all practical purposes is 560 mph. It is only when speeds become appreciable with respect to the speed of light that a relativistic analysis is necessary.

Time

Measurements such as time, length, and mass take on new meaning in a relativistic world, leading to several interesting consequences. These consequences deal with observations made from different frames of reference. As an initial example, consider time and simultaneity. Simultaneous events are defined as those that occur at the same time. In order to see how the concept of time and simultaneity changes with a relativistic interpretation, a thought experiment can be performed. Einstein was famous for performing *gedanken* (German for thought) experiments in which he pondered the possible results of hypothetical situations. Consider a train traveling at a very fast speed down a track. In a train car a light is located exactly halfway between the front and back of the car. A person in the car turns on the light and observes that light strikes the front and back walls of the car simultaneously. This is what would be expected, since the distances to the front and back walls are the same and the speed of light travels forward and backward at constant speed. Now consider what a person standing on the ground outside the train car observes as the car passes (it is assumed that the train car is transparent so that an outside observer can see into the car with no problem). The light is turned on and travels forward and backward at a constant speed of c. Because the train is moving passed the observer, the front and back walls have moved a small distance forward between the time when the light is turned on and when it strikes the wall. Therefore, the distance light must travel is a little longer before it hits the front wall and a little shorter before hitting the back wall. The observer will say that the light strikes the back wall first and then the front wall. What was interpreted as simultaneous events inside the car are not simultaneous when observed from outside the car.

The gedanken experiment for the train car indicates that time is not absolute, but relative, depending on the relative motion between the observer and what is being observed. We generally consider that a strobe light flashes with the same frequency whether it is observed when stationary or moving. The special theory of relativity predicts that the frequency decreases when it is moving. The decreasing frequency of a strobe when it moves is an example of **time dilation.** Time dilation is the lengthening of time between events that occurs when observations are made between reference frames in relative motion.

To illustrate time dilation, consider again the hypothetical train moving down the tracks and an observer on the ground

Figure 15.6
A light beam traveling between two mirrors travels along the solid line according to an observer inside the train car. An observer outside the car would see the light take a path as indicated by the dashed lines.

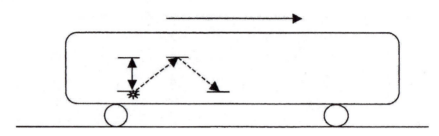

observing the train as it passes. Inside the train is a fictitious light clock. This light clock keeps time by sending out a beam of light toward a mirror where it is reflected back to its source (Figure 15.6). One second on the light clock is given by the time interval it takes for the beam to travel from its source and return to the mirror. For an observer on the train, 1 second is given by the time interval as it travels in a straight vertical direction to the mirror and back. This is no different than if the train was at rest. The observer watching the train would see light travel at an angle. It would not only move vertically up and down but also have a horizontal component because of the motion of the train. Since the distance the light travels in making its round-trip is greater in this situation, the time interval for one tick on the clock is lengthened. This means that for the observer outside that time has slowed down. With geometry, it can be shown that if the time for one tick of the clock is t_o for an observer on the train, then the time, t, for one tick for the observer outside the train is given by

$$t = \frac{t_o}{\sqrt{1 - \dfrac{v^2}{c^2}}}$$

In this equation, v is the speed of the train and c is the speed of light. The denominator in this equation appears in a number of other equations involving relativistic measurements such as length and mass. The value of $1/\sqrt{1 - v^2/c^2}$ is referred to as the Lorentz factor. Putting numbers into this equation illustrates time dilation. Assuming the train can travel at relativistic speeds and that t_o is 1 second, Table 15.1 gives various values for the Lorentz factor, symbolized by γ. The values for γ would equal the values for t. Table 15.1 shows that when the train is traveling at half the speed of light, that 1 second for the train-bound observer increases by about 15% for the earthbound observer. If the train could approach the speed of light and travel at 99% of c, time would be slowed downed by a factor of 7. This means that when 7 hours have passed on Earth, only 1 hour will have elapsed for an identical clock on the train observed from Earth. A passenger on the train will also measure a time interval of 7 hours. It is only when observations are made between different reference frames that relativistic effects must be considered.

The previous example shows that the time interval between two events is dependent on the reference frame of the observer. While the observer on the train would have

Table 15.1
Time Dilation

v	0	0.1c	0.2c	0.3c	0.4c	0.5c	0.6c	0.7c	0.8c	0.87c	0.9c	0.99c
γ	1	1.005	1.021	1.048	1.051	1.155	1.250	1.400	1.667	2.028	2.294	7.089

a clock that gave ticks with a frequency of 1 second, an earthbound observer would believe the same clock had a slower frequency. To distinguish between the two different reference frames, the frame in which an observer is at rest with respect to the events being observed is called the **proper reference frame.** In the previous example, an observer on the train would be in the proper reference frame when observing the light clock, and 1 second would be the proper time. The word "proper" is a relative term and shouldn't be interpreted to mean "correct." Proper means it is a result of viewing events in the same frame of reference. In this example, it is assumed that the train is moving and an observer on the ground is motionless. It is just as valid for an observer on the train to consider that the train is motionless and that an earthbound observer is moving past the train. In this case, an observer on the train would measure a light clock outside the train as running slower. The same dilation effect would apply for the train observer observing a clock on the earth, as for the earthbound observing the clock on the train. In situations such as this, the results are said to be symmetric or that symmetry exists.

An experiment to test the time dilation effects of special relativity was performed in 1971 by placing very precise cesium atomic clocks on commercial jet airliners traveling eastward and westward around the Earth. The motion of the clock traveling westward opposed the Earth's rotation, and the clock traveling eastward had motion in the same direction as the Earth. Special relativity predicted that the clocks should show contrasting differences with the atomic clock at the U.S. Naval Observatory, due to time dilation effects. After the three-day trip around the Earth, the clocks confirmed the results predicted, which were on the order of one ten-millionth of a second. Measurements on particles called muons created by cosmic rays in the upper atmosphere and in particle accelerators also displayed time dilation. These particles have extended lifetimes that allow muons to be detected at the earth's surface. The extended lifetime of muons is consistent with the time dilation predicted by special relativity. As technology has advanced in recent years, the effects of relativity have been incorporated into this new technology. One example where relativistic effects are important deals with advances in global positioning systems (GPSs). These systems are widely used for all forms of commercial transportation and in recent years have been available for the general consumer. Hand held GPS units costing several hundred dollars are capable of fixing a location on the earth within 10 m. Global positioning systems depend on a network of 24 satellites orbiting the Earth at an altitude of approximately 20,000 km. Each of these satellites contain an atomic clock that sends out signals to receivers. Because the satellites are orbiting the Earth, there is a time dilation effect when signals are received from them. These time dilation effects (along with gravitational general relativity effects) must be taken into account when fixing a position on Earth.

The time dilation effect with physical objects such as ticking clocks and physical processes such as radioactive decay extends to biological processes. Therefore, the beating of a heart and aging would represent such metabolic processes. This means that when a person on Earth observes a person in a rocket streaking by at $0.87c$, that the rocket person's watch not only runs two times slower (a speed of $0.87c$ gives a time dilation factor of approximately 2), but that person's heart also beats at half the rate. Compared to the person on the rocket, the person on Earth ages at twice the rate compared to the person in the rocket. This has led to the classical problem of the twin paradox in the discussion of relativity. The twin paradox consists of two identical twins. One twin remains on Earth while the other twin boards a rocket capable of flight at relativistic speeds. Assume the twin leaves the Earth and travels at $0.99c$. At this speed the time dilation factor is slightly greater than 7. Also assume that rocket travels for 21 Earth years for the trip out into space and takes another 21 years to return. If the twins were 30 years old when the rocket took off, the earthbound twin would have aged 42 years when the rocket returned, making this twin an elderly 72 years old. The rocket twin experiences a time dilation effect of approximately a factor of 7. This means each 7 years on earth is observed as a 1-year interval for the rocket twin. While the earthbound twin has aged 42 years, the rocket twin returns to Earth, has aged only 6 years, and looks like a 38-year-old person. All of this is consistent with time dilation. The apparent paradox arises when the situation is considered as symmetrical. That is, rather than assuming that the rocket travels at $0.99c$ away from a stationary Earth, it is just as valid to consider

the rocket stationary and assume that the Earth travels away from the rocket at $0.99c$. In this case, the earthbound twin only ages six years and the twin on the rocket has aged 42 years. Special relativity, general relativity (considered later in this chapter), or some combination of both can be used to explain the paradox. Special relativity is based on inertial reference frames. In the twin paradox, the Earth can be considered an inertial reference frame, but two reference frames are needed for the rocket: one as it travels away from Earth and another for the return trip. The two reference frames needed for the rocket traveler means that the situation for the rocket traveler and earthbound person are not symmetrical. A complete analysis of the twin paradox is beyond the scope of this chapter, but the analysis reveals that the when the two reference frames for the rocket are considered, the rocket traveler is always younger upon the return to Earth. Another factor to consider is that the theory of special relativity requires inertial reference frames. The rocket experiences several accelerations and decelerations in making its round-trip journey. These accelerations make the reference frame in the rocket noninertial. Therefore, the theory of special relativity cannot be applied to the situation. Using the general theory also results in the rocket twin aging less than the earthbound twin does. The results of the twin paradox have been used in a number of science fiction plots to account for space travelers returning to the future. The scenario generally has space travelers returning to Earth and, due to time dilation, several hundred years have past since they left the planet, for example, *Planet of the Apes*.

The discussion of time in this section has demonstrated that time is not absolute, but relative. Relativity requires the addition of time to the three-dimensional coordinates that define space. A more complete description of an event involves space-time. In this system, an event is defined by both its location and time. Space-time, the speed of light, and the vastness of the universe make the distinctions of past, present, and future relative. For example, consider an supernova (the violent explosion of a giant star in its latter stages of existence) that takes place at a location 100 light-years away from the Earth. When earthlings record the explosion as occurring in the year 2005, it means that is when we observe the supernova. Since the light from the event takes 100 years to reach the Earth, the supernova occurred in 1905, and earthlings are actually looking into the past when they see the event. Inhabitants on fictitious Planet X midway between the Earth and the event would see the explosion after 50 years, in 1955, and could predict our future by saying we will observe a supernova in the year 2005. Of course, this prediction could not be shared with us since a prediction requires an announcement before the event occurs. If a signal is sent as soon as possible, the best that Planet Xers can achieve is having the message arrive just after the supernova has already been observed on Earth. Using the same argument, inhabitants on Planet X observing the Earth would observe that Eisenhower was president, space travel had not commenced on Earth, and many of the readers of this book had not yet been born.

Length Contraction

In addition to time dilation, objects moving relative to an observer will appear to be shortened in the direction of their motion. The concept of length contraction can be illustrated by considering a trip to a planet 5 light-years away in a rocket capable of a speed of $0.87c$. The latter value is chosen so that the Lorentz factor is 2. The relative speed of the rocket to both the Earth and the planet is 0.87c. According

to a person on Earth this speed is equal to the total distance covered, 10 light-years, divided by time it takes to make the trip, which would be about 11.5 years. From the reference frame of a person in the rocket, the Earth and planet move by the rocket at the same speed of $0.87c$. The time it would take to cover the round-trip according to an Earth clock would be 5.75 years due to time dilation. In order for the relative speed of the Earth and the planet to be $0.87c$ with respect to the rocket, the distance between the Earth and the planet would have to be 2.5 light-years, which is half of the original distance.

The amount of contraction in the direction of motion is given by the equation. $L = L°/\gamma \ or \ \ L = L_0\sqrt{1-(v^2/c^2)}$ In this equation, the terms are the same as those used for time dilation, except that L is the contracted length and L_0 is the **proper length.** The proper length is the length observed when the object is at rest with respect to the observer. Using values from Table 15.1, a meterstick moving (in a lengthwise orientation) at a relative speed of $0.87c$ would appear as though it was only 0.5 m long. The contraction of an object according to special relativity is a distortion of space-time and not an illusion. The contraction is real in the sense that observations of the world around us determine reality and these observations are made from particular frames of reference. To say that an object is contracted means that length is a relative concept. It would be misleading to interpret the proper length as the real length. The proper length is no more real than the length observed from any other reference frame.

Mass, Momentum, and Energy

In addition to time dilation and length contraction, the special theory of relativity predicts an increase in mass of an object as its relative speed to an observer increases. The increase in mass is given by γm_o, where m_o is the **rest mass** or **proper mass** of the object. Einstein's special theory of relativity requires that for linear momentum to be conserved, the Lorentz factor must also be applied to the momentum equation. The momentum of an object is properly given by the equation $p = \gamma m_o v$ rather than $p = m_o v$, as presented in chapter 3. The relativistic momentum of an object only needs to be considered for objects moving at relativistic speeds. The increase in mass and momentum is illustrated when subatomic particles are accelerated to high speeds in particle accelerators. Electrons accelerated to near the speed of light in particle accelerators acquire masses that are approximately 40,000 times greater than their rest mass. A proton accelerated to 99.99% the speed of light will have a relativistic mass 71 times greater than its rest mass.

In addition to the increase in mass of an object in relative motion, Einstein derived a connection between mass and energy. Einstein showed that an object moving with relative speed v has a total energy equal to

$$E = \frac{m_o c^2}{\sqrt{1 - \dfrac{v^2}{c^2}}} = mc^2$$

The equation $E = mc^2$ is probably the most famous equation in science. In this equation, m_o is the rest mass and m is the relative mass. The equation reduces to $E_o = m_o c^2$ when an object is at rest, that is, when $v = 0$. The rest energy is denoted by E_o. The energy equation demonstrates that energy and mass are equivalent. When an object accelerates to a relative speed, it acquires kinetic energy. The object's total energy equals its rest energy and kinetic energy.

Using the equation $E_o = m_o c^2$ shows that a tremendous amount of energy is associated with a small amount of mass. One gram of matter has an energy equivalent of 9×10^{13} J. This would be enough energy to light a 100 W bulb for 30,000 years. Changes in mass accompany both fission and fusion nuclear reactions. In the former, a heavy nucleus, such as uranium, splits into smaller nuclei, and in the process a small amount of mass is converted to energy. This is the basis of nuclear power. In nuclear fusion, light nuclei come together to form a heavier nucleus. This process also results in a conversion of mass to energy. Nuclear fusion is the process that fuels the Sun.

General Relativity

The general theory of relativity extended the special theory to include mass and gravitation and extended Einstein's idea of spacetime. Einstein's special theory focused on inertial reference frames. Accelerating reference frames could be interpreted with special relativity, but the mathematics was difficult. Einstein realized that the special theory was limited. Newton's theory of gravitation did not account for time effects. The force of gravity between objects had to be communicated at a finite speed limited by the speed of light. Newton's definition of mass also posed problems. Mass can be defined as **gravitational mass** or **inertial mass.** The former was a static measurement that was a measure of the attraction between two objects as given by the inverse square law, $F = Gm_1 m_2 / r^2$. Specifically, the mass of an object was related to the attraction between the Earth and the object. In contrast, inertial mass was a dynamic measurement given by the equation $F = ma$. Even though the gravitational and inertial masses of an object were considered equivalent, there was no reason why they should be. Einstein wrestled with these basic problems, and this led him to the development of the general theory of relativity in the decade following publication of his special theory.

Einstein's thoughts on gravitation were supposedly inspired while he was sitting in the patent office in Bern in 1907 and observed through the window a painter falling off a scaffold. This led Einstein to one of the basic tenets of general relativity known as the **equivalence principle.** Einstein's equivalence principle expanded upon the equivalence principle of Galileo and Newton. This equivalence principle equated inertial and gravitational mass. According to this basic version, also known as the weak equivalence principle, all objects regardless of their mass fall with constant acceleration in a gravitational field. Einstein's equivalence principle reformulated the classical interpretation of gravity in terms of the curvature of space-time, and thereby he was able to incorporate gravity into his special theory. According to Einstein's equivalence principle, acceleration and gravity are equivalent. This means the laws of physics are the same in a uniform accelerating reference frame and in a uniform gravitational field. Einstein's equivalence principle can be illustrated by considering a situation of acceleration in a remote region of space where the effects of gravity from objects with mass (galaxies, stars, etc.) are ignored. Einstein used thought experiments of this type in formulating his general theory. First consider two elevators (as Einstein did): one at rest on Earth and one at rest in remote space in the absence of a gravitational field (Figure 15.7). If a person on the Earth elevator drops a ball, the ball will fall to the floor of the elevator with an acceleration of g. A ball dropped in the space elevator, at rest and in the absence of a gravitational field, floats next to the person. A similar result occurs if the elevator moves at a constant

Figure 15.7
Accelerating upward in an elevator is equivalent to the effects of gravity

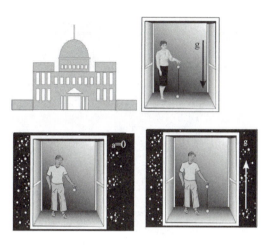

speed. As long as the space elevator and its contents all move at a constant speed, there are no gravitational effects. A similar result is approximated when a spacecraft heads off toward the Moon and cruises at a constant speed. Away from the Earth and Moon, the spacecraft acts as an elevator in remote space, and the effects gravity are removed. Now consider what happens if the space elevator accelerates upward at a rate equal to g. When the ball is released, the floor of elevator accelerates up to it, and the ball strikes the floor. To an observer in the space elevator, the motion of the ball is exactly the same as what a person in the Earth elevator observes when dropping the ball. There is no way for an observer to determine whether she is in an elevator on the Earth's surface or accelerating in remote space at a rate of g. Using this type of reasoning, Einstein postulated that the effects of a uniform accelerating reference frame and gravitational field are the same.

As in special relativity, Einstein surmised that the equivalence principle holds for all laws of physics, including

Figure 15.8
In an accelerating elevator, a light beam will appear to travel in a straight line for an outside observer using distant stars as the frame of reference. For someone in the elevator, the light will travel in a parabola, as indicated by the dotted line, as the elevator accelerates upward.

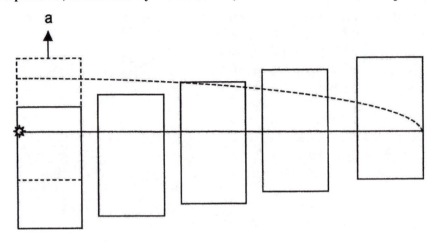

electromagnetic phenomena. Expanding upon the elevator thought experiment gives insight into the nature of light. Consider a beam of light that travels from one wall of the space elevator to the opposite wall while the elevator is accelerating upward. For an observer outside the elevator, the path of the light beam would be straight with respect to the distant stars and would strike the opposite wall at a point closer to the bottom of the elevator (Figure 15.8). An observer in the accelerating elevator would see the light take a parabolic path across the elevator and strike the opposite wall at the same point observed by the outside observer. The equivalence principle means that since light is bent in an accelerating reference frame it would also be bent in a gravitational field. The first major experimental test of general relativity involved observation of the bending of light by the gravitational field of the Sun. In 1919, an expedition was mounted to photograph the Sun during a total solar eclipse. According to the general theory, light from stars that were positioned behind the Sun would be visible due to the bending of light around the Sun by its gravitational field. Normally these stars would not be visible due to the brightness of the Sun obscuring them. The shift in the position of the stars could be ascertained by comparing photographs taken during the solar eclipse and comparing these to photographs taken when the light from didn't pass by the Sun (Figure 15.9).

The bending of light by gravity objects in the universe gives rise to what is known as a gravitational **lens.** Just as a piece of glass can bend light, stars and galaxies can bend light. This results in multiple images from selected regions of space as light from an object is bent in multiple paths as it passes a massive object. One of the most dramatic results from the general theory and the bending of light is the prediction of **black holes.** A black hole can be considered a region of the universe so massive that light and radiation cannot escape from the region. One method of their formation is thought to be from the collapse of a massive star during

Figure 15.9
The bending of light in a gravitational field was demonstrated in 1919 by observing the position of stars during a solar eclipse, and when starlight did not pass the Sun. When the Earth is in position 1, the light from the star bends around the Sun and is observed in the apparent position. Six months earlier, when the Earth is in position 2, the star is observed in its actual position.

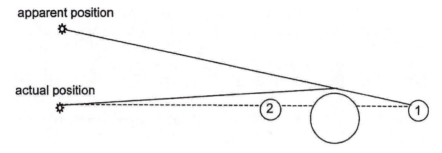

Figure 15.10
Precession of Mercury's orbit provides evidence for general relativity

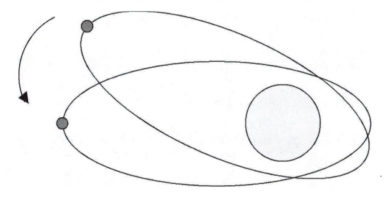

the last stages of its life cycle. The action of a black hole can be understood by considering how objects travel on Earth. When a ball is thrown up in the air, it naturally returns to Earth. In order for the ball to leave the Earth, it has to obtain escape velocity. The escape velocity is the minimum velocity an object needs to escape the gravitational field of the Earth. It is approximately 40,000 km/h (25,000 mph). The escape velocity depends on the strength of the gravitational field. For example, the escape velocity on Jupiter is about 214,000 km/h (128,000 mph). A black

hole is so massive that the escape velocity exceeds the speed of light. While light cannot escape from a black hole, the black hole affects the region around it to give indirect evidence of its presence, for example, pulling matter from a visible object into it.

Another astronomical observation that supports the general theory is the precession of the orbit of Mercury. The orbit of Mercury changes over time, as shown in Figure 15.10. Changes in the orbit of Mercury were predicted using Newton's laws applied to Mercury's gravitational interaction with other bodies in the solar system, but there was an unaccounted error in the predictions. Mercury is the planet closest to the Sun and, therefore, would experience the greatest relativistic effects predicted by the general theory. The orbit of Mercury is a type of clock that will be affected by the gravitational field of the Sun. The precession of Mercury's orbit is 570 seconds per century (1 second is 1/360 of a degree). Approximately 40 seconds of this 570 couldn't be accounted for by using Newtonian physics. Calculations based on general relativity accounted for this discrepancy.

Rather than consider an attractive force of gravity, Einstein's concept of gravitation is put in terms of the curvature of space-time. There is a mutual relationship between space-time and mass. Mass causes distortions in space-time, and space-time dictates how matter moves through it. The curvature of space can be pictured as an elastic sheet such as the top of a trampoline. When the trampoline has nothing on its top, it is flat. When a person stands on top of the trampoline, it is compressed, creating a depression under the weight of the person. Water poured onto the trampoline near the person will run down into the depression. In a similar manner, mass tends to bend the space around it. The universe can be considered a continuum of space-time with hills and valleys around regions of matter. Thus, a gravitational field results from the bending of space-time by mass. Objects entering this local space can be considered to have their motion changed from a natural state of free fall in a region of flat space to motion along curved spaced.

The curvature of time is another result of the general theory of relativity. The curvature of time means that time is dilated in the presence of gravitational field. This means that time passes more slowly for someone on the first floor of a building compared to someone on an upper floor. The slowing of time in a gravitational field affects both physical and biological clocks. Therefore, people living in Denver, the mile-high city, age more slowly than people living at sea level. The time dilation on the Earth's surface is about 1 second every 32 years. In 1958, Rudolph Mössbauer (1929–) discovered a process where gamma rays were emitted and absorbed by atomic nuclei with little loss of energy. The process was named the Mössbauer effect in honor of its discoverer, who received the 1961 Nobel Prize in physics for the work. The process involved the emission and absorption of gamma radiation by atoms held tightly in a crystalline lattice. The Mössbauer effect made it possible to use atomic nuclei as atomic clocks. The Mössbauer effect was used in 1960 by Robert V. Pound (1919–) and doctoral student Glen A. Rebka to test general relativity. Pound and Rebka looked at the change in frequency of gamma radiation as it was emitted and absorbed at the bottom and top of a 75-foot elevator shaft in the Jefferson Tower at Harvard University. According to general relativity, the frequency of radiation decreases as its leaves a gravitational field. A lower frequency equates to a longer wavelength. The increase in wavelength when moving from a region of stronger to weaker gravity is termed the **gravitational red shift.** The shift occurs because as radiation works against gravity, it loses energy. A loss of energy results in a lower frequency and longer wavelength. Pound and Rebka were able to measure the gravitational red shift with the Mössbauer effect as gamma radiation from iron, Fe-57, traveled up the Harvard tower. The change in frequency measured agreed with the results predicted from the general theory.

Summary

This chapter introduced the theory of relativity and a few of the results that this theory predicts. Relativity stretches the imagination and forces us to "think outside the box." Einstein's theory of relativity builds upon and expands Newtonian physics so that phenomena such as electromagnetic radiation and relativistic motion can be understood. Although relativity is a more comprehensive theory than Newtonian physics, we continue to rely upon the simplicity and intuitive nature of the latter for common understanding. Newtonian physics works perfectly well for putting people on the Moon, building cities, and analyzing motion at our tortoiselike (compared to light) pace. Practical applications of relativity do exist, and as technology continues to develop, a relativistic interpretation will become increasingly necessary. Today an understanding and appreciation for relativity is still in its infancy. Our knowledge of relativity has just scratched the surface but has contributed greatly to understanding the universe. Speculation concerning aspects such as the fate of the universe, space travel, wormholes, and multiple universes has provided ample material for science fiction, but today's science fiction may be tomorrow's reality. Galileo and Newton created a physics that sufficed for 200 years; we can only guess how long Einstein's physics will serve us. The wonderful aspect of nature is that it is open to understanding using science and imagination. As Einstein himself stated: "The most incomprehensible thing about the world is that it is at all comprehensible."

16

Physics Experiments

Introduction

Science begins with careful observation using all of our senses. Most people associate observation with the sense of vision, but our senses of smell, taste, hearing, and touch are also used to observe. Observations accompanying physics often occur through senses other than sight. For example, the roughness of a surface as related to friction can be determined by feeling the surface, and hearing is particularly important when studying sound. Nearly all of our daily observations are accepted at face value, but frequently we make observations that don't fit our expectations. Many of these observations are quickly dismissed, but some tend to pique our curiosity and focus our attention. We might question whether we can believe our senses and then attempt to verify a strange observation. If we trust our senses, we might begin to ask questions to explain our observations. Many paths can be taken when trying to explain observations, but the one used by scientists is termed the scientific method. In this chapter, we will take a brief look at the scientific method, discuss some elements of experimentation, and conclude with a series of physics activities and experiments.

The Scientific Method

As stated in the previous section, questions naturally arise from observations made in daily activities. Questions may also arise from something read, a discussion, or an experience. There are various methods used to seek the answers to questions. Simply consulting another person, looking in a book, or searching the Internet is often sufficient to supply the answer. Scientific questions are questions to which the scientific method can be applied. The scientific method is a systematic procedure used to answer questions in order to develop logical explanations for the world (universe) around us. The general scientific method follows a common pattern for answering a question. The procedure consists of developing a question into a testable **hypothesis,** developing an experiment to test the hypothesis, performing the experiment, and using the results to come to some conclusion about the original question. As described in this fashion, the scientific method seems as though it's a straightforward procedure for seeking an answer. In reality, the scientific method often follows a circuitous path. Applying the scientific method is like driving in different locations. In familiar locations, there is little problem of getting from one place to another. Several routes may be used, but overall steady progress is made in arriving at a destination. Conversely, driving in an unfamiliar area, such as a large city, often consists of wrong turns, backtracking, and running into dead ends. Even with a road map and directions, the path taken can be less than direct. In the rest of this section some of the key elements of the scientific method will be briefly discussed.

It is important when using the scientific method to formulate a scientific question about a subject and then use this question as a guide toward further knowledge on the subject. Posing a question may seem trivial, but great scientists are often characterized by having asked the right question. This often means looking past the obvious and focusing on questions that haven't been asked. Alternatively, perhaps everyone else has ignored the obvious and overlooked a very basic question. Scientific questions must be clearly defined. Students starting projects often use terms loosely in posing questions. For example, a question such as "Which battery is better?" is unclear. Questions arise concerning what is meant by "better" (cheaper, longer life, more powerful), under what conditions, and the type of battery. An alternate question might be "How does the life of rechargeable batteries on a single charge compare to nonrechargeable alkaline batteries in audio equipment?" Formulating an appropriate question leads the researcher to developing a testable hypothesis.

A hypothesis is a possible explanation that answers the question. It is often presented as a concise statement of what the outcome of an experiment will be. A hypothesis is often defined as being an educated guess. Before a hypothesis is stated, as much background information as possible about the subject should be gathered. In fact, it may be that an answer to the question will come from the preliminary research. Books, journal articles, the Internet, and experts on relevant subjects can be used to guide the researcher in refining the question and developing a hypothesis. Researching the subject allows the researcher to draw tentative conclusions based on the work of others. This is why the hypothesis is only a possible explanation. Researching the subject can help identify important **variables** in formulating the hypothesis. Variables are conditions that change and may affect the results of an experiment. For example, in a battery study comparing the life of batteries it may be important to know how batteries operate at different temperatures. If the purpose were to compare battery use of flashlights used both indoors and outdoors, then temperature would be an important variable to consider. This in turn would affect how the hypothesis was formulated, and ultimately how the experiment was designed.

Once a hypothesis is formulated, an experiment is designed and conducted to test the hypothesis. Experimentation is what distinguishes physics, and other experimental sciences, from other disciplines. In simple experiments, the researcher designs the experiment to examine the effect of one **independent variable** on a **dependent variable.** The independent variable is systematically changed variable while attempting to hold all other variables constant. Again, using the battery example, if the experiment is designed to test battery life in flashlights at different temperatures, then temperature would be the independent variable and battery life the dependent variable. It would be important to define in exact terms the variables "temperature" and "battery life" precisely and how these would be measured. The researcher would then attempt to **control** all other variables, thereby conducting a controlled experiment. The type of flash-light could be controlled by using the same or identical flashlights. The same type and brand of batteries would be used in different trials. Although the researcher makes every attempt to control for every variable except the independent variable, it is impossible to control for all variables. In the flashlight example, even using the same flashlight in different trials means that one set of batteries would be placed in a flashlight with a bulb that was slightly more used. On the other hand, the researcher might use several identical flashlights and conduct the experiment, but there would still be unknown differences between the flashlights that couldn't be controlled. A researcher can never control for all extraneous variables but must make every effort to control for those believed to significantly affect the independent variable.

The battery example demonstrates why it is important to identify the important variables when formulating the hypothesis and designing the experiment. A few variables will generally be identified as important to control; many variables will be ignored. For example, variables such as air pressure, relative humidity, position of flashlight during the experiment, color of batteries, and so forth would seemingly have no affect on battery life. It is important to remember that an experiment without proper controls is generally useless in reaching any valid conclusions concerning the hypothesis.

In conducting an experiment, it is important to keep careful and accurate records. The standard method for doing this is using a laboratory notebook. Official laboratory notebooks are sold at bookstores, or a notebook of graph paper may be used. A simple laboratory notebook can be made by placing sheets of graph paper in any three-ring binder. The first several pages of the notebook should be left blank to allow for a table of contents. Your name and phone number should be on the inside cover along with any emergency information. The latter is important in situations where an accident might occur and authorities may have to be contacted. The laboratory notebook is used to maintain a comprehensive record of the experiment. Entries should be dated, and important relevant information on the experiment should be recorded. Information such as equipment used, notes, times, observations, questions, calculations, experimental conditions, and so on should be written in the laboratory notebook. By going back through the laboratory notebook, it should be possible to retrace the steps taken during the experiment. This is important in analyzing why things worked or didn't work as expected. It is also important to use the laboratory notebook to identify areas where the experiment can be improved or modified. Results recorded in the laboratory notebook should provide the basis for future experiments. What you record in the notebook should be written neatly in ink. Changes should be crossed out with a single line, enabling the crossed out material to be read.

Once the experiment has been conducted and data collected and analyzed, a conclusion can be drawn concerning the original hypothesis. The experiment does not establish absolute truth concerning the hypothesis. A researcher should resist the temptation to state that the hypothesis was proven or not proven. In science, absolute truth can never be established. Results of an experiment may either support or not support the hypothesis. Ambiguous results can cause the researcher to modify the hypothesis and redesign the experiment. Even when the results are not ambiguous, an experiment is often repeated. One of the benchmarks of science is reproducibility. When an experiment is conducted and certain conclusions reached, another researcher should be able to repeat the same experiment and obtain similar results.

When analyzing the results of an experiment, it is important to consider sources of errors and how these sources affected the results. Two primary types of errors occur in experiments: **random error** and **systematic error.** The researcher has no control over random error. Random error involves the variability inherent in the natural world and in making any measurement. As its name implies, random error varies in a random manner. An attempt is made to control for random error by taking multiple measurements. For example, say we wanted to know the mass of an average penny. Naturally, older pennies might be expected to have smaller masses due to wear, and newer pennies more mass (we will ignore the fact that penny composition changed in 1982). If a large random sample of pennies was used, it would be expected that there would be a good mixture of pennies. A large sample of pennies should include pennies of many ages. When measured on a scale, some pennies would have more than the true mass of a penny, others less, but it is assumed that these would tend to balance out and give a fairly accurate measure of a penny's mass.

Systematic error is error that occurs in the same direction each time. For example, if a scale had a dirty pan, each measurement would be increased by the mass of dirt on the pan. Another example of introducing systematic error in an experiment might be reading the meniscus level of mercury in a barometer at an angle. Looking down from above on the mercury when measuring its volume would systematically underestimate the volume. Systematic error can be eliminated using proper techniques, calibrating equipment, and employing standards. An example of accounting for systematic error is checking a thermometer to see if it measures $0°C$ in a freshwater-ice mixture and $100°C$ in boiling freshwater. If both measurements are $2°C$ high, then it can be assumed that a systematic error of $+2°C$ exists in the temperature measurements made with this thermometer.

Physics Experimentation

The most important aspect of conducting physics activities and experiments is safety. The excitement of physics is often portrayed to the public through dramatic demonstrations. The caveat "don't try this at home" is often spoken in jest but should be taken seriously. Individuals performing public demonstrations have practiced them and honed their skills in science classrooms and museums. They know what conditions and safety precautions are needed to conduct the demonstration safely. Physics demonstrations, like magic tricks, are often "hyped" for the audience. Nevertheless, if you are not thoroughly familiar with a physics activity and its possible hazards, do not attempt it. This is especially true when working with electricity.

Even when doing simple experiments it is important to practice safety. It is a good habit to wear safety glasses, avoid loose fitting clothing, have extinguishers and other safety equipment available if fire is involved, and when in doubt, ask questions. The simple activities listed in the following section can be performed safely at home if directions are followed. Before doing any activity or experiment written by someone else, it is important to read and think it through before actually conducting it. Gather items to conduct the activity, and do the activity in a location that is safe and won't damage property. Safety begins and ends with common sense. If a procedure is unclear, advice should be sought from an adult.

One of the difficulties in doing physics at home is obtaining materials. Much of your own equipment can be built or obtained from local stores for building materials, auto supplies, aquariums, and gardening. Hardware stores carry a number of items useful in doing physics experiments. A small collection of inexpensive instruments may be assembled by searching the shelves of various stores, second-hand shops, and garage sales. A simple stopwatch or wristwatch with a stopwatch feature can be found at any discount stores for a few dollars. Eyedroppers, tongs, tweezers, syringes, funnels, and measuring cups can be used without modification. Rubber or plastic tubing can be obtained from hardware or automotive stores. Alcohol thermometers can be found for less than $2; they are usually encased in a housing from which they can be removed. Mercury thermometers should not be used. A postage scale can substitute for a balance, a small camping stove for a burner, Pyrex cookware for beakers, and a warming plate can be used as a hot plate. If it is safe, gas or electric ranges, microwaves, and regular ovens, and refrigerators may be used to conduct experiments. Motors can be salvaged from old appliances, but care should be taken to observe electrical safety. Old skateboards and skates are a good source of wheels. Pulleys, springs, and weights can be obtained from the fishing section of discount stores or sporting shops.

Physics Activities

This section contains physics investigations that may be conducted at home. Most are not full experiments but should serve to provide useful activities that could be developed into more thorough investigations. Each illustrates basic concepts explained in the previous chapters. The purpose of this section is to pique interest in doing physics rather than reading about physics. It is not the intention to provide a comprehensive set of activities, but to present a few activities that illustrate basic concepts. Most activities involve only a few simple steps, while some are more complicated. Again, it should be emphasized that the most important aspect of these activities is to be safe, have fun, and learn. Before actually trying to do the activity, make sure to read and understand the activity fully. Assemble all materials before starting and do the activity in a safe place.

Activity 1: Simple Pendulum 1

The study of a simple pendulum is a basic experiment performed by many students just beginning a study of physics. A simple pendulum consists of a mass attached to a length of string or thin wire. The mass is assumed to be concentrated at a single point at the end of the string, and the string itself is assumed to have no mass. A simple pendulum is assumed to be frictionless, with no air resistance as the mass swings back and forth. A simple pendulum can be made using a thin length of string (dental floss is ideal) and a mass such as a metal nut or washer. The nut should be attached to one end of the string by making a loop in the string and threading it through the whole in the nut and looping it back on itself. Using a loop rather than tying the nut makes it easier to remove the nut and replace it with a different mass. Hang the free end of the string from any suitable support. This can be a coat rack in a closet, a knob, or a pencil taped to a table so that it overhangs the edge. It is important to have an area that allows a free-swinging pendulum. The goal of this activity is to determine what variables affect the period of a pendulum.

The period is the dependent variable, and one other variable is examined, the independent variable, while other variables are controlled. The period is the round-trip time it takes for the pendulum to make one round-trip swing from its starting position back to its starting position. It is best to use a stopwatch; watches with a stopwatch feature are available at discount stores for just a few dollars. If a stopwatch is not available, a watch with a second hand will suffice. In order to time the period, the total time for a number of swings should be divided by the number of swings. Observe at least 10 swings to make a good measurement of the period. If 10 periods takes 22 seconds, the period is 2.2 seconds. Further details of this activity are omitted here to enable you to perform the experiment. They will be addressed in the next activity, Pendulum 2. This will allow you to perform the experiment with a minimum of bias, in a true scientific fashion.

Activity 2: Simple Pendulum 2

In the activity Pendulum 1, you were challenged to determine what variables affect the period of a pendulum. The important variables to explore are the amount of mass on the end of the pendulum, the length of the string, and the amplitude of the pendulum. Using a loop at one end of the string enables masses to be easily changed and may have alerted you that mass was one variable to explore. This variable can be examined by using different size washers or adding multiple washers for different trials. The trials with different masses must take place by controlling for other important variables. When examining the variable "mass," the length and amplitude of the pendulum should be controlled. Since the same string (wire) is used for all trials, the type of pendulum is controlled. The amplitude is the angular displacement of the mass from the vertical position. Unless you employed a protractor or some other means to measure the displacement angle, this variable has to

be estimated. One way to control for amplitude is to hang the pendulum near a vertical object such as a wall and always release the weight from the wall. The mass should not affect the period of the pendulum.

The amplitude of the pendulum affects its period, but not appreciably for angles less than about 20°. This is referred to as the small angle assumption. The period of the pendulum increases as the amplitude increases. For a pendulum that is 1 meter long, the period increases by 0.01 second between amplitudes of 10° and 20°. At an amplitude of 50°, the period would be almost 0.1 second greater than at 10°. The small change in period with amplitude can't be detected without a stopwatch, and even with a stopwatch, it will be difficult to detect. When performing trials with the pendulum it is best to keep the displacement at a constant small angle.

The one variable that affects the period of a simple pendulum measurably is its length. The length of a pendulum is the distance between the point of suspension and the center of the suspended mass. A simple pendulum's period increases with its length. The formula for the period, T, of a simple pendulum is $T = 2\pi\sqrt{l/g}$, where l is its length in meters and g is the acceleration due to gravity, 9.8 m/s². If the pendulum distance is measured in feet, the value of g is 32 ft/s², and if measured in inches the value of g is 384 in/s².

Using the formula, data collected with your pendulum can be compared to the period calculated from the formula. The difference between the calculated and observed values is a measure of the experimental error. The error also incorporates the assumptions and variables that can't be controlled, for example, friction and wind resistance. The equation for a simple pendulum gives a period of 2.0 seconds for a 1 meter (3.28 yard or 39.4 inch) long pendulum. Science has taken advantage of the regular motion of pendulums. Pendulum clocks rely on the regular motion of a mass called a "bob" to keep accurate time. By measuring the period of a standard pendulum 1 meter long, the acceleration due to gravity at various points on the Earth's surface was measured during scientific expeditions throughout the eighteenth and nineteenth centuries. Typically, scientists would carry calibrated pendulums aboard ships to make measurements when they went ashore. Scientists on land expeditions set up stations at different locations on the globe. Results of these expeditions showed that the acceleration due to gravity was not constant, but varied. This eventually led to the discovery that the Earth was not a perfect sphere, but an oblate spheroid with a greater equatorial diameter.

This activity illustrates many of the concepts of the scientific method. The important independent variables considered for the pendulum were weight, length, and amplitude. Many other variables affect the pendulum's, but these were not considered significant and not controlled. For example, variables such as temperature, humidity, atmospheric pressure, shape of the pendulum, type of string or wire, and latitude can all be considered to affect the pendulum's period. These were ignored in a simple experiment but need to be considered to make more precise measurements. For example, fluctuations in temperature would cause the expansion and contraction of the pendulum, and this would affect the period. Atmospheric conditions would produce different air resistance as the pendulum swings, and different types of string or wire would stretch as more mass is added to them. These few examples demonstrate that perfect control in any experiment is never possible.

Activity 3: Reaction Time

The time it takes to physically react to an event is important in many instances. For example, an object in the road might cause a driver to suddenly apply the brakes. Reaction times vary depending on factors such as the person's age and physical state. In this activity, you can measure reaction time by using the equations for acceleration due to gravity. Two people are required for this activity. A qualitative demonstration of reaction time can be made by challenging someone to catch a falling dollar bill (or any other type of currency bill). One person extends an arm with the thumb and forefinger about an inch apart in a pinching position. This person is the catcher. The other person takes a flat, unwrinkled bill and holds it at one end, letting it hang between the catcher's thumb and forefinger. The dollar bill should be centered so that the thumb and forefinger are on opposite sides of the head on the bill, for example, George Washington for a one-dollar bill. The person holding the bill releases it without warning and the goal of the catcher is to catch it by only closing the two fingers together. The catcher can't move his arm down as the bill falls, but can only close his fingers to catch the bill. In order to prevent "cheating," the arm can be rested on a flat surface so that it can't be moved down. A person responding to the unknown time of release of the bill can't react fast enough to catch it. Although it seems like a simple task, human reaction time is too slow to catch the bill. The reaction time in this activity includes the perception time, which is the time it takes to observe the event of the bill's release and mentally process the decision to react, and the movement time, which is the time it takes for the fingers to close.

Rather than just trying to catch a bill, a quantitative measure of reaction time can be made by replacing the bill with a ruler and measuring the distance the ruler falls before it is caught. When using the ruler, have the catcher position the two fingers around the 1 inch mark. In this activity, the dropped ruler will drop a certain distance before the person reacts, closes the fingers, and catches the ruler. The distance the ruler drops can be measured by observing where it is caught. For example, if it is caught at the 6.5 inches, it dropped 5.5 inches. One inch has to be subtracted from the position where it is caught to account for the fact that it was at the 1 inch mark at the start. The following table can be used to measure the catcher's reaction time. This table is generated using the equation for the distance and the acceleration due to gravity: distance = $1/2(gt^2)$. This equation can be rearranged to give reaction time: $t = \sqrt{2d/g}$. If the distance the ruler falls, d, is measured in centimeters, then the appropriate value for g is 980 cm/s².

Under optimal conditions, the reaction time for a driver is about 0.200 s. In addition to the time it takes to react, the driver must depress the pedal and the vehicle's brake must respond to the brake pressure. The distance a vehicle travels during perception, reaction, and the mechanical engagement of the brakes is referred to as the thinking distance during the braking. The distance traveled after the brakes actually engage is the braking distance. The thinking distance in feet is about equal to the speed at which a vehicle is traveling in miles per hour. At 30 mph, the thinking distance is about 30 ft and the total stopping distance (thinking distance + braking distance) is 75 ft, while at 60 mph, thinking distance is 60 ft and total stopping distance is 240 ft. Both

Inches Fallen	Centimeters Fallen	Reaction Time (seconds)
4.0	10.2	0.144
4.5	11.4	0.153
5.0	12.7	0.161
5.5	14.0	0.169
6.0	15.2	0.176
6.5	16.5	0.184
7.0	17.8	0.191
7.5	19.1	0.198
8.0	20.3	0.204
8.5	21.6	0.210
9.0	22.9	0.216
9.5	24.1	0.222
10.0	25.4	0.228

thinking and braking distances vary greatly depending on conditions. As previously mentioned, thinking time will depend on the age and condition of the driver as well as factors such as whether it is day or night, the amount of daylight if it is day, and the type of car. The braking distance depends on the condition of the tires and road surface as well as weather conditions, for example, dry versus wet pavement.

Activity 4: Projectile Motion

One of the most fundamental areas of study in mechanics is that of projectile motion. Projectile motion was important in ancient warfare as armies launched objects using catapults and other contraptions to attack an enemy's fortress. Arrows shot from bows also required an intuitive knowledge of projectile motion. Although ancient ammunition has been replaced by artillery shells and missile warheads, their motion is still described by projectile motion. Projectile motion is also readily apparent in many sports, such as when a batter hits a fly ball or a golfer hits a tee shot.

A simple way to study projectile motion is with a garden hose. A garden hose can be used to determine what launch angle gives the greatest down-range distance to a projectile. To determine the angle that produces the greatest distance, simply turn on a garden hose with a spray nozzle set to produce a steady stream of water. By pointing the nozzle at various angles and noting how far the water travels, you can get a general idea of the angle that gives the greatest distance. To quantify this, a large protractor can be made by transferring the angles from a regular protractor to piece of cardboard as, shown below. The cardboard protractor can be covered with clear plastic or placed in a large zip-lock bag to waterproof it.

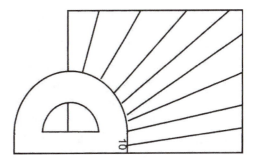

When using the large protractor, align the bottom of it parallel to the ground and to the nozzle. You should discover that the maximum range occurs when the nozzle points at a 45° angle. The angle may be slightly less than 45° due to friction.

Activity 5: Coefficient of Restitution

The coefficient of restitution is a measure of the elasticity of a collision. An elastic collision occurs when kinetic energy is conserved. For an object such as a ball, dropped on a floor, the coefficient of restitution is given by the equation v_s/v_a, where v_s is the speed of the object just after striking the floor and v_a is the approach speed or speed just before striking the floor. In a perfectly elastic collision, $v_s = v_a$, and the coefficient of restitution would be 1. Because it is difficult to measure the speed of dropped objects just before and after striking a surface, the rebound height of a dropped object is used to determine the coefficient of restitution. If an object is dropped from a height h and rebounds to a height h_r, then the coefficient of restitution is given by $\sqrt{h_r/h}$. The coefficient of restitution for dropped objects will depend primarily on the object's composition, the height from which it is dropped, and the type of floor surface. The coefficient of restitution for different balls on various surfaces can be measured. Coefficients of restitution for several common balls on a hard surface should be approximately 0.70 for a tennis ball, 0.55 for a baseball, 0.75 for a basketball (properly inflated), and 0.85 for a superball. The effect of temperature on the coefficient of restitution can be tested by placing tennis balls or baseballs in a freezer and comparing values obtained with balls at normal room temperature. An example of a situation with a very high coefficient of restitution is a small steel ball bearing bouncing on a hard steel surface. This is illustrated with the common toy consisting of several small steel balls that hang vertically, barely touching each other. When the end ball is released, a series of highly elastic collisions causes the balls to stay in periodic motion for an extended period.

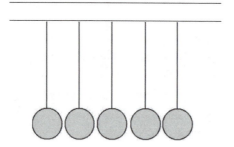

Activity 6: Momentum

To examine the conservation of momentum and build on the previous activity, two different size superballs can be used to project the smaller superball to a significant height. Do this activity outside or in a large room such as a gymnasium to ensure you have ample vertical height. Place the smaller ball on top of the larger ball and drop the combination vertically from waist height. The superball on top will be launched with a speed of three times its speed just before the collision with the floor and be projected to an appreciable height. Using laws of conservation of momentum and energy, it can

be shown that the height the small superball should reach is almost nine times the height from which it is dropped. The actual height the small ball reaches will be less than nine times its dropped height because the collisions are not perfectly elastic.

Activity 7: Center of Mass Balancing Act

The center of mass of an object is the point where it can be assumed that all of the mass of the object is concentrated. For example, for a solid sphere of uniform density, the center of mass would be a point at the geometric center of the sphere. Center of gravity refers to the point where the weight of an object is concentrated. For our purposes, center of mass and center of gravity can be assumed synonymous, although in special circumstances, such as large objects over which the force of gravity varies, they can differ. For an object to be stable, its center of mass must be aligned vertically with a point of support. If the center of mass of an object and the point of support are not aligned, a torque produces a rotation, and the object topples over. A number of toys that show objects that appear to balance in awkward positions are based on this principle. This activity will show you how to construct several balancing arrangements.

To balance a clothespin on the end of your finger, use a traditional wooden clothespin (the type that is a piece of wood with a slot, not the spring type) and a leather belt. Thread the belt through the clothespin so that it divides the belt into approximately equal halves. With a little adjustment, you should be able to balance one of the clothespin's prongs on the end of your forefinger. If your finger is slippery, wet it by licking it. Another variation of this arrangement is to take a loop of string and balance a ruler off the end of a table with a hammer, as shown in the diagram.

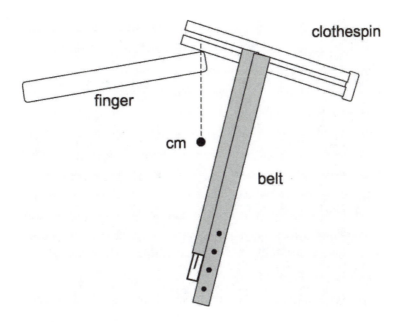

You can progressively move the ruler out and will find that most of the ruler can be made to overhang the table. Be careful with the hammer and use a pillow or towels to protect the floor in case the hammer falls.

In the above activities, the center of mass of the combined objects (belt and clothespin or ruler and hammer) is not within in the objects but lies outside the objects. The center of mass of the belt and clothespin is

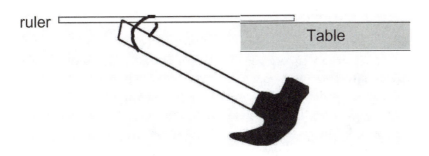

approximated in the figure at the point indicated "cm."

Another tricky feat is to balance a quarter on the edge of a drinking glass. Two identical forks are position so that their tines line up and the handles point outward and away from each other. A quarter is placed horizontally in the middle opening made by the tines. The

quarter should only be placed far enough into the tines so that the forks stay together. The outer edge of the quarter is rested on the lip of a glass. This balancing act may take a little practice and it works best using a glass with a flat, rather than rounded, edge. Experiment with several different drinking glasses to get the quarter to balance.

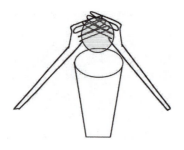

Activity 8: Cartesian Diver

An activity that demonstrates buoyancy is the construction of a Cartesian diver. To make a Cartesian diver, obtain a clear 2 liter soda bottle with its cap and a medicine dropper. Any clear plastic bottle that can be squeezed is acceptable, but a 2 liter soda bottle is about the right size and readily available. Fill the soda bottle completely with water. The medicine dropper must be filled with water so that when it is placed in the soda bottle it just floats. In order to do this take a large drinking glass or mug and fill it almost full with water. Draw water into the medicine dropper and then place the dropper vertically in the drinking glass to check its buoyancy. Add or remove drops of water from the dropper as necessary to establish neutral buoyancy. Once the right buoyancy is determined, carefully place the dropper in the soda bottle and screw the cap back on, making sure the bottle stays full of water. The dropper, which is the diver, should sink when the bottle is squeezed and float back to the surface when pressure on the bottle is released. The sinking and floating is due to changes in the density of the diver as pressure on the bottle changes. Squeezing the bottle increases the pressure inside the bottle, forcing water into the diver. This compresses the air within the dropper. The net result is that the diver becomes denser and sinks. Releasing the bottle relieves the pressure, and the air within the diver expands,

forcing water out of it, making it less dense and causing it to rise to the surface.

Activity 9: Air Pressure and Gas Laws

Caution: This experiment should be done with adult supervision. Wear safety glasses or goggles.

To do this activity, assemble the following items: safety glasses, hardboiled egg, large glass juice bottle with a mouth slightly narrower than an egg, rubbing alcohol, cotton, aluminum pie pan, and tongs. On Earth, partial vacuums are used for all kinds of applications. The most common is the vacuum cleaner; another is using a drinking straw. The word "suck" is used when dealing with vacuums or straws, but this is misleading. When a liquid moves up a straw, a partial vacuum is created and atmospheric pressure pushing on the fluid is forced up the straw.

To do this activity, first boil and peel an egg. The egg should be slightly wider than the mouth of the bottle. Wet the cotton ball with a little alcohol. You don't need a lot. **Put the cap back on the rubbing alcohol container and remove it from the area.** Now place the cotton ball in the pie pan and light it. Using the tongs, grab the flaming cotton ball and drop it in the bottle. Quickly place the egg, narrow end up, in the opening. In a few seconds the flame should go out and atmospheric pressure should push the egg into the bottle.

To get the egg out, position the egg with its narrow end in the opening. Tilt the bottle back and blow into it as if you're blowing a trumpet. The egg should slide out.

When asked why the egg is forced into the bottle, most people respond that a partial vacuum is created because the oxygen is removed during combustion. While it is true

that oxygen is reduced, it must be remembered that carbon dioxide is created and replaces the oxygen. The egg is forced into the bottle because of the relationship between temperature and pressure. The flame heats up the air in bottle. Since the bottle is initially open to the atmosphere, the pressure inside the bottle is the same as the atmospheric pressure. Once the opening of the bottle is blocked with the egg, the flame goes out due to lack of oxygen. The gas molecules comprising the air inside the bottle are no longer heated and are less energetic. This translates into a drop in pressure inside the bottle, and in pops the egg. The drop in pressure is due to cooling not removal of oxygen.

To get the egg out it wouldn't make sense to suck on the bottle's mouth. The pressure inside the bottle is the same as atmospheric pressure. Extracting more air would just lower the pressure inside the bottle even further. To get the egg out you have to increase pressure inside the bottle so that it is higher than the atmospheric pressure. This is accomplished by blowing into the bottle. Give the bottle a good blow and out slides the egg. It may take a little practice, but you should be able to get the egg out with a good blow.

Activity 10: Balloon in the Bottle

Wear safety glasses for this activity. As an alternative or addition to Activity 9, this activity can be performed. Prepare a baby bottle for this activity by removing the nipple from the top lid. The screw-on lid should now have a hole in place of the nipple. Take a balloon and inflate it several times to stretch it. Stretch the balloon evenly over the hole in the lid so that the balloon's opening is centered. Put the lid close to a microwave oven. In the bottle, place several tablespoons of water. About 3 cm (1 inch) of water on the bottom of the bottle is suf-

ficient. Put on gloves. Place the bottle in a microwave for a minute or however long it takes to boil the water in the bottle for at least 30 seconds. After the water has boiled for at least 30 seconds, quickly remove the bottle from the microwave oven and screw on the lid. Once the lid is screwed on, make sure the balloon is upright and not folded over to one side. You may have to move it slightly to the upright position.

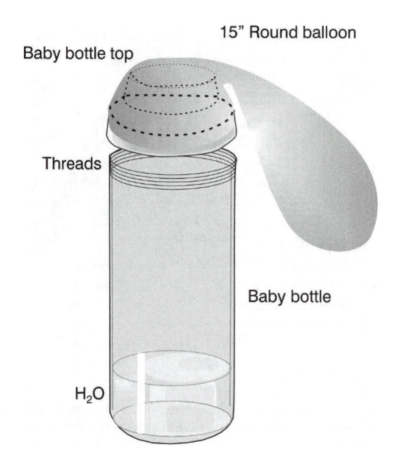

In a short period, usually less than a minute, the balloon will be pushed into the bottle, similar to the egg being pushed into the bottle. In this case, the baby bottle fills with water vapor when it is boiling. When the bottle is removed from the oven it is

filled with steam and it immediately begins to cool. Quickly placing the lid on the bottle traps the steam. As the bottle continues to cool, the steam condenses. This decreases the pressure inside the bottle and the balloon is forced into it. The same principle is used to seal jars when making jelly or canning.

Activity 11: Heat Capacity of Water

Caution: This experiment should be done with adult supervision. Do this activity outdoors or in a well-ventilated area.

Water has a very high heat capacity. Substances with high heat capacities experience smaller temperature changes when heat is exchanged between them and their surroundings as compared to substances with smaller heat capacities. In this activity, the high heat capacity of water enables a piece of fabric to stay below its kindling temperature, and in the process the fabric dries. Prepare a mixture of about 1/2 cup water and 1/2 cup of 70% rubbing alcohol. Once you are done with the alcohol, remove it from the area. Cut a small, approximately 10 cm (3 inch) square out of out of an old 100% cotton sock, towel, or piece of clothing. Soak the fabric in the water-alcohol mixture. Ring out the fabric, dry your hands, and place the fabric in the pie pan. Light the fabric with a long fireplace match or barbecue lighter. Alternately, hold the fabric with tongs and light it, then place it in the pie pan. The alcohol in the fabric burns, but the fabric does not ignite. The water absorbs much of the heat and keeps the temperature below the ignition temperature of the fabric. This activity can also be done with paper such as a dollar bill.

Another demonstration of water's high heat capacity is to boil water in a paper cup. Use a small 3 ounce paper Dixie cup and fill it almost to the top with water. Obtain a coffee cup to pour the water into when finished. Place the coffee cup next to a gas burner. Using tongs, hold the paper cup over the flame provided by a camp stove or gas burner. The water in the cup will boil, but the paper will not ignite. After the water boils, you can dump the water directly into the mug.

Activity 12: Fluid Flow

The acceleration of fluids is caused by pressure gradients. Fluids move from a region of high pressure to a region of low pressure. As fluids move around an obstacle, pressure changes perpendicular to the streamlines occur, and these can cause horizontal accelerations. The activities here are often erroneously attributed to the Bernoulli effect. One common demonstration of fluid flow is the inward movement of a shower curtain when water flows out of a showerhead. This is often interpreted as a decrease in pressure inside the shower stall due to an increased flow of water from the showerhead. Another interpretation involves buoyancy effects due to the heating of the air from the warm water. Turning on the shower with cold water doesn't seem to effect the inward movement, so the buoyancy effect is minimal. Physicists have found that the inward movement is actually due to a vortex created in the shower when the water is turned on. The vortex lowers the pressure inside the shower, and the shower curtain is pushed in by the higher pressure outside the shower.

Several activities with table tennis or Styrofoam balls can demonstrate fluid flow effects. Attach a table tennis ball with tape to a length of thread or floss that is 1 to 2 feet long. Suspending the ball near a flow of water from a faucet causes it to move toward the water. The flow of water lowers the pressure immediately around it, causing the ball to move toward it. Another demonstration of the lowering of pressure can be seen using two table tennis balls on strings. The balls are suspended side by side about 2 inches apart. Use a straw to blow air between the two balls. The two balls will move together due to the lowering of pressure between the two balls. A free table tennis ball can be suspended in midair by using a pistol-type hair dryer. Point the hair dryer so it blows air directly upward. Initially put the hair dryer on its lowest setting and place the ball in the upward flowing air. You can increase the speed of the dryer if it has multiple settings, and the ball will rise higher. As the ball rises, you can tilt the dryer so that the ball is not directly above the hairdryer, but at an angle. A greater lift can be accomplished using a leaf blower.

A final activity with a table tennis ball can be done using a moderate size funnel. Place the table tennis ball in the funnel and with the stem pointing down try to blow the table tennis ball out of the funnel. The ball should stay in the funnel. The blown air moves faster at the bottom of the ball and lowers the pressure at this point. The higher pressure on the top of the ball keeps the ball inside the funnel. To move the ball out of a funnel, air can be blown across the large opening at the top of the funnel. This can also be shown using two small, short drinking glasses and a table tennis ball, although for this activity a Styrofoam ball works better. Place the ball in one glass and try to move the ball into the other glass without touching it. The feat can be accomplished by blowing a strong burst of air across the top of the glasses.

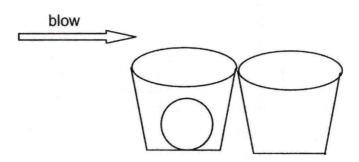

Activity 13: Surface Tension

Surface tension is the energy needed to increase the surface area of a liquid per unit area. It is a measure of the intermolecular attractions of a liquid. Because of hydrogen bonding, water has a high surface tension. To demonstrate this, take a small glass and a box of paperclips. Fill the glass to the very top with water. Once you believe the glass is completely full, start to place the paper clips into the glass one at a time. Count how many paper clips you can place in the glass before water spills out. Another variation of this activity is to guess the number of water drops that can be placed on a penny using a medicine dropper. In both variations of this activity, a much larger number of paper clips or drops than expected will be found. As many as several hundred paper clips may be placed in the glass and 25–30 drops placed on a penny. The reason for this is that the intermolecular forces and surface tension of water cause the water molecules to form into a spherical tightly bound shape.

Activity 14: Electrostatics

Water is a polar molecule. The oxygen atom in the water molecule carries a partial negative charge, and the hydrogen atoms carry a partial positive charge. This polarity can be observed by using a charged comb. Turn on a faucet so that a very small continuous stream of water is coming out of the faucet. Open the faucet the minimum amount to get a continuous stream. Take a comb and run it several times through your hair. Your hair should be dry and you should be able to hear little discharge sparks as you comb your hair. Alternately, the comb can be rubbed on a piece of fur or even a pet's fur. Bring the charged comb up to the stream of water. Notice how the water is attracted to the comb. Move the comb around and observe how the water "dances" in response to the comb's movement. The positive hydrogen end of the polar water molecules is attracted to the negatively charged comb, while the negative oxygen is repelled.

Another electrostatic activity is to take a small florescent bulb into a dark room and stroke the bulb with fur. Small florescent bulbs made for camping lights are ideal for this activity. A plastic ruler rubbed with a piece of wool (a wool sock works well) or a comb charged by combing dry hair will attract pepper sprinkled on a sheet of paper, but not salt. A mixture of salt and pepper can be separated using a charged plastic ruler or comb.

Electric Current Activities

The next series of activities involves electricity. The purchase of a few basic items facilitates the construction of simple circuits and simplifies doing these electrical activities. Basic items are bulbs, bulb holders, and a battery holder. These may be available from a local hardware or large building supply store, but the best source is an electronics supply store such as RadioShack®.

A couple of miniature lamp holders for E-10 screw-base lamps (RadioShack part # 272–357), a battery holder that holds two D batteries (RadioShack part # 270–0386), and several miniature lamps can be obtained for about five dollars. The miniature bulbs should have a voltage rating of approximately 3 volts. Several types of wire can be used to construct circuits. The wire should be insulated and have a small diameter. Solid wire is easier to work with than stranded wire and has the added advantage of being able to hold its shape when bent. Wire is sized by gauges, the higher the gauge the smaller the diameter. A gauge size of 20 or 22 is good for electrical activities. The ends of wires should be stripped; strip only about 1/4 to 1/2 inch of the insulation off the ends of the wire. Wires can be connected by twisting their ends together, or a package of screw-type alligator clips can be obtained for about a dollar, and the clip can be attached to the ends of the wire.

Activity 15: Series and Parallel Circuits

Simple circuits can be constructed with batteries, flashlight bulbs, and wire. The circuit diagrams described in this section will be illustrated using the standard symbols given in chapter 13. Series and parallel circuits with two bulbs can be constructed according to the schematics shown below and the brightness compared to the circuit with a single bulb. Observations should indicate that the brightness of two bulbs connected in series decreases, while two bulbs connected in parallel retain the same brightness as a single bulb. When two bulbs are connected in series, the resistance of the circuit (neglecting the resistance of the wire) doubles. The same current flows through each bulb, and there is a voltage drop across each bulb. The total voltage drop in the circuit is 3.0 volts because two D cells, each rated at 1.5 volts, are used as the power source. Since the bulbs are identical, the voltage drop across each bulb is about 1.5 volt. If a single D cell is used to light a single bulb, the brightness will be about the same as that when the two D cells are used to illuminate two bulbs in series. Bulbs connected in parallel cause the current to be split between the two branches. The voltage drop across each bulb is the same, in this case 3 volts. Therefore, the bulbs glow with the same brightness as a single bulb illuminated by the two D cells.

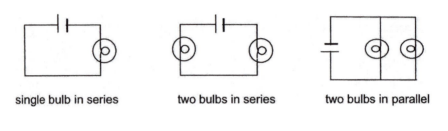

single bulb in series   two bulbs in series   two bulbs in parallel

Activity 16: Conductors and Insulators

By setting up the circuit shown below, different materials can be tested for their ability to conduct electricity. By touching the bare ends of each wire to an object, the ability of the object to conduct can be assessed. The brightness of the bulb can be used to qualitatively assess the conductivity of materials. Good conductors, including metals such as iron, will cause the bulb to glow brightly. Pencil lead from a mechanical pencil causes a dull glow, indicating that graphite (pencil lead is graphite) is a relatively poor conductor. Moving the two wire ends apart along the length of the pencil lead causes the bulb to dim due to increased resistance as the length of graphite through which current flows increases. Some objects may appear to be made of conducting metals, but testing indicates they don't conduct electricity. Some common materials to test include rubber bands, aluminum foil, chalk, rocks, wood, and charcoal. Materials that don't conduct electricity are called insulators.

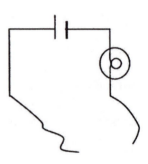

Activity 17: Electromagnet

A simple electromagnet can be made by coiling wire around a steel nail. For the electromagnet circuit, a switch is added to turn the electromagnet on and off. A simple switch can be made using a small block of wood with a couple of thumbtacks. The thumbtacks serve as terminals of the switch. Make sure the tacks have bare metal tops that are not painted. Press or tap the tacks partially into the block of wood about 2 inches apart. Place wires from the circuit under each tack and a paper clip bent open under one of the tacks. Press down firmly on the tacks to secure the wires and end of the paper clip. Bend the unsecured end of the paperclip upward so that it doesn't make

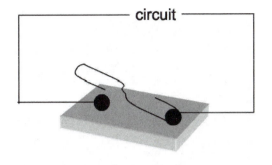

a simple switch

contact with the tack beneath it. The switch can be placed in the circuit. Pressing the paper clip down to make contact with the top of the tack closes the circuit and releasing the paper clip opens it. The switch can be tested with a simple single light bulb circuit.

To make the electromagnet, use a steel nail at least 3 inches long. Coil wire along the length of the nail. Don't wrap the last 1/2 inch of the nail's pointed end. Strip the ends of the

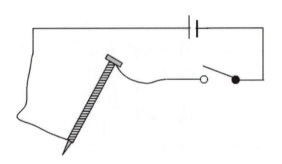

wire coiled around the nail and place it in a circuit with the cell and switch. Obtain paper clips, tacks, or other small metal objects to test the electromagnet. The electromagnet can be activated by closing the switch. Do not keep the switch closed for more than about a minute because the wire may overheat. You can experiment by determining the number of objects the magnet can pick up if you vary the number of coils around the nail or use only one battery instead of two.

Activity 18: Magnetic Field

An electrical current generates a magnetic field, and a moving magnetic field produces electric current. To observe how a magnetic field affects the space around a current-carrying wire, use the circuit below and a compass. Position the compass under the wire on one side of the circuit with the compass needle aligned with the wire. Closing the switch should cause the needle to deflect 90°. Align the compass needle under the wire on the opposite side of the circuit and again observe the deflection when the switch is closed. In this case, the needle should deflect 90° in the opposite direction. The deflection is consistent with the right-hand rule. If the circuit wires are held closely together in a parallel alignment, the magnetic fields created by the current moving in opposite directions in the opposite

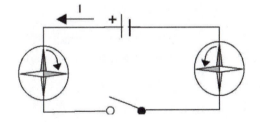

sections of wire will cancel each other out, and there will be no deflection.

Activity 19: Electric Motor

A simple electric motor can be built using the items used in the previous activities, plus a strong rectangular magnet. Magnets can be obtained from RadioShack, toy stores, or scavenged. A good source for a magnet is a hide-a-key compartment. Several small magnets can be stacked together to produce a stronger magnetic field. The first step in building a motor is winding the armature. To do this take about 18 inches of wire and start wrapping it around a circular object about 1 inch in diameter. Good objects to use are a size C battery or a film canister; it will be assumed you use a battery. Make about seven or eight tight loops, leaving 2–3 inches of free wire at each end. Excess wire should be trimmed with wire cutters, but don't trim excess wire until necessary. Slip the coil off the battery. To keep the several loops together in a tight coil use the two free wire ends to wrap the loops. Make sure the binding wraps are directly opposite each other. This will give a nice symmetric armature. There should be about 1 inch or slightly more of free wire extending out from the binding wraps. This is a good time to trim the wire so that both extensions are the same length. Strip each extension. Finally, for one extension, use some nail polish or typing correction fluid such as Liquid Paper to coat the top half of one of the extensions. The goal is to insulate the top half of this extension.

A cradle to hold the armature is easily made with two large paperclips. Unwrap two clips according to the diagram. The base of the clip can be inserted into clay, a piece of hard Styrofoam, or a block of wood in which two small holes have been created by tapping a small brad nail into the wood and removing. Space the paper clips so that the middle of the armature extensions rests securely in the paperclip loop. The height of the armature should be enough to just clear the top of the magnets when the armature spins.

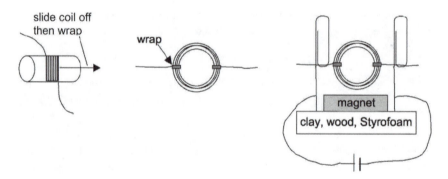

Once the motor is assembled and connected to the power source, give the armature a gentle spin to get it started. It should rotate at several rotations per second. You will probably have to adjust the exact positioning of the magnet to obtain the best results. Don't run the motor for more than a minute or two to avoid overheating. Notice that the paper clips heat up the longer you run the motor. Once you have determined the correct positioning of the components and have the motor running smoothly, you can experiment by turning the magnet upside down and placing it on top instead of underneath. Note which direction the motor spins when the magnet is place in a different position. You can also change the magnetic field strength by using multiple magnets.

The armature acts as an electromagnet when power is applied to it. One face of the coil forms a north pole and the other a south pole. The interaction between the armature electromagnet and the permanent magnet causes it to spin. The magnetic field of the permanent magnet exerts a torque on the coil when current flows through the coil. The insulated side of one extension turns off the current every half rotation. Inertia of the armature allows it to move through the succeeding half rotation, at which time current again flows through the armature. Insulating half of the extension creates a commutator that allows the motor to rotate in one direction. If the top half of one extension was not insulated it would turn one-half turn and wobble to a stop. You can test this by removing the insulation and observing what happens when you try to start the motor. The insulation can be easily reapplied using correction fluid or nail polish.

Activity 20: Tyndall Effect/Rayleigh Scattering

To demonstrate Rayleigh scattering, all that is needed is a flashlight, a tall drinking glass with smooth sides, and a little bit of milk. Fill the glass almost to the top with water. Looking down from above, view the light as it is shined through the water in the

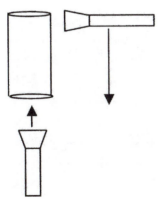

glass from the side. Continue to view the light as the flashlight is moved down the side of the glass and shined through the bottom.

Repeat your observations several times, each time adding a few drops of milk to the water until you have added approximately a teaspoon of milk. If you have several similar glasses, you can fill each with water and add the milk incrementally to each one and in this way continue to compare observations for different concentrations of milk throughout the activity.

As more milk is added to the glass, you should notice that the color of the light beam through the water changes from a blue hue to orange. The change in color is due to the scattering of different wavelengths that compose visible light. As the wavelength decreases, the amount of scattering increases. Other factors that affect the observed color are the size of particles light encounters and the distance through which light travels. The color of the sky and Sun depends on scattering of light. Light from the Sun is yellow, as observed during most of the day. This also indicates that yellow light passes through the atmosphere with relatively minimal scattering. The sky is blue because blue light, with a short wavelength, experiences the most scattering by atmospheric particles such as nitrogen and oxygen molecules. The color of the Sun may change as the size of particles changes or the light passes through more of the atmosphere. The Sun may appear orange or red when the atmosphere is polluted. A polluted atmosphere contains larger particles, which have the ability to scatter longer wavelengths, for example, yellow. Red skies at sunrise or sunset are due to light traveling through a greater distance and, therefore, encountering a greater number of particles, such as pollutants or sea salts, that have the ability to scatter long wavelengths. The longest wavelengths, red and orange, are least likely to be scattered and define the color of the Sun under these conditions.

The change in color of the light beam through the water depends on the number of milk particles and the distance through which the light travels. The orange colors are not visible at small concentrations of milk, even when viewing light that has traveled through the entire height of water. As more milk is added to the glass, the orange hue becomes stronger; it may even be observed when the flashlight shines through the middle portion of the glass.

Activity 21: Transmission of Sound

The travel of sound through objects can be easily experienced by using string and metal objects such as coat hangers, cooling racks, or eating utensils. Attach two strings about 2 feet long to opposite sides of the metal object. Wrap the free ends two or three turns around the forefingers of each hand. Now place you fingers in your ears with the object hanging down. Have some-

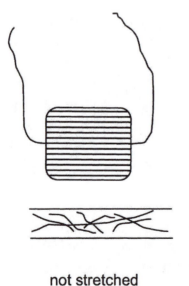

not stretched

one gently stroke the metal object with a spoon and listen to the sound. The vibrations imparted to the object travels through the string and your fingers to your ears. The sound should be much fuller and pleasant sounding than listening to the same sound that travels through air.

Activity 22: Entropy and the Rubber Band

According to the second law of thermodynamics, an increase in entropy occurs during a natural spontaneous process. An interesting demonstration of entropy and energy changes can be experienced with just a rubber band. Take a 3 or 4 inch rubber band and hold it to the top of your lip to sense its temperature. Next, grab a small section of rubber band with both hands. Using your thumb and forefinger, quickly stretch the rubber band and place it against your upper lip. Note the difference in temperature. When the rubber band is stretched, it should feel warm. Stretching the rubber band is a nonspontaneous process, so the change in Gibbs free energy, ΔG, is positive. The fact that the rubber band feels warm also indicates that energy is given off when the rubber band is stretched; therefore, the change in enthalpy, ΔH, is negative. One version of the second law of thermodynamics states: $\Delta G = \Delta H - T\Delta S$, where ΔS is the change in entropy. In this example, a positive ΔG and a negative ΔH would mean that ΔS is negative. Entropy is a measure of disorder of a system. A negative entropy change means that the system becomes less disordered, or more ordered. For a rubber band, the order

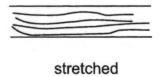

stretched

involves an ordering of its molecules as they change from a more random orientation to an orientation in the direction of the stretching.

Activity 23: Ray Tracing

The interaction of light with mirrors and lenses can be explored using a light box and simple mirrors and lenses. A light box can be made out of a shoebox with a flat bottom and a flashlight. In one end of the shoebox cut a hole that is the same size as the end of a flashlight. In the opposite end of the box cut a hole that is about 2 inches high and 3 inches wide. Cut several rectangular pieces of cardboard or construction paper that are slightly larger than the rectangular hole made in the shoebox. Cut slits in the rectangular pieces. Make pieces with one, three, and five slits. Keep the width of the slits as small as possible, preferably no more that 1/16 inch wide. The spacing between slits should be about 3/8 inch. The pieces with the slits will cover the opening in the shoebox, and a flashlight will be placed in the round opening. Put the lid on the shoebox and tape the single-slit rectangle over the opening. When the flashlight is turned on and placed in the round opening, a single, well-defined ray should emerge from the box. Using the box in a darkened room makes it much easier to observe the ray.

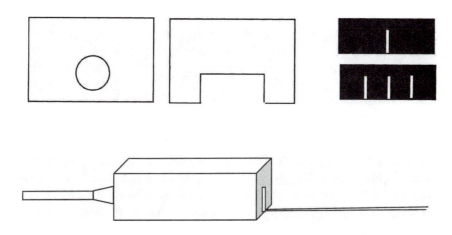

A simple activity with the ray box is to look at the angle of incidence and the angle of reflection. Use a flat upright mirror. If a wall mirror is available, you can place a table or chair to provide a flat surface next to the mirror. It is important to have a flat working area in front of the mirror. Use a sheet of plain white paper and draw a line across the width of the paper at its centerline. Position the paper against the mirror so that the centerline of the paper is in the

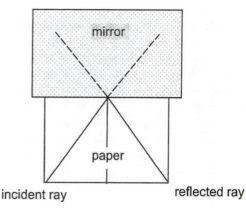

middle of the mirror and perpendicular to it. Use your ray box to project a ray at an angle to the centerline and so that it strikes the mirror right on the centerline. Mark the incident and reflected rays using a pencil at points along the rays. Use a straight edge to connect the points and measure the incident and reflected angles. Once the rays have been measured check the angles with a protractor. The centerline on the paper is the normal. You can also take the paper and see that the reflected lines in the mirror are extensions of the incident and reflected rays.

Activity 24: Lenses

Replace the one-slit rectangle with a five-slit rectangle. In a darken room place the light box on a flat surface; the floor is a good surface to study lenses. When you turn on the flashlight, five parallel beams should emerge from the light box. Obtain lenses from old eyeglasses, cameras, binoculars, and so forth. Place a lens so that the parallel rays from the light box pass through it and observe how the lens affects the rays. More than one lens can be placed in the path of the light. Use a drinking glass with water to see how this affects the rays.

Activity 25: Concave and Convex Mirrors

The easiest way to observe the reflection from concave and convex mirrors is to use a shiny spoon or metal ladle. The back of the spoon acts as a convex mirror and gives an upright smaller image. The front of the spoon is a concave mirror. The image from the inside of the spoon will be upside down and smaller. Although an upright image is possible for a concave mirror, the size of the spoon is too small to ever obtain an upright image.

Summary

This chapter has introduced the basic concepts of experimentation and offered various activities to illustrate some of the concepts presented throughout the book. This set of limited activities should serve to give the reader practice in conducting a scientific investigation. In addition to the activities presented in this chapter, there are a numerous Web sites devoted to virtual labs, simulations, and demonstrations. Some of these are listed in the Selected Bibliography and Internet Resources section in this book. The reader is encouraged to explore the listed sites and continue to experiment and develop scientific skills.

17

Physics as a Career and Physics Jobs

Introduction

Compared to other scientific disciplines, physicists are rare. While it is difficult to categorized scientists in discrete disciplines because many jobs utilize training and education in several areas, the job title of physicist is rare. The number of scientists directly employed in specific scientific areas in 2003 is reported in Table 17.1, demonstrating that physicists are greatly outnumbered by the other science disciplines listed. Another indicator of the number of scientists in various science fields can be inferred from the number of graduates with different degrees. The National Science Foundation's publication *Science and Engineering Degrees: 1966–2001*, published in March 2004, presents statistics on the number of graduates in science and engineering disciplines. The statistics for 2001, the last year available, are shown in Table 17.2. Tables 17.1 and 17.2 provide a general view of the number of physicists compared to other disciplines but do not reveal the number of individuals working in the disciplines and related fields. For example, over 1 million people are employed in the chemical industry. Similarly, numerous people are employed in areas related to biology. The number of people in the health and medical industry would greatly inflate the numbers in biology compared to the other disciplines. Although there are many fewer physics jobs when compared to the other major disciplines, education and a knowledge of physics is required for practically all technically related jobs. This chapter will introduce physics as a career. It is rare when you find someone who calls him- or herself a physicist, but many jobs depend heavily on a knowledge of physics.

Table 17.1
Approximate Number of Scientists in Four Major Areas

Discipline	Number
Biology	70,000
Chemistry	90,000
Geology	35,000
Physics	13,000

Table 17.2
Graduates in Selected Science Areas in 2001

	Bachelors		Masters		Doctorate	
	Male	Female	Male	Female	Male	Female
Astronomy	145	195	57	36	145	41
Physics	2,701	756	1,090	286	1,036	155
Biological sciences	25,005	37,084	2,732	3,755	3,132	2,547
Chemistry	5,047	4,775	1,184	825	1,349	628

Physics-Related Fields

Many jobs exist in the physical sciences area. These combine physics with other disciplines. A partial list of physics related jobs and a brief description of each follows:

Astronomy. Astronomy is often considered a subdiscipline of physics. Two broad, but related, areas of astronomy are observational and theoretical astronomy. Astronomers make observations and conduct research on celestial phenomenon using equipment that includes various types of telescopes and spectrometers (optical, radio, x-ray, etc.). This work is conducted from both ground-based observatories and remotely from instruments mounted on rockets, balloons, satellites, and space probes. Astronomers study a range of subjects related to the universe, including the evolution of stars and stellar systems, **cosmology,** extraterrestrial life, celestial mechanics, and planet formation. Some astronomers are involved in applied areas related to navigation, communications, weather, and climatology.

Biophysics. Biophysicists combine biology and physics to understand the physical aspects of organisms and life systems. This may entail a study of the dynamics of blood flow through arteries and veins, the transmission of electrical impulses in nerves and muscles, the absorption of light by plant pigments, and the mechanics of animal movement.

Computer Software Design. Software designers modify and write programs to perform a variety of tasks. The technical training and logical thinking involved in physics is especially well suited for writing computer code and designing software. Many software jobs are filled by computer science graduates, but surprisingly, more physics graduates are employed in jobs related to computer software than any other area.

Engineering. All engineering disciplines rely heavily on an understanding of physics, and the areas of physics stressed depend on the type of engineering. Electrical engineers are more likely to use principles in electricity and magnetism as they work with subjects such as power transmission, electrical system design, and electrical equipment construction. Mechanical engineers and civil engineers are more likely to utilize knowledge in mechanical subjects such as **statics** and **dynamics**. The former design mechanical parts used in machinery, and the latter design structures, for example, bridges, foundations, and so forth. One type of engineering that applies the principles of physics to a broad range of practical problems is engineering physics. Engineering physicists apply fundamental physics to problems in engineering.

Geophysics. Geophysicists apply physics to study the Earth. One type of geophysicist is a seismologist. Seismologists gather information from earthquakes through the study of earthquake waves to locate and identify (determine the magnitude) earthquakes. The natural propagation of earthquake waves through the earth has been extended by geophysicists to identify geologic formations and pinpoint areas of commercial interest. Using sound to probe the Earth, geophysicists explore the Earth for geologic formations with high potential for oil and gas and commercial minerals. Many geophysicists work with geologists, engineers, and scientists in natural resource companies (oil, coal, metals).

Health Physics. Health physicists maintain a safe working environment in those areas where radiation is of concern. This may include hospitals, nuclear power plants, research facilities, and military installations. Health physicists monitor individuals and the physical environment for radiation levels and educate personnel on radiation hazards. They may also conduct research regarding inspection standards, decontamination procedures, and standard operating procedures. Health physicists are often required to deal with the general public and government regulatory agencies regarding potential radiation hazards.

Industrial Physics. Industrial physics combines a knowledge of physics with practical laboratory work for use in industrial research, design, and development. The area increases the marketability of physics graduates, who often find it difficult to obtain a job in industry directly out of college without additional work in engineering. Industrial physicists are broadly trained and must apply this knowledge to address unique problems in various industries. For example, an industrial physicist might examine the insulating and wicking abilities of different fabrics in the clothing industry, or the resistance of various flooring material to impact. Industrial physicists apply physics to the practical social and economic aspects of industrial problems.

Material Science. Material scientists are concern with the structure and properties of materials, how materials are made, and their performance. Material scientists often use their knowledge of materials in the quest to change a material or to combine materials into composites to perform a certain function. Material scientists may be categorized according to the type of material they work with, for example, metallurgists (metals) and ceramic engineers. Material scientists are responsible for producing a wide array of materials used everyday: Teflon cookware, carbon-fiber golf clubs, ceramic dental fillings, silicon semiconductors, polymer plastics, and so forth.

Medical Physics. Medical physicists use physics in the treatment of patients. They often work with a wide array of diagnostic equipment in areas such as radiology, ultrasound, magnetic resonance imaging, x-rays, and positron emission tomography and with therapeutic equipment such as dialysis machines. Medical physicists also work with individual patients' treatments involving pacemakers and monitors. Medical physicists set up and calibrate equipment, modify existing equipment, and design new equipment. In their role, medical physicists consult with other medical staff on patient treatment involving electronic equipment.

Meteorology. Meteorologists study the Earth's atmosphere and how it interacts with the Earth. They report on weather conditions and use current conditions and a variety of computer models to make forecasts. Meteorologists apply physics to the atmosphere by examining mass transport of air and water over the surface of the Earth. The importance of energy balances and heat transport applied on a global scale enable meteorologists to make weather predictions and address climate variability.

Nuclear Plant Operation. Nuclear operators control equipment associated with power generation in nuclear power plants. Training and requirements for these positions are governed by the Nuclear Regulatory Commission (NRC). The NRC licenses nuclear plant operators in the United States. The Navy depends on personnel, both enlisted persons and commissioned officers, for its fleet of nuclear submarines and aircraft carriers. Many nuclear operators receive their education in the military before moving into the private sector.

Physical Oceanography. Physical oceanographers apply physics to the study of the oceans. They are chiefly concerned with ocean circulation and study surface currents, waves, tides, and large-scale **thermohaline** circulation. Physical oceanographers work closely with other types of oceanographers (geological, chemical, biological) to answer interdisciplinary questions involving subjects such as fisheries productivity, beach erosion, climate change, and oil spill impacts.

Radiology. Several job titles fall under this category. They range in training from radiologic technicians to physicians who specialize in radiology. Radiologic technicians set up and operate a wide array of diagnostic equipment. The most common example of radiologic technician work is performing x-rays at clinics and hospitals. Other routine procedures performed by radiologic technicians include sonograms, mammograms, and CAT scans. Radiologic technicians may also be involved in radiation therapies and be called radiation therapists. These individuals are more concerned with treatment as opposed to diagnosis. Radiation therapists work directly with physicians and patients using radiation treatments. Depending on the situation, several job titles may be describe similar work. For example, a radiologic technician, nuclear medicine technologist, radiation therapist, and medical physicist may perform very similar or very different tasks, depending on the situation.

Teaching. Physics teaching at the secondary level is one area that students with an undergraduate physics degree and a teaching certificate can perform without additional graduate education. Because of the rigors of obtaining a physics degree and the relatively low number of qualified teachers, physics teachers are in high demand at many school districts across the country.

Physics Work

Physicists study the structure and behavior of matter, energy transfer, and the interaction of energy and matter. Theoretical physicists work on the frontiers of science and explore basic questions related to the origin of the universe, the fundamental structure and behavior of matter, and the nature of radiation and energy. Basic research is performed to advance scientific knowledge, and often the immediate impact of discoveries is unclear. Physicists work in any number of subdivisions within the field: elementary particle physics, nuclear physics, molecular physics, high-energy physics, solid-state physics, optics, acoustics, plasma physics, fluid mechanics.

Theories formulated by theoretical physicists are tested by collaborating with experimental physicists using a variety of equipment and instruments such as lasers, particle accelerators, telescopes, and spectrometers. The immense cost and sophisticated equipment involved in physics research dictates that much of the support for this research comes from national governments. The United States directly funds national labs such as Fermi National Accelerator Laboratory (Fermilab) outside Chicago, Brookhaven on Long Island, and Los Alamos in New Mexico. The federal government also assists labs managed by universities, for example Stanford Linear Accelerator Center and Columbia Plasma Physics Laboratory. In Europe, the major lab is CERN, outside Geneva, Switzerland. Even though large national labs are the most prominent sites of physics research, most physics research still occurs in small or medium size labs.

Applied physicists use basic research to refine processes and develop new products. Applied research is specifically targeted to improve or produce a product or process. For example, knowledge in solid-state physics is the basis for the modern computer and electronics industry. Advances in optics have been applied to improve human vision and observe the far reaches of the universe with products such as the Hubble telescope. Optics combined with knowledge of energy and radiation has led to a plethora of remote sensing devices to monitor our planet on a continual real-time basis. This can be illustrated by sitting at a computer and accessing Web sites maintained by federal agencies (NASA, NOAA, USGS, NWS) that enable citizens to observe all types of information related to the environment, for example, weather, tides, ocean temperatures, landforms, and pollution levels. Applied research is supported by private companies that can benefit directly from discoveries made in this area. Some large corporations have invested in their own labs to support their work. Some of these such as IBM's Thomas Watson Research Center and AT&T Research have established an international reputation through their work. Research leading to several Nobel Prizes has taken place in these labs.

Education

Although more than 500 colleges and universities offer a bachelor's degree in physics, work as a physicist with an undergraduate degree is limited. For this reason, many physics graduates have limited job opportunities in the field directly out of college. Many physics graduates undertake further education that builds upon their physics education to obtain employment in related areas, as noted in the previous section. Physics graduates who desire to work in research should plan to obtain a doctoral degree in physics. Approximately 175 institutions offer a doctorate program in physics. Degree opportunities in the related area of astronomy are even more limited than in physics. Only about 70 colleges and universities offer an undergraduate degree in astronomy; about 40 of these have separate astronomy programs, and 30 are administered within a physics program. About 40 schools have a doctoral program in astronomy. An undergraduate degree in physics is highly recommended for a number of fields of graduate study in which undergraduate education is limited. In addition to astronomy, these include geophysics, meteorology, physical oceanography, biophysics, and nuclear engineering.

Students planning to major in physics should take as many mathematics, science, and computer courses as possible in high school. Undergraduate programs in physics consist of a basic preparation in mathematics, science, and electives during the first two years for upper level (third- and fourth-year) work. Physics majors during the first two years will take mathematics that includes several calculus courses, a course on differential equations, and perhaps a course in linear algebra. General physics with lab and general chemistry with lab are typically required during the first two years. Specialized physics course during the first two years are limited, although there may be a general course in mechanics or modern physics. The last two years of a physics program is devoted to a more comprehensive study of specific subdivisions of physics. The requirements for the last two years vary between schools, but it can be expected that many schools will require courses in the major areas that define physics: mechanics, optics, thermodynamics and heat, electromagnetism, quantum physics, and modern physics. Courses in these areas are supplemented with specialized courses consistent with the curriculum and faculty of the individual schools and may include work in any number of areas: acoustics, atomic physics, solid-state physics, astrophysics, chemical physics, statistical mechanics, and so forth.

Elective courses should be used by students to enhance their physics education. These courses are often neglected by students, who assume that mastering the technical material is sufficient to become a successful scientist. For example, one of the most highly valued skills sought in graduates by potential employers is the ability to communicate clearly in writing and speech. A technical writing or speech course helps students develop these skills. Another benefit of elective courses is to help the student develop career goals. This is especially important in physics, where many jobs exist in related areas. Unlike the chemical industry, there is no physics industry that provides abundant employment opportunities for physicists. If a student wants to work in sales for an instrument company or work in a non-English-speaking country, courses in marketing or foreign language would be beneficial.

Another option that the student should consider is completing an internship as part of the undergraduate program, or summer employment in a physics-related area. Some colleges require an internship as part of their programs. There are many advantages to completing an internship. An internship provides a professional experience that exposes the student to practical applications of physics. The internship can be used to network with professionals to assist in obtaining employment upon graduation or being admitted into a graduate program. An internship also exposes students to advanced equipment and processes. For instance, work as an intern in a national energy lab would acquaint the student with equipment and instruments often unavailable to undergraduates. Internships can be found through the college's physics department or career service department. Students may also contact the human resources department of large companies, individual physicists, or faculty members to learn about internships. The Internet is an invaluable tool for exploring internship opportunities. Many government agencies such as NASA, national energy laboratories (FERMI, Brookhaven, etc.), and universities have summer programs offering internship opportunities.

Physics Organizations

A number of organizations are devoted to physics and related areas. These organizations provide a wealth of information on physics and related topics. The American Institute of Physics (AIP, home page: www.aip.org/index.html) is a federation of 10 member societies. Some of these member societies are devoted to the general physics community, such as the American Physical Society, while others focus on specific areas of physics, for example, Acoustical Society of America and Optical Society of America. The American Institute of Physics, through its member societies, provides services that support the physics community. This includes promoting physics through its publications, providing career and professional development services, providing resource material to educators, offering scholarship and fellowship opportunities, coordinating regional and national meetings, issuing reports on relevant topics, and recognizing significant work in physics with awards. AIP publishes several magazines: *Physics Today, The Industrial Physicist,* and *Computing in Science and Engineering. Physics Today* contains articles of general interest and information intended to keep readers informed on current events in physics. The American Physical Society has resources for students considering a career in physics: www.aps.org/jobs/guidance/student. cfm. Especially useful is a book published by AIP and authored by John Rigden, *Landing Your First Job: A Guide for Physics Students,* August 2002.

The American Astronomical Society consists mostly of research astronomers. The society conducts most of the same functions mentioned above for the American Institute of Physics, but with a limited focus on areas connected to astronomy. Table 17.2 indicates that relatively few individuals receive astronomy degrees each year. There are only about 6,000 astronomers employed in North America. The American Astronomical Society has a brochure on its Web site providing information on a career in astronomy. The address for this is www.aas.org/education/students.htm#careers.

The American Association of Physics Teachers's (AAPT) mission is the dissemination of physics knowledge, particularly through teaching. While its primary focus is toward physics education in high school, the organization deals with all levels of physics. AAPT provides a wealth of resources for those involved in physics teaching. Its monthly publication *The Physics Teacher* contains articles on classroom and laboratory activities, book reviews, teaching methods, physics equipment and instruments, and numerous other subjects related to the teaching of physics. Students contemplating a career in physics teaching can join the organization at a reduced rate. AAPT has a publication entitled *Planning for Graduate Education in Physics* on its Web site: www. aapt.org/ (look under Teacher Resources).

Job Outlook

Employment of physicists and astronomers is expected to lag behind general employment through the year 2012. Many physics jobs in research are directly tied to federal expenditures for research funds in physics and astronomy. While these funds are expected to grow in the next decade, their growth will not match that of the past. Most employment opportunities for research physicists in the next decade will come through the replacement of retiring physicist rather than increased opportunities. Many private laboratories are also reducing basic research, and this will further reduce the need for research scientists. Competition for physics and astronomy positions requiring a doctorate will be high over the next decade.

Although competition for research positions will be high, individuals with physics degrees possess knowledge in science and mathematics that make them highly qualified for many other occupations. Bachelor degree physics majors may find employment opportunities in computer science, engineering, information technology, environmental science, and education. Physics teachers are in high demand in many school districts.

Summary

The job title of *physicist* is rather limited in society. I challenge you to go to the classified section of your Sunday newspaper and find "physicist" listed in the employment section. This chapter has attempted to broaden your view with respect to physics and the job opportunities available in the area. While relatively few individuals are called physicists in their jobs, a person can use a physics education as an entry to many fields. Physics requires excellent quantitative skills, the ability to analyze, skill in working with equipment and making measurements, and the ability to critically analyze and communicate information. These skills are highly valued by employers and can be used by physics graduates to gain employment in many areas. Some of those areas have been outlined in this chapter.

Glossary

Absolute Zero the lowest possible temperature, equal to $-273.15°$ C or 0 kelvin (K); temperature at which molecules possess the minimum kinetic energy and the entropy of a system is zero.

Acceleration rate of change of a velocity vector with respect to time.

Activity the intensity of a radioactive material, generally measured in disintegrations per unit time, with 1 disintegration per second equal to 1 becquerel.

Adhesion attraction between the surfaces of two different bodies.

Alpha decay nuclear process in which an alpha particle is emitted.

Alpha particle radioactive particle consisting of two protons and two neutrons, equivalent to the nucleus of a helium atom.

Amorphous solid with a random particle arrangement.

Ampere unit of electrical current equal to flow of 1 coulomb per second, abbreviated as amp or A.

Angular momentum measure of the amount of angular motion of an object, moment of inertia times the angular velocity.

Anion negatively charged ion.

Anode the positively charged electrode in an electrolytic cell where oxidation takes place.

Antinodes location of maxima of displacement in a standing wave system.

Antiparticle particle with the same mass but having properties complementary to matter, for example, positively charged electrons (positrons).

Applied physics broad area of physics in which principles of physics are used to solve a specific problem to serve societal needs, or the use of physics in applications.

Armature wire coil of an electromagnetic device such as a motor.

Assimilate to incorporate into human tissue.

Atomic mass unit mass equal to 1/12 the mass of the most abundant isotope of carbon (C-12), abbreviated amu.

Atomic number number of protons contained in the nucleus of an atom of a particular element.

Avogadro's law law that states equal volumes of gases at the same temperature and pressure contain an equal number of particles.

Background radiation low-level naturally occurring radiation present at the Earth's surface due to cosmic rays and radioactive materials in Earth's crust.

Baryons family of heavy subatomic particles that includes protons and neutrons, from the Greek *barys* meaning "heavy."

Becquerel   SI unit for activity equal to 1 disintegration per second, abbreviated Bq.

Bernoulli principle   conservation of energy law applied to fluids, for a nonviscous incompressible fluid exhibiting steady flow it states that the sum of pressure, kinetic energy, and potential energy is constant.

Beta decay   nuclear process in which a beta particle is emitted

Beta particle   subatomic particle equivalent to an electron.

Binding energy   the amount of energy holding the nucleus of an atom together, equivalent to mass defect times velocity of light squared (Δmc^2).

Black hole   an area of space with a gravitational field so intense that escape velocity is equal to or greater than the speed of light.

Bose-Einstein condensate   phase of matter just above absolute zero in which atoms lose their individual identity.

Bosons   class of particles that includes alpha particles, photons, and pions.

Boyle's law   law that states the volume of an ideal gas is inversely related to its pressure.

Breeder reactor   nuclear reactor that produces fissionable material in the process of producing energy from the original source of fissionable material.

Brownian motion   random motion of small particles suspended in a fluid.

Buckminsterfullerenes   C-60, allotropes of carbon consisting of a spherical arrangement of carbon atoms, named after inventor Buckminster Fuller.

Calorie   unit of energy equivalent to amount of energy required to raise the temperature of 1 gram of water from $14.5°C$ to $15.5°C$, equivalent to 4.18 joule.

Calorimeter   instrument used to make thermodynamic measurements on heat exchange.

Calorimetry   experimental technique used to determine the specific heat capacity of a substance or heat exchange in chemical and physical processes.

Cathode   negatively charged electrode in an electrolytic cell, where reduction takes place.

Cation   a positively charged ion.

Center of mass   point in an object or a system of objects where mass can be assumed to be concentrated.

Centrifugal force   center fleeing force, the apparent outward force on an object moving in a circular path due to inertia of the object.

Centripetal acceleration   acceleration in circular motion directed toward the center of curvature.

Centripetal force   force in circular motion directed toward the center of curvature, the force required for circular motion.

Charles law   law that states the volume of an ideal gas is directly related to its absolute temperature.

Classical physics   era of physics that consists of knowledge prior to the twentieth century.

Closed system   system in which matter is confined to the system and does not flow across a boundary into the surroundings.

Cohesion   intermolecular attractive force between particles within a substance.

Complementarity principle   principle attributed to Niels Bohr that states an experiment will only define one aspect of complementary concepts; applies to the wave and particle nature of light.

Compression   in a longitudinal (sound) wave the area where an increase in pressure occurs, in contrast to rarefaction.

Compton effect   increase of wavelength of an electromagnetic wave after interacting with electrons in a material.

Concave mirror   spherical mirror in which the reflecting surface is on the inside of the spherical section defining the mirror.

Conductance measure of the ability of a material to conduct electricity, the reciprocal of resistance.

Conductor material that conducts electricity (electrical conductor) or heat (thermal conductor).

Conservative property a property of a physical system that remains constant even though the system may change, for example, energy, momentum.

Constructive interference addition of waves to produce a single wave in which individual waves have reinforced each other.

Control rods long cylindrical tubes used in nuclear reactor to regulate the fission process.

Convex mirror spherical mirror in which the reflecting surface is on the outside of the spherical section defining the mirror.

Correspondence principle principle credited to Niels Bohr in 1923 that states predictions of quantum mechanics are the same as those of classical mechanics when applied to macroscopic systems.

Cosmic radiation radiation from an extraterrestrial source.

Coulomb unit of electrical charge equal to the mount of charge transferred in 1 second by a current of 1 ampere.

Covalent crystal crystal in which atoms are held together by covalent bonds in a rigid three-dimensional network, for example, diamond.

Critical mass quantity of fissionable material necessary to give a self-sustaining nuclear reaction.

Crystalline solid in which atoms are arranged in a definite three-dimensional pattern.

Curie point temperature above which a ferromagnetic material loses permanent magnetism.

de Broglie wavelength wavelength of a particle given by the equation $1 = h/mv$,

where h is Planck's constant, m is the mass of the particle, and v is the particle's velocity.

Decoupling phenomenon in which the interaction and differentiation of elementary particles disappears, important in theories of cosmology.

Destructive interference addition of waves to produce a single wave in which individual waves have cancelled each other.

Deuterium isotope of hydrogen with one neutron, heavy hydrogen.

Diffraction bending of wave fronts of a propagating wave around obstacles.

Diffuse reflection reflection from a rough surface where light is scattered in numerous directions.

Displacement change in position of a particle.

Doppler effect change in frequency (pitch) of sound as emitting source moves in relative motion to the receiver; the frequency increases as source and receiver approach each other.

Dose amount of energy imparted to matter by ionizing radiation.

Dose equivalent measure of amount of radiation absorbed that takes into consideration the type of radiation, measured in rems or the SI unit sieverts (Sv).

Dynamics study of motion with specific emphasis on the forces causing motion.

Echolocation use of sound by organisms for positioning and locating objects.

Eddy currents current induced in small circles when a conducting plate is moved through a magnetic field.

Elastic limit stress level that an object can sustain and still return to its original shape and dimensions.

Electrolyze to bring about a chemical change by passing an electric current through an electrolyte.

Electric potential amount of work per unit of electric charge required to move a charge from a reference point, usually taken as infinity, to another point in an electric field, measured in volts.

Electrode   a terminal that conducts current in an electrochemical cell.

Electrolysis   process in which an electric current passes through an electrochemical cell, forcing a nonspontaneous reaction to occur.

Electrolyte   a substance that ionizes in water or a substance that when melted conducts current.

Electromagnetism   phenomenon associated with the interaction of electricity and magnetism.

Electron volt   unit of energy equal to work required to move one electron through a potential difference of 1 volt.

Electrostatic   electricity associated with stationary sources of charge.

Electroweak force   in the standard model the unification of the weak force and electromagnetic force.

Energy   having the potential to do work.

Entropy   a measure of the disorder of a system.

Equivalence principle   concept that the laws of physics give corresponding results in all inertial reference frames.

Escape velocity   minimum speed that an object must have to escape a gravitational field; for Earth the escape velocity is about 11.2 kilometers per second (25,000 miles per hour).

Experimental physics   broad area of physics in which ideas are tested through experiments.

External force   a force applied from outside a system.

Far infrared   long wavelength electromagnetic infrared radiation; the wavelength for infrared is not firmly established but generally falls in the range between 20 and 1,000 microns.

Fermat's principle   the path of a light beam between two points will be the one that takes the least time to traverse.

Ferromagnetic   characteristic of substances such as iron, nickel, and cobalt that have high magnetic permeability.

Field   the concept that a physical quantity permeates space, and every point in space can be associated with the quantity, for example, a gravitational field.

First harmonic   standing wave pattern having the longest wavelength and lowest frequency, for a string it would be the frequency associated with vibration between two nodes at attached ends.

First law of thermodynamics   conservation of energy; the change in energy of a system results from heat flow into or out of system or work done on or by the system.

Forced convection   heat transfer when the motion of the transfer medium, such as air or water, is supplied externally, for example, by a motor or pump.

Frequency   in any phenomenon exhibiting periodic motion, the rate at which the phenomenon (wave) repeats.

Fuel cell   type of electrochemical cell in which fuel and oxidant are continually supplied to create electrical energy, for example, hydrogen (fuel) and oxygen.

Fundamental   lowest-order vibration mode of a standing wave, also called fundamental frequency.

Gamma   type of electromagnetic radiation consisting of high-energy photons or high-energy x-rays.

Gamma ray   electromagnetic radiation consisting of photons with energies greater than 100 thousand electron volts and wavelengths ranging between about 10^{-10} and 10^{-14} meters.

Grand unified theory (GUT)   consisting of gravitational, electromagnetic, weak nuclear, and strong nuclear interactions into one field theory.

Geometric optics   area of optics that considers light to travel in straight lines, enabling ray tracing to describe results of reflection and refraction.

Geosynchronous orbit that has a period of one sidereal day, such that the object stays above a fixed point on Earth (geostationary).

Gravitational mass mass defined by interaction of an object with a gravitational field as given by the inverse square law.

Gravitational red shift change in wavelength of electromagnetic radiation to a longer wavelength in a gravitational field as predicted by general theory of relativity.

Gravitrons hypothesized particle with no mass or charge that mediates the gravitational force.

Gray SI unit for energy absorbed from ionizing radiation equivalent to 1 joule per kilogram.

Ground a large conducting body and connector that provides a zero potential to provide a path for excess charge.

Grounded a conducting body connected to a large body (such as the Earth) that provides a zero potential.

Hadrons class of particles involved in the strong force, consisting of baryons and mesons.

Half-life time it takes for the activity of a radioactive substance to be reduce to half of its original activity.

Heat transfer of energy from one object to another due to a difference in temperature; form of energy associated with motion of particles comprising an object.

Heat change energy flow into or out of a system.

Heat of fusion energy required to melt a unit mass of a substance to liquid at the melting point of the substance.

Heat of vaporization energy required to vaporize a unit mass of a substance to gas at the boiling point of the substance.

Heisenberg uncertainty principle quantum mechanical concept that both the position and momentum of a particle can't be determined simultaneously; furthermore the more precisely position is determined, the less precisely momentum is known, and vice versa.

Holography method of producing a three-dimensional image using interference patterns recorded on a film or plate, which is then illuminated by light.

Hooke's law in mechanics, a principle that states strain is directly proportional to stress as long as the limit of elasticity is not exceeded.

Huygen's principle concept on wave propagation that says every point on a primary wave front can be thought of as a source of secondary waves.

Ideal gas gas that obeys the ideal gas law.

Ideal gas law statement of relationship between pressure, temperature, moles, and pressure of gas under ambient conditions, $PV = nRT$, where R equals the ideal gas law constant.

Impulse instantaneous change in momentum or change in momentum over a time period, equal to average force times the time period over which it acts.

Impulsive wave wave caused by a sudden displacement in the medium through which it travels, for example, tsunamis.

Index of refraction ratio of speed of light in a vacuum to speed of light in a given medium, also called refractive index.

Inductor a component of an electrical system that operates by induction, often a small wire coil that transfers current.

Inertia property of matter to remain at rest or in state of constant motion unless acted upon by an external force.

Inertial mass mass as defined by Newton's second law, $F = ma$.

Inertial reference frame reference frame in which the law of inertia holds, a frame of reference that moves with constant velocity with respect to another inertial reference frame.

Insulator substance that does not conduct electricity (or heat or sound when referring to these phenomena).

Internal energy the total kinetic and potential energy of a system.

Internal force a force that acts between components within a system.

Inverse square rule general rule that applies when the intensity of property varies inversely with the square of a defined distance, for example, intensity of light, gravitational attraction between two masses.

Ionic solid a solid composed of anions and cations arranged in a crystalline network, such as NaCl.

Ionizing radiation radiation possessing sufficient energy to dislodge electrons from atoms in tissue, forming ions with the potential to cause cell damage.

Isolated system a system in which neither mass nor energy is exchanged with the surroundings.

Isotope different forms of the same element with different mass numbers due to a difference in number of neutrons in the nuclei of atoms.

Kinematics study and description of motion without regard to its cause.

Kinetic energy energy associated with motion, $(1/2)mv^2$.

Kinetic molecular theory model that defines behavior of an ideal gas and assumes the average kinetic energy of gas molecules is directly proportional to the absolute temperature.

Latent heat energy transfer to a substance without a change in the substance's temperature; associated with phase changes such as melting (freezing) and boiling (condensing).

Lattice point position in a unit cell in a crystalline structure occupied by an atom, molecule, or ion.

Leptons class of elementary particles associated with the weak force, includes relatively small particles such as electrons and neutrinos.

Lever arm perpendicular distance from the axis of rotation to the line of the force vector, also called moment arm.

Liquid crystals materials that have properties of both solids and liquids, used extensively in digital displays.

Longitudinal wave a wave in which the individual particles oscillate in same direction in which the wave propagates, for example, sound wave.

Long-range order general term to describe a regular repeating structure associated with crystals.

Lorentz-Fitzgerald contraction in relativity, the shortening of an object in its direction of motion when observed from a reference frame moving relative to the reference frame of the observed object.

Luminiferous ether hypothesized substance proposed to fill the universe and used to account for observations made using classical physics.

Magnetic declination angular deviation of a compass from true geographic north or south; angular displacement between the geographic pole and magnetic pole.

Magnetic domain small area in a ferromagnetic material in which electrons spin to produce a small magnetic region.

Magnetic inclination dip of a compass needle.

Magnetosphere asymmetrical region surrounding the Earth that contains the Earth's magnetic field; the magnetosphere traps electrically charged particles from outer space and contains the Van Allen radiation belts.

Mass quantity of a physical object that gives it inertia.

Mass defect the decrease in mass when a nucleus is formed from its constituent nucleons.

Mass number number of nucleons equal to the sum of protons and neutrons in the nucleus of an atom.

Mechanical advantage the ratio of load or resistance to the force applied when a machine is used to perform work; the factor by which applied force is increased in a machine.

Mechanics area of physics that deals with the action of forces on mater at rest or in motion, includes statics and dynamics.

Meson class of subatomic particles involved in strong interactions; mesons have masses intermediate between leptons and baryons and can have negative, neutral, or positive charge.

Metallic solid type of solid characterized by delocalized electrons and metal atoms that occupy lattice points.

Moderator a substance such as graphite or deuterium used to slow down neutrons in a nuclear reactor.

Modern physics refers to physics after approximately 1900; incorporates relativity and quantum mechanics.

Modulus of elasticity Young's modulus, defined as the ratio of stress to strain

Molecular solid solid that contains molecules at lattice points.

Moment of inertia a measure of the resistance of an object to a change in angular acceleration.

Momentum physical quantity in mechanics conserved in the absence of external forces; measure of a moving object's inertia, linear momentum equals mass times velocity.

Muon a mu meson, a fundamental unstable subatomic particle with a mass 207 times that of an electron.

Mutual induction process in which the changing current in a coil causes (induces) a current in another coil.

Near infrared electromagnetic radiation with a wavelength just beyond that of visible light up to a wavelength of approximately 1,000 nanometers.

Nebula diffuse mass of gas and dust particles in interstellar space.

Neutrino stable subatomic particle with no charge and very small mass.

Newton basic unit of force in SI system equal to 1 kilogram-meter per second^2; force required to accelerate 1 kilogram to 1 meter per second^2.

Node a stationary point or area of zero amplitude in wave motion.

Noninertial reference frame reference frame in which the law of inertia doesn't hold; a frame of reference that is accelerating with respect to another inertial reference frame.

Nonionizing radiation radiation possessing insufficient energy to dislodge electrons from atoms in tissue.

Normal force force a surface exerts on object in contact with it; reaction force applied by a surface on a mass resting on that surface.

Nuclear fission splitting of the nucleus of an atom of a heavier element into smaller nuclei, resulting in the production of energy.

Nuclear fusion combination of nuclei of lighter elements into heavier nuclei, resulting in the production of energy.

Nucleon a proton or neutron; the number of nucleons in an atom is the sum of its protons and neutrons.

Ohm measure of electrical resistance; a conductor with a current of 1 ampere and a voltage drop of 1 volt has a resistance of 1 ohm.

Ohm's law expresses the relationship between current, voltage, and resistance and states current is directly proportional to the electromotive force and inversely proportional to the resistance, $V = IR$.

Open system a system in which both matter and energy are exchanged with the surroundings.

Optics   area of physics concerned with the study of light.

Orbital velocity   minimum velocity needed for an object to revolve around another body; velocity needed to put object in orbit, for Earth approximately 8 kilometers per second (18,000 miles per hour) for an orbit of 500 kilometers (300 miles).

Oxidation   electrochemical process involving the loss of electrons.

Pascal   basic SI unit of pressure equal to 1 newton per square meter.

Pascal's principle   pressure applied at any point in a confined fluid will be transmitted undiminished throughout the fluid.

Pauli exclusion principle   in quantum mechanics, the principle that states no two particles can occupy the same quantum state simultaneously.

Peak voltage   maximum displacement from the reference voltage in an alternating current sine wave.

Period   time interval it takes for two consecutive in-phase parts of a wave, such as wave crests, to travel past the same point; the time it takes for a repetitious phenomenon to repeat itself.

Periodic waves   regularly repeating disturbance that travels through a medium.

Photoelectric effect   transfer of energy when light strikes a metal, causing an electron or electrons to be dislodged from the metal.

Photon   fundamental particle of electromagnetic radiation (light).

Physical optics   area of optics that deals with diffraction, refraction, and polarization where light interacts with objects having physical dimensions similar to those of the wavelength of light.

Plane polarization   polarization process resulting in light vibrating in one plane.

Plasma   state of matter in which atoms or molecules of gas have been ionized to produce equal numbers of free electrons and positive ions.

Polarized light   light in which the vibrations are confined to one planar direction by passing the light through a substance called a polarizer.

Polyatomic ion   ion consisting of more than one atom.

Population inversion   in laser action, when many particles (molecules, atoms, ions) achieve a metastable excited state.

Positron   subatomic particle with the mass of an electron and a charge of +1.

Potential energy   energy by virtue of the position or configuration of a system.

Pressure   force per unit area.

Primary cell   an electrochemical cell that cannot be recharged.

Primary coil   coil to which current is supplied in an electrical device such as a transformer.

Proper length   in special relativity, the length of an object measured from within the same reference frame as the object.

Proper reference frame   reference frame of observer.

Proper time   in special relativity, the time measured when the clock is at rest with respect to the observer.

Protostar   early stage in formation of a star, preceding the onset of nuclear fusion.

Quantum chromodynamics   theory of nuclear physics involving how quarks of different colors combine to form larger particles.

Quantum electrodynamics   quantum theory involving the interaction of electrons, muons, and photons with electromagnetic radiation.

Quantum mechanics   branch of physics developed in the first several decades of the twentieth century to explain behavior of matter and electromagnetic radiation at the atomic level.

Quark   hypothesized particle with fractional charge; there are six quarks with either a charge of 1/3 or 2/3 thought to compose hadrons.

Quark confinement idea that a free independent quark cannot exist.

Rad unit of an absorbed dose of radiation equal to 100 ergs of energy absorbed by 1 gram of matter.

Radioactive when a substance contains unstable atomic nuclei that spontaneously emit particles and energy.

Radioactive tracers radioactive substance used to monitor movement and behavior of chemicals in biological or physical process.

Refraction bending of a wave as it passes from one medium to another, for example, bending of light as it moves from air to water.

Rarefaction in a longitudinal (sound) wave, the area where decompression or a decrease in pressure occurs.

Rayleigh scattering scattering of electromagnetic radiation by particles much smaller than those of the incident light; the particles absorb and scatter the incident light without a change in wavelength.

Real image in optics, where the image produced results from rays actually passing through the image.

Reduction electrochemical process involving the gain of electrons.

Relativistic examining physical phenomenon where the theory of relativity is needed, that is, when speeds are significant with respect to the speed of light.

Rem unit to measure biological effects of an absorbed dosage of radiation equivalent to the effect produced by 1 rad of high-level x-rays.

Resistance property that gives the opposition of passage of electrical current through a substance; resistance results in current being converted to electrical energy or heat, measured in ohms.

Resistivity intrinsic property of a material that gives its resistance as related to its cross-sectional area and length; resistivity of a material equals its resistance times cross-sectional area divided by length.

Rest mass classical concept of mass in which mass is considered stationary in a reference frame, as opposed to relativistic mass, where mass increases when an object is moving.

Right-hand rule general method to determine direction of vector forces and fields in mechanics and electricity.

Rotational equilibrium state for an object when the external torques sum to zero resulting, in no rotation of the object.

Scalar quantity a physical quantity with a magnitude but no direction, for example, speed.

Scintillation a flash or spark of light produced by the absorption of a photon or particle.

Second law of thermodynamics law stating how energy flows in the universe; the second law of thermodynamics says that the entropy of the universe is constantly increasing, and a spontaneous process requires the entropy of the universe to increase.

Secondary coil coil in which current is induced in an electrical device such as a transformer.

Self-induction process in which a change in current in a coil is caused (induced) by current itself.

Semiconductor a material with properties intermediate between conductors and insulators, for example, crystalline silicon used in computer chips.

Sensible heat heat transfer characterized by a change in temperature; heat that can be "sensed" with a thermometer.

Sidereal measurement with respect to the distance stars; sidereal time is measured with respect to stars, for example, a sidereal day is 23 hours 56 minutes 4 seconds, as opposed to 24 hours.

Snell's law law that describes how light is refracted through a medium and states that the sine of the angle of incidence divided by the sine of the angle of refraction is equal to the index of refraction.

Solar wind   continuous stream of high-speed charged particles coming from the Sun.

Sound   energy created by a longitudinal wave traveling through an elastic medium and associated with sense of hearing.

Sound intensity   measure of sound energy per unit area per unit time, measured in decibels.

Specific heat capacity   the amount of heat required to raise the temperature of a substance by $1°C$.

Specular reflection   reflection in which an incident beam is reflected at a single angle equal to the angle of incidence.

Speed   scalar quantity equal to the distance traveled divided by the time of travel.

Spherical aberration   blurring of an image due to reflection from the edge of a mirror or the passage of light through the edge of lens due to variations of focal length at different points on the optical device and its deviation from a perfect sphere.

Streamline   path of a particle in laminar fluid flow.

Stefan-Boltzmann law   law that states that the energy radiated from the surface of a blackbody per unit time is proportional to absolute temperature to the fourth power.

Stellar aberration   apparent motion of light from a star (or other heavenly bodies) due to the finite speed of light; causes elliptical motion of star and depends on Earth's motion and angle of observation.

Stress   internal forces interacting throughout an object caused by external forces and producing strain.

Subcritical mass   when the quantity of radioactive fuel is insufficient to produce a self-sustaining chain reaction.

Supercritical mass   when the quantity of radioactive fuel is more than sufficient to produce a self-sustaining chain reaction.

Superconductor   substance that conducts electricity with zero resistance, a process that is approached near absolute zero in certain metals and ceramic materials.

Supernova   rare, spectacular explosion of giant star in its late stages of life, resulting in a brief appearance of a bright light source in the universe and that emits a large amount of energy.

Superposition principle   states that net wave displacement at point in space at a particular time is the result of adding individual waves at that point.

Surface tension   property of liquid causing it to contract to the smallest possible surface area due to unbalance of forces at the liquid's surface.

Surroundings   remainder of universe outside a system.

System   that part of the universe defined for physical analysis.

Tangential   pertains to an instantaneous quantity at a point measured at some distance from the axis of rotation in rotational motion, for example, tangential velocity, tangential acceleration, tangential force.

Theoretical physics   general area of physics that attempts to explain nature by proposing physical models to explain nature and make predictions.

Theory of Everything (TOE)   general term given to a broad overarching theory to explain all physical phenomenon under one theory that unifies the four fundamental interactions.

Thermal energy   energy from kinetic energy associated with the random motion of particles making up a substance; energy by virtue of a substance possessing an absolute temperature.

Thermodynamics   area of physics concerned with the transformations of energy, heat, and work.

Time dilation   slowing down of time when a clock in a moving reference frame is compared to a clock in a stationary reference frame.

Torque moment of force that produces rotation; vector product of force times the perpendicular distance force acts from the axis of rotation.

Total internal reflection phenomenon when light moving through a medium strikes an interface at an angle greater than the critical angle and is reflected back into the medium.

Transmutation nuclear transformation in which the atomic nucleus of one element is converted into the nucleus of another element.

Transverse wave wave in which particle displacement is perpendicular to the direction of wave motion.

Tribolelectricity static electricity produced by friction between two objects.

Turbulent fluid flow characterized by small-scale random eddies and fluctuations in the fluid.

Ultraviolet catastrophe term given to the prediction that power curves derived from blackbody radiation approach infinity at higher frequencies using classical physics; was resolved using quantum theory.

Unified field theory theory that unites different theories into a more general theory, generally applied to theories attempting to unite the four fundamental interactions.

Uniform circular motion motion at a constant speed in a circular path.

Unit cell most basic repeating unit of a lattice structure.

Valence electron electron in the outer shell of an atom; valence electrons are involved in chemical reactions.

Vector quantity physical quantity characterized by both a magnitude and a direction, for example, velocity, force.

Velocity vector quantity equal to the change in displacement with respect to time.

Virtual image in optics, when the image produces light rays that appear to emerge from the image, but light does not come from the actual image position.

Wavelength distance between two points in phase, such as two crests or two troughs, on a wave.

Weight force force experienced by an object due to gravitational attraction of another object, for example, force by which an object is attracted to Earth.

White dwarf type of star formed in the latter stages of stellar evolution, with low luminosity, small size, high density, and emitting white light.

Wiens's law law giving the relationship between the wavelength of peak emission of blackbody radiation and temperature; states the wavelength (in meters) equals 0.002898/absolute temperature in kelvins.

Work product of force times distance through which force acts; transfer of energy from one body to another, causing displacement.

X-ray crystallography method used to determine the structure of a crystal by examining a diffraction pattern produced by x-rays that pass through the crystal.

Zero-point motion quantum mechanical motion present in a system in its lowest energy state.

Zeroth law of thermodynamics law that states if two systems are in thermal equilibrium with another system then they are in thermal equilibrium with one another.

Selected Symbols, Abbreviations, and Units

Symbol or Abbreviation	Quantity
Ω	ohm (electric resistance)
τ	torque
A	ampere (current)
a	acceleration
C	coulomb (charge)
F	farad (capacitance)
H	henry (inductance)
Hz	hertz (frequency)
J	joule (energy, work)
K	kelvin (temperature)
g	gravitational acceleration
m	meter (length)
mol	mole (quantity)
N	newton (force)
p	momentum
Pa	pascal (pressure)
s	second
S	entropy
T	tesla (magnetic field)
V	volt (electric potential)
W	watt (power)
Wb	weber (magnetic flux)

Prefixes and Multiples for Metric Units

f	femto (10^{-15})
p	pico (10^{-12})
n	nano (10^{-9})
μ	micro (10^{-6})
m	milli (10^{-3})
d	deci (10^{-1})
c	centi (10^{2})
k	kilo (10^{3})
M	mega (10^{6})
G	Giga (10^{9})

SI Units

Physical Quantity	Unit	Symbol
Length	meter	m
Mass	kilogram	kg
Time	second	s
Electric current	ampere	A
Temperature	kelvin	K
Quantity of substance	mole	mol
Luminous intensity	candela	cd

Selected Physical Constants Data

Quantity	Symbol	Value
Astronomical unit	AU	1.4959787×10^{11} km
Atmospheric pressure at sea level		1.013×10^5 Pa
Atomic mass unit	amu	1.67377×10^{-27} kg
Boltzmann's constant	K	1.3807×10^{-23} J/K
Density of air at STP		1.29 kg/m3
Density of water at 4.0°C		1.000×10^3 kg/m^3
Earth's mass		5.974×10^{24} kg
Earth's mean orbital speed		30 km/s (18 miles/s)
Earth's mean radius		6.371×10^3 km
Electron radius		2.81794×10^{-15} m
Electron rest mass	m_e	9.10956×10^{-31} kg
Elementary charge	e	1.60×10^{-19} C
Faraday constant	F	9.65×10^4 C/mol
Gas law constant	R	8.3145 J/K−mol
Gravitational acceleration at Earth's surface	g	9.8067 m/s^2
Gravitational constant	G	6.67×10^{-11} N-m/kg^2
Latent heat of fusion of water	ΔH_f	3.35×10^5 J/kg
Latent heat of vaporization of water	ΔH_v	2.26×10^6 J/kg
Moon's mass		7.348×10^{22} kg
Moon's mean radius		1.738×10^3 km
Neutron rest mass	m_n	1.67942×10^{-27} kg
Permeability of free space	μ_0	$4\pi \times 10^{-7}$ T-m/A
Permittivity of free space	ε_0	8.85×10^{-12} C^2/N-m^2
Planck's constant	h	6.626176×10^{-34} J-s
Proton rest mass	m_p	1.67261×10^{-27} kg
Rydberg constant	R_∞	1.097×10^7 m^{-1}
Specific heat capacity of water		4.186×10^3 J/kg-C°
Speed of light in vacuum	c	2.998×10^8 m/s
Speed of sound in air at 20°C		343 m/s (1,100 ft/s)
Stefan-Boltzmann constant	σ	5.67×10^{-8} W/m^2-K^4
Sun's mass		1.989×10^{30} kg
Sun's mean radius		6.970×10^8 m
Wien's constant	σ_w	2.898×10^{-3} m-K

Table of Elements

Element	Symbol	Atomic Number	Atomic Mass	Element	Symbol	Atomic Number	Atomic Mass
Actinium	Ac	89	227	Cobalt	Co	27	58.93
Aluminum	Al	13	26.98	Copper	Cu	29	63.55
Americium	Am	95	243	Curium	Cm	96	247
Antimony	Sb	51	121.8	Dubnium	Db	105	262
Argon	Ar	18	39.95	Dysprosium	Dy	66	162.5
Arsenic	As	33	74.92	Einsteinium	Es	99	254
Astatine	At	85	210	Erbium	Er	68	167.3
Barium	Ba	56	137.3	Europium	Eu	63	152.0
Berkelium	Bk	97	247	Fermium	Fm	100	253
Beryllium	Be	4	9.012	Fluorine	F	9	19.00
Bismuth	Bi	83	209.0	Francium	Fr	87	223
Bohrium	Bh	107	264	Gadolinium	Gd	64	157.3
Boron	B	5	10.81	Gallium	Ga	31	69.72
Bromine	Br	35	79.90	Germanium	Ge	32	72.59
Cadmium	Cd	48	112.4	Gold	Au	79	197.0
Calcium	Ca	20	40.08	Hafnium	Hf	72	178.5
Californium	Cf	98	249	Hassium	Hs	108	265
Carbon	C	6	12.01	Helium	He	2	4.003
Cerium	Ce	58	140.0	Holmium	Ho	67	164.9
Cesium	Cs	55	132.9	Hydrogen	H	1	1.008
Chlorine	Cl	17	35.45	Indium	In	49	114.8
Chromium	Cr	24	52.00	Iodine	I	53	126.9

Element	Symbol	Atomic Number	Atomic Mass	Element	Symbol	Atomic Number	Atomic Mass
Iridium	Ir	77	192.2	Rhenium	Re	75	186.2
Iron	Fe	26	55.85	Rhodium	Rh	45	102.9
Krypton	Kr	36	83.80	Rubidium	Rh	37	85.47
Lanthanum	La	57	138.9	Ruthenium	Ru	44	101.1
Lawrencium	Lr	103	257	Rutherfordium	Rf	261	261
Lead	Pb	82	207.2	Samarium	Sm	62	150.4
Lithium	Li	3	6.941	Scandium	Sc	21	44.96
Lutetium	Lu	71	175.0	Seaborgium	Sg	106	263
Magnesium	Mg	12	24.31	Selenium	Se	34	78.96
Manganese	Mn	25	54.94	Silicon	Si	14	28.09
Meitnerium	Mt	109	268	Silver	Ag	47	107.9
Mendelevium	Md	101	256	Sodium	Na	11	22.99
Mercury	Hg	80	200.6	Strontium	Sr	38	87.62
Molybdenum	Mo	42	95.94	Sulfur	S	16	32.07
Neodymium	Nd	60	144.2	Tantalum	Ta	73	180.9
Neon	Ne	10	20.18	Technetium	Tc	43	99
Neptunium	Np	93	237	Tellurium	Te	52	127.6
Nickel	Ni	28	58.69	Thallium	Ti	81	204.4
Niobium	Nb	41	92.91	Thorium	Th	90	232.0
Nitrogen	N	7	14.01	Thulium	Tm	69	168.9
Nobelium	No	102	253	Tin	Sn	50	118.7
Osmium	Os	76	190.2	Titanium	Ti	22	47.88
Oxygen	O	8	16.00	Tungsten	W	74	183.9
Palladium	Pd	46	106.4	Ununnilium	Uun	110	269
Phosphorus	P	15	30.97	Unununium	Uun	111	269
Platinum	Pt	78	195.1	Unumbium	Uub	112	277
Plutonium	Pu	94	242	Uranium	U	92	238.0
Potassium	K	19	39.10	Vanadium	V	23	50.94
Praseodymium	Pr	59	140.9	Xenon	Xe	54	131.3
Promethium	PM	61	147	Ytterbium	Yb	70	173.0
Protactinium	Pa	91	231	Yttrium	Y	39	88.91
Radium	Ra	88	226	Zinc	Zn	30	65.39
Radon	Rn	86	222	Zirconium	Zr	40	91.22

Conversion Tables

Length

	inch	cm	ft	m	mile	km
1 inch 5	1	2.54	8.33×10^{22}	2.54×10^{22}	1.58×10^{25}	2.54×10^{25}
1 cm 5	0.394	1	3.28×10^{22}	0.01	6.21×10^{26}	10^{25}
1 ft 5	12	30.5	1	0.305	1.89×10^{24}	3.05×10^{24}
1 m 5	39.4	100	3.28	1	6.21×10^{24}	10^{23}
1 mile 5	6.34×10^{4}	1.61×10^{5}	5.28×10^{3}	1.61×10^{3}	1	1.61
1 km 5	3.94×10^{4}	10^{5}	3.28×10^{3}	10^{3}	0.621	1

Area

	in^2	cm^2	ft^2	m^2	acre
1 in^2 5	1	6.45	6.94×10^{23}	6.45×10^{24}	1.59×10^{27}
1 cm^2 5	0.155	1	1.08×10^{23}	10^{24}	2.47×10^{28}
1 ft^2 5	144	929	1	9.29×10^{22}	2.30×10^{25}
1 m^2 5	1.55×10^{3}	104	10.8	1	2.47×10^{24}
1 acre	6.27×10^{6}	4.047×10^{7}	4.36×10^{4}	4.047×10^{3}	1

Mass

	g	oz	lb	kg
1 g =	1	3.53×10^{22}	2.20×10^{23}	10^{23}
1 oz =	28.4	1	6.25×10^{22}	2.83×10^{22}
1 lb =	454	16	1	0.454
1 kg =	10^3	35.3	2.20	1

Force

	newton	pound	dyne
1 newton =	1	0.2248	10^5
1 pound =	4.448	1	4.448×10^5
1 dyne =	10^{25}	2.248×10^{26}	1

Energy

	joule	calorie	kW-h	electron volt	BTU
1 joule =	1	0.239	2.778×10^{27}	6.242×10^{18}	9.481×10^{24}
1 calorie =	4.186	1	1.163×10^{26}	2.613×10^{19}	3.698×10^{23}
1 kW-h =	3.6×10^6	8.60×10^5	1	2.247×10^{25}	3,410
1 electron volt =	1.602×10^{219}	3.83×10^{220}	4.45×10^{226}	1	1.519×10^{222}
1 BTU =	1,056	252	2.93×10^{24}	6.59×10^{21}	1

Speed

	$\dfrac{ft}{s}$	$\dfrac{cm}{s}$	$\dfrac{m}{s}$	mph	$\dfrac{km}{h}$
$1\dfrac{ft}{s}=$	1	30.5	0.305	0.681	1.10
$1\dfrac{cm}{s}$	3.28×10^{22}	1	0.01	2.24×10^{22}	3.60×10^{22}
$1\dfrac{m}{s}=$	3.28	100	1	2.24	3.60
1 mph 5	1.47	44.7	0.447	1	1.61
$1\dfrac{km}{hr}$	0.911	27.8	0.278	0.621	1

Power

	watt	kilowatt	calorie second	horsepower
1 watt 5	1	0.001	0.2389	1.341×10^{23}
1 kilowatt	1,000	1	238.9	1.341
$1\dfrac{calorie}{second}=$	4.184	4.184×10^{23}	1	5.611×10^{23}
1 horsepower 5	745.7	0.7457	178.2	1

	$kPa = 10^3 \frac{N}{m^2}$	$\frac{pound}{in^2}$	atmosphere	cm Hg	bar
1 kilopascal (kPa) 5	1	0.145	$9.869\ 3\ 10^{23}$	0.750	0.0100
$1\frac{pound}{in^2} =$	6.895	1	0.0680	5.172	0.0689
1 atmosphere 5	101.3	14.7	1	76.0	1.013
1 cm Hg 5	1.333	0.1934	0.01316	1	0.01333
1 bar 5	100	14.504	0.987	75.00	1

	in^3	cm^3	ft^3	m^3	gal	L
1 in^3 5	1	16.4	5.79×10^{24}	1.64×10^{25}	4.33×10^{23}	1.64×10^{22}
1 cm^3 5	6.10×10^{22}	1	3.53×10^{25}	10^{26}	2.64×10^{24}	10^{23}
1 ft^3 5	1.73×10^3	2.83×10^4	1	2.83×10^{22}	7.48	28.3
1 m^3 5	6.10×10^4	10^6	35.3	1	264	10^3
1 gal 5	231	3.79×10^3	1.34×10^{21}	3.79×10^{23}	1	3.79
1 L 5	61.0	10^3	3.53×10^{22}	10^{23}	0.264	1

	$\dfrac{\text{ft}^3}{\text{s}}$	$\dfrac{\text{m}^3}{\text{s}}$	$\dfrac{\text{gal}}{\text{s}}$	$\dfrac{\text{L}}{\text{s}}$	MGD	$\dfrac{\text{m}^3}{\text{d}}$
$1\dfrac{\text{ft}^3}{\text{s}}$	1	2.83×10^{22}	7.48	28.3	0.646	2.45×10^3
$1\dfrac{\text{m}^3}{\text{s}}$	35.3	1	264	103	22.8	8.64×10^4
$1\dfrac{\text{gal}}{\text{s}}$	1.34×10^{21}	3.79×10^{23}	1	3.79	8.64×10^{22}	327
$1\dfrac{\text{L}}{\text{s}}$	3.53×10^{22}	10^{23}	0.264	1	2.28×10^{22}	86.4
1 MGD 5	1.55	4.38×10^{22}	11.6	43.8	1	3.79×10^3
$1\dfrac{\text{m}^3}{\text{d}}$	4.09×10^{24}	1.16×10^{25}	3.06×10^{23}	1.16×10^{22}	2.64×10^{24}	1

Nobel Prize Winners in Physics*

*Years in which the prize was split equally are indicated, with a horizontal line separating how the prize was split. The horizontal line also indicates work performed. In years for which several individuals are listed, the prize was shared for common work. For example, in 2002 Davis and Koshiba shared one-half the prize for work on detection of cosmic neutrinos, while Giacconi received half the prize for work on astrophysics and cosmic x-ray sources.

Year	Winners	Work
1901	Röntgen, Wilhelm Conrad	Discovery of x-rays
1902	Lorentz, Hendrik Antoon Zeeman, Pieter	Effect of magnetism on radiation
1903	Becquerel, Henri Curie, Marie Curie, Pierre	Spontaneous radioactivity
1904	Rayleigh, John William Strutt	Discovery of argon, density of gases
1905	Lenard, Philipp	Cathode rays
1906	Thomson, Sir J.J.	Electrical conductivity of gases
1907	Michelson, Albert Abraham	Spectroscopic/metrological investigations
1908	Lippmann, Gabriel	Photographic reproduction of color
1909	Braun, Ferdinand Marconi, Guglielmo	Wireless telegraphy
1910	Waals, Johannes Diederik van der	Equation of state of fluids
1911	Wien, Wilhelm	Laws on radiation of heat
1912	Dalén, Nils	Automatic regulators for gas
1913	Kamerlingh Onnes, Heike	Matter at low temperature
1914	Laue, Max von	X-ray diffraction with crystals
1915	Bragg, Lawrence Bragg, William	Crystal structure using x-rays

Year	Winners	Work
1917	Barkla, Charles Glover	Characteristic x-ray spectra of elements
1918	Planck, Max	Energy quanta
1919	Stark, Johannes	Splitting of spectral lines in electric field
1920	Guillaume, Charles Édouard	Anomalies in nickel and steel alloys
1921	Einstein, Albert	Photoelectric effect
1922	Bohr, Niels	Atomic structure and radiation
1923	Millikan, Robert Andrews	Elementary electric charge
1924	Siegbahn, Karl Manne Georg	X-ray spectroscopy
1925	Franck, James Hertz, Gustav	Impact of electron on atom
1926	Perrin, Jean	Discontinuous structure of matter, sedimentation equilibria
1927	Compton, Arthur Holly Wilson, C.T.R.	Compton effect Invention of cloud chamber
1928	Richardson, Owen Willans	Richardson's law, electron emission of hot metals
1929	Broglie, Louis-Victor	Wave nature of electrons
1930	Raman, Chandrasekhara Venkata	Raman effect, light diffusion
1932	Heisenberg, Werner	Quantum mechanics
1933	Dirac, Paul Adrien Maurice Schrödinger, Erwin	Wave equations in quantum mechanics
1935	Chadwick, James	Discovery of neutron
1936	Anderson, Carl David Hess, Victor Francis	Discovery of positron Discovery of cosmic rays
1937	Davisson, Clinton Joseph Thomson, George Paget	Crystal diffraction of electrons
1938	Fermi, Enrico	Neutron irradiation and discovery of new elements
1939	Lawrence, Ernest Orlando	Invention of cyclotron
1943	Stern, Otto	Discovery of proton magnetic moment
1944	Rabi, Isidor Isaac	Magnetic resonance of atomic nuclei
1945	Pauli, Wolfgang	Exclusion principle of electrons
1946	Bridgman, Percy Williams	High-pressure physics
1947	Appleton, Edward Victor	Upper-atmosphere physics
1948	Blackett, Patrick M.S.	Nuclear and cosmic physics with cloud chamber
1949	Yukawa Hideki	Prediction of mesons
1950	Powell, Cecil Frank	Photographic method for meson studies

Year	Winners	Work
1951	Cockcroft, John Douglas Walton, Ernest Thomas Sinton	Transmutation of nuclei
1952	Bloch, Felix Purcell, E.M.	Nuclear magnetic resonance methods
1953	Zernike, Frits	Phase-contrast microscopy
1954	<u>Born, Max</u> Bothe, Walther	<u>Statistical treatment of wave function</u> Coincidence method
1955	<u>Kusch, Polykarp</u> Lamb, Willis Eugene, Jr.	<u>Electron magnetic moment</u> Hydrogen structure
1956	Bardeen, John Brattain, Walter H. Shockley, William B.	Semiconductors and discovery of transistor
1958	Cherenkov, Pavel Alekseyevich Frank, Ilya Mikhaylovich Tamm, Igor Yevgenyevich	Discovery and interpretation of Cherenkov effect
1959	Chamberlain, Owen Segrè, Emilio	Antiproton
1960	Glaser, Donald A.	Bubble chamber
1961	<u>Hofstadter, Robert</u> Mössbauer, Rudolf Ludwig	<u>Structure of nucleons</u> Absorbance and emission of photons
1962	Landau, Lev Davidovich	Condensed matter and liquid helium
1963	Jensen, J. Hans D. <u>Goeppert-Mayer, Maria</u> Wigner, Eugene P.	<u>Nuclear shell structure</u> Fundamental symmetry principles
1964	Basov, Nikolay Gennadiyevich Prokhorov, Aleksandr Mikhaylovich Townes, Charles Hard	Maser-laser principles
1965	Feynman, Richard P. Schwinger, Julian Seymour Tomonaga Shin'ichiro	Quantum electrodynamics
1966	Kastler, Alfred	Hertzian resonance in atoms
1967	Bethe, Hans Albrecht	Energy production in stars
1968	Alvarez, Luis W.	Elementary particles and resonance states
1969	Gell-Mann, Murray	Quark model and elementary particles
1970	<u>Alfvén, Hannes</u> Néel, Louis-Eugène-Félix	<u>Magneto hydrodynamics</u> Ferromagnetism and anti-ferromagnetism
1971	Gabor, Dennis	Development of holography
1972	Bardeen, John Cooper, Leon N. Schrieffer, John Robert	Theory of superconductivity

Year	Winners	Work
1973	Esaki, Leo <u>Giaever, Ivar</u> Josephson, Brian D.	<u>Tunneling in superconductors</u> <u>Supercurrents and Josephson effects</u>
1974	Hewish, Antony Ryle, Sir Martin	Radioastronomy and pulsars
1975	Bohr, Aage N. Mottelson, Ben R. Rainwater, James	Atomic nucleus structure
1976	Richter, Burton Ting, Samuel C.C.	Discovery of J/psi particle
1977	Anderson, Philip W. Mott, Sir Nevill F. Van Vleck, John H.	Electronic structure of magnetic and noncrystal-line solids
1978	<u>Kapitsa, Pyotr Leonidovich</u> Penzias, Arno Wilson, Robert Woodrow	<u>Helium liquefaction</u> Cosmic background radiation
1979	Glashow, Sheldon Lee Salam, Abdus Weinberg, Steven	Unification of electromagnetic and weak interaction
1980	Cronin, James Watson Fitch, Val Logsdon	Violation of symmetry principles in mesons
1981	Bloembergen, Nicolaas <u>Schawlow, Arthur Leonard</u> Siegbahn, Kai Manne Börje	<u>Laser spectroscopy</u> Electron spectroscopy
1982	Wilson, Kenneth Geddes	Phase transitions
1983	<u>Chandrasekhar, Subrahmanyan</u> Fowler, William A.	<u>Stellar evolution</u> Element formation in universe
1984	Meer, Simon van der Rubbia, Carlo	Discovery of W and Z particles
1985	Klitzing, Klaus von	Quantizied hall effect
1986	Binnig, Gerd <u>Rohrer, Heinrich</u> Ruska, Ernst	<u>Scanning tunnel microscopy</u> Electron microscopy
1987	Bednorz, J. Georg Müller, Karl Alex	Superconductivity in ceramics
1988	Lederman, Leon Max Schwartz, Melvin Steinberger, Jack	Discovery of muon neutrino
1989	Dehmelt, Hans Georg <u>Paul, Wolfgang</u> Ramsey, Norman Foster	<u>Ion trap technique</u> Atomic clocks

Year	Winners	Work
1990	Friedman, Jerome Isaac Kendall, Henry Way Taylor, Richard E.	Deep elastic scattering and quark discovery
1991	Gennes, Pierre-Gilles de	Order transitions in liquid crystals
1992	Charpak, Georges	Multiwire proportional chamber detector
1993	Hulse, Russell Alan Taylor, Joseph H., Jr.	Discovery of binary pulsar
1994	<u>Brockhouse, Bertram N.</u> Shull, Clifford G.	<u>Neutron scattering</u> Neutron diffraction.
1995	<u>Perl, Martin Lewis</u> Reines, Frederick	<u>Discovery of tau lepton</u> Detection of the neutrino
1996	Lee, David M. Osheroff, Douglas D. Richardson, Robert C.	Superfluidity in He-3
1997	Chu, Steven Cohen-Tannoudji, Claude Phillips, William D.	Laser cooling and trapping of atoms
1998	Laughlin, Robert B. Stormer, Horst L. Tsui, Daniel C.	Fractional quantum Hall effect
1999	Hooft, Gerardus 'T Veltman, Martinus J. G.	Quantum structure and electroweak interaction
2000	Alferov, Zhores I. <u>Kroemer, Herbert</u> Kilby, Jack St. Clair	<u>Semiconductor heterostructures</u> Invention of integrated circuits
2001	Cornell, Eric A. Ketterle, Wolfgang Wieman, Carl E.	Bose-Einstein condensation of alkali atoms
2002	Davis, Raymond, Jr. <u>Koshiba, Masatoshi</u> Giacconi, Riccardo	<u>Detection of cosmic neutrinos</u> Astrophysics and cosmic x-ray sources
2003	Abrikosov, Alexei A. Ginzburg, Vitaly L. Leggett, Anthony J.	Superconductivity and superfluids
2004	Gross, David J. Politzer, David H. Wilczek, Frank	Asymptotic freedom and the strong interaction
2005	<u>Glauber, Roy J.</u> Hall, John L. Hänsch, Theodor W.	<u>Quantum theory of optical coherence</u> laser spectroscopy

Selected Bibliography
and Internet Resources

Abbott, David, ed. *Physicists (Biographical Dictionary of Scientists)*. New York: Peter Bedrick Books, 1984.

Asimov, Isaac. *Understanding Physics*. New York: Barnes & Noble Books, 1993.

Atkins, P. W. *The Second Law*. New York: Scientific American Books, 1984.

Bloomfield, Louis A. *How Things Work: The Physics of Everyday Life*. New York: John Wiley & Sons, 1997.

Cohen, I. Bernard. *The Birth of a New Physics*. New York: W. W. Norton & Co., 1985.

Cullerne, J. P., and Valerie Illingsworth. *The Penguin Dictionary of Physics*. London: Penguin Books, 2000.

Culver, Roger B. *Facets of Physics: A Conceptual Approach*. St. Paul, MN: West Publishing Co., 1993.

Cutnell, John D., and Kenneth W. Johnson. *Physics*, 4th ed. New York: John Wiley & Sons, 1998.

Einstein, Albert. *Relativity: The Special and General Theory* trans. Robert W. Lawson. New York: Routledge Classics, 2001.

Fritzsch, Harald. *Quarks: The Stuff of Matter*. New York: Basic Books, 1983.

Gardner, Martin. *Relativity Simply Explained*. Mineola, NY: Dover Publications, 1997.

Halliday, David, Robert Resnick, and Jearl Walker. *Extended Fundamentals of Physics*, 6th ed. New York: John Wiley & Sons, 2000.

Henderson, Harry. *Nuclear Physics*. New York: Facts on File, 1998.

Heuvelen, Alan Van. *Physics: A General Introduction*. Boston: Scott Foresman & Co., 1986.

Hewitt, Paul. *Conceptual Physics,* 8th ed. Reading, MA: Addison Wesley, 1998.

Hobson, Art. *Physics: Concepts and Connections,* 2nd ed. Upper Saddle River, NJ: Prentice Hall, 1999.

Kahan, Gerald. $E = mc^2$: *Picture Book of Relativity*. Blue Ridge Summit, PA: Tab Books, 1983.

Lampton, Christopher F. *Particle Physics: The New View of the Universe*. Hillsdale, NJ: Enslow Publishers, 1991.

Nave, Carl R. *Hyperphysics* (CD). Atlanta, GA: Rod Nave, Georgia State University, 2004.

Ne'eman, Yeval, and Kirsch Yoram. *The Particle Hunters*. Cambridge: Cambridge University Press, 1996.

Sambursky, Shmuel, ed. *Physical Thought from the Pre-Socratics to the Quantum Physicists: An Anthology*. New York: Pica Press, 1975.

Schwarz, Cindy. *A Tour of the Subatomic Zoo,* 2nd ed. New York: Springer, 1997.

Sutton, Christine. *The Particle Connection*. New York: Simon and Shuster, 1984.

Walker, James S. *Physics,* 2nd ed. Upper Saddle River, NJ: Pearson Education, 2004.

Weaver, Jefferson Hane. *The World of Physics: Volume I The Aristoleian Cosmos and the Newtonian System, Volume II The Einsteinian and the Bohr Atom, Volume III The Evolutionary Cosmos and the Limits of Science*. New York: Simon and Shuster, 1987.

Weinberg, Steven. *The Discovery of Subatomic Particles,* rev. ed. Cambridge: Cambridge University Press, 2003.

Internet Resources

The following resources contain URLs for physics resources groups in general categories with a brief note on content.

General Physics Lessons and Activities

http://directory.google.com/Top/Science/Physics/. Google's physics directory.

www.physicsclassroom.com/. Physics classroom constructed by Mathsoft Engineering & education, Inc.

www.sciencejoywagon.com/physicszone/. Science Joy Wagon site on physics topics; costs $5.00 for annual access.

http://scienceworld.wolfram.com/physics/. A physics dictionary and abbreviated encyclopedia by Eric Weisstein's World of Science by Wolfram Research, Inc.

http://webphysics.davidson.edu/Applets/Taiwan-Univ/indexPopup.html. Interactive activities, simulations, and virtual labs from Fu-Kwun Hwang at the Department of Physics, National Taiwan Normal University.

www.exploratorium.edu/snacks/index.html. Exploratorium site with activities from one exploratorium interactive science museum in San Francisco.

www.ied.edu.hk/has/phys/phydemo_lr/mechanic/index.htm. Various physics demonstrations by HKIEd HAS Center.

http://Galileo.phys.Virginia.EDU/classes/109N/more_stuff/Applets/. Physics applets by Michael Fowler at the University of Virginia.

http://jersey.uoregon.edu/vlab/. Virtual labs from the Physics Department at the University of Oregon.

www.nasa.gov/home/index.html. Home page for NASA.

www.hazelwood.k12.mo.us/~grichert/sciweb/applets.html. Virtual labs and simulations from two teachers, Richert and Pipkin, from the Hazelwood School District.

Mechanics

http://wildcat.phys.nwu.edu/ugrad/vpl/mechanics/. Virtual mechanics laboratories by the Department of Astronomy and Physics at Northwestern University.

www.jracademy.com/~jtucek/physics/physics.html. Interactive lessons dealing mainly with mechanics.

http://liftoff.msfc.nasa.gov/academy/rocket_sci/orb-mech/vel_calc.html. Calculates the orbital velocity and period of a satellite at any altitude above the Earth's surface; from NASA.

http://monet.physik.unibas.ch/%7Eelmer/pendulum/. Several activities on pendulums by Franz-Josef Elmer.

http://it.gse.gmu.edu/projects/ziop/student.html. Lessons on Newton's laws of motion by The Stepping Stones to Technology Grant and George Mason University, Graduate School of Education.

www.learner.org/exhibits/parkphysics/. Amusement park physics from Annenberg/CPB.

Conversions and Units

http://physics.nist.gov/cuu/Units/. Home page of the National Institute of Standards and Technology.

www.digitaldutch.com/unitconverter/. Interactive converter between units by Digital Dutch.

Optics

www.ecok.edu/academics/schools/ms/cp/mathsci_chemphy_physics_sites 3.asp. Links to optics sites from East Central University, Oklahoma.

http://idisk.mac.com/kwillims/Public/Optics.html. Links to optics sites by Emily Scott, ECU Senior.

Modern Physics

http://phys.unsw.edu.au/~jw/time.html. Devoted to special relativity and time dilation from University of New South Wales, Sydney, Australia.

http://phys.educ.ksu.edu/vqm/index.html. Simulations on quantum mechanics from Physics Education Research Group at Kansas State University.

www.ecok.edu/academics/schools/ms/cp/mathsci_chemphy_physics_sites 2.asp. Links to modern physics sites from East Central University, Oklahoma.

www.pbs.org/wgbh/nova/time/. Time travel and relativity from PBS WGBH Science Unit.

www.superstringtheory.com/. Devoted to superstring theory.

http://particleadventure.org/particleadventure/. Particle physics provided by the Particle Data Group at Lawrence Berkeley National Laboratory.

http://hepunx.rl.ac.uk/neutrino-industry/. Devoted to neutrino research.

www.fnal.gov/pub/inquiring/physics/index.html. Fermi lab Web site. http://public.web.cern.ch/public/. Home page for CERN.

http://my.unidata.ucar.edu/content/staff/blynds/tmp.html. Devoted to temperature, heat, and thermodynamics from "About Temperature" by Beverly T. Lynds, UCAR in Boulder, Colorado.

http://lectureonline.cl.msu.edu/~mmp/applist/blackbody/black.htm. Applet to see how spectrum changes with temperature by W. Bauer.

Nobel Prize Winners

http://nobelprize.org/physics/. Nobel site devoted to winners of Nobel prizes.

http://home.gwu.edu/~reevesme/superlinks.html. Superconductivity information provided by research group of Mark reeves at The George Washington University.

Orders of Magnitude

http://micro.magnet.fsu.edu/primer/java/scienceopticsu/powersof10/. Virtual tour from 10 million light years away from Earth to the quark level by Michael W. Davidson and Florida State University.

Index

About the Author

RUSTY L. MYERS is professor of environmental science at Alaska Pacific University where he has taught for the past 25 years. He teaches undergraduate and graduate courses in chemistry, environmental science, and statistics. Dr. Myers' research includes work on air quality, water quality, science education, and science and the humanities. He has received several national awards for teaching, including the 1994 Alaska Professor of the Year award from the Carnegie Foundation for Higher Education.